Alessandro Boselli

A Panchromatic View of Galaxies

Related Titles

Greggio, L., Renzini, A.

Stellar Populations
A User Guide from Low to High Redshift

2011
ISBN: 978-3-527-40918-1

Phillipps, S.

The Structure and Evolution of Galaxies

2005
ISBN: 978-0-470-85506-5

Morison, I.

Introduction to Astronomy and Cosmology

2008
ISBN: 978-0-470-03334-0

Salaris, M., Cassisi, S.

Old Stellar Systems
Decoding the Fossil Record of Galaxy Formation

2013
Softcover, ISBN: 978-3-527-41059-0; Hardcover, ISBN: 978-3-527-41076-7

Stiavelli, M. S.

From First Light to Reionization
The End of the Dark Ages

2009
ISBN: 978-3-527-40705-7

Roos, M.

Introduction to Cosmology

2003
ISBN: 978-0-470-84910-1

Alessandro Boselli

A Panchromatic View of Galaxies

WILEY-VCH

WILEY-VCH Verlag GmbH & Co. KGaA

The Authors

Dr. Alessandro Boselli
Physics of Galaxies group
Laboratoire d'Astrophysique de Marseille
38, rue Joliot-Curie
13388 Marseille
France

■ All books published by Wiley-VCH are carefully produced. Nevertheless, authors, editors, and publisher do not warrant the information contained in these books, including this book, to be free of errors. Readers are advised to keep in mind that statements, data, illustrations, procedural details or other items may inadvertently be inaccurate.

Library of Congress Card No.: applied for

British Library Cataloguing-in-Publication Data:
A catalogue record for this book is available from the British Library.

Bibliographic information published by the Deutsche Nationalbibliothek
The Deutsche Nationalbibliothek lists this publication in the Deutsche Nationalbibliografie; detailed bibliographic data are available on the Internet at http://dnb.d-nb.de.

© 2012 WILEY-VCH Verlag GmbH & Co. KGaA, Boschstr. 12, 69469 Weinheim, Germany

All rights reserved (including those of translation into other languages). No part of this book may be reproduced in any form – by photoprinting, microfilm, or any other means – nor transmitted or translated into a machine language without written permission from the publishers. Registered names, trademarks, etc. used in this book, even when not specifically marked as such, are not to be considered unprotected by law.

Typesetting le-tex publishing services GmbH, Leipzig
Cover Design Adam-Design, Weinheim
Printing and Binding Fabulous Printers Pte Ltd, Singapore

Printed in Singapore
Printed on acid-free paper

ISBN Print 978-3-527-40991-4

ISBN oBook 978-3-527-64025-6
ISBN ePub 978-3-527-64026-3
ISBN ePDF 978-3-527-64027-0
ISBN Mobi 978-3-527-64028-7

to Felice

Contents

Preface *XIII*

1 **Introduction** *1*
1.1 Galaxies *1*
1.2 A Multifrequency Approach *4*
1.3 The Purpose of this Book *10*

Part One Emitting Sources and Radiative Processes in Galaxies *15*

2 **X-ray** *17*
2.1 Continuum *17*
2.1.1 Discrete Sources *18*
2.1.2 X-ray Emission in Active Galaxies *20*
2.1.3 Hot Gas *21*

3 **UV-Optical-NIR** *25*
3.1 Continuum: Stellar Emission *26*
3.2 Emission Lines *28*
3.2.1 Hydrogen Lines *32*
3.2.2 Metals *34*
3.3 Absorption Lines *35*
3.3.1 Hydrogen Lines *37*
3.3.2 Other Elements *38*
3.4 Molecular Lines *39*
3.4.1 H_2 Near-Infrared Emission Lines *39*
3.4.2 H_2 UV Absorption Lines *39*

4 **The Infrared** *41*
4.1 Continuum: Dust Emission *42*
4.2 Emission Lines *44*
4.2.1 PAHs *44*
4.2.2 Cooling Lines in PDR *45*
4.2.3 H_2 Lines *47*
4.2.4 Dust Absorption of Ly_α Scattered Photons *49*

5	**Millimeter and Centimeter Radio** 51
5.1	Continuum 51
5.1.1	Free–Free Emission 52
5.1.2	Synchrotron Emission 53
5.1.3	Dust Emission 53
5.2	Emission Lines 54
5.2.1	Molecular Lines 54
5.2.2	HI 55
5.3	Absorption Lines 57
5.3.1	HI 57

Part Two Derived Quantities 59

6	**Properties of the Hot X-ray Emitting Gas** 61
6.1	X-ray Luminosity 61
6.2	Gas Temperature 61

7	**Dust Properties** 63
7.1	The Far-IR Luminosity 63
7.2	Dust Mass and Temperature 65

8	**Radio Properties** 71
8.1	Determining the Contribution of the Different Radio Components 71
8.1.1	Synchrotron vs. Free–Free Radio Emission in the Centimeter Domain 71
8.1.2	The Emission of the Cold Dust Component at $\lambda \leq 1.5$ mm 72
8.2	The Radio Luminosity 74

9	**The Spectral Energy Distribution** 77
9.1	The Emission in the UV to Near-Infrared Spectral Domain 79
9.1.1	UV, Optical, and Near-IR Colors 81
9.1.2	Fitting SEDs with Population Synthesis Models 83
9.2	The Dust Emission in the Infrared Domain 84
9.2.1	Mid- and Far-Infrared Colors 86
9.3	The Thermal and Nonthermal Radio Emission 90

10	**Spectral Features** 91
10.1	Galaxy Characterization through Emission and Absorption Lines 91
10.1.1	Classification of the Nuclear Activity 92
10.1.2	Classification of Post-Starburst and Post-Star-Forming Galaxies 92
10.1.3	Line Diagnostics 95
10.2	Gas Metallicity from Emission Lines 101
10.3	Stellar Age and Metallicity from Absorption Lines 103

11	**Gas Properties** 107
11.1	Gas Density, Mass, and Temperature 107
11.1.1	The Atomic HI Mass 108
11.1.2	The Molecular H_2 Mass 115

12	**Dust Extinction** *125*	
12.1	Galactic Extinction *126*	
12.1.1	Extinction Curve *127*	
12.2	Internal Attenuation *132*	
12.2.1	Attenuation of the Emission Lines *133*	
12.2.2	Attenuation of the Stellar Continuum *134*	
13	**Star Formation Tracers** *143*	
13.1	The Initial Mass Function *143*	
13.2	The Star Formation Rate *144*	
13.3	The Birthrate Parameter and the Specific Star Formation Rate *146*	
13.4	The Star Formation Efficiency and the Gas Consumption Time Scale *147*	
13.5	Hydrogen Emission Lines *147*	
13.6	UV Stellar Continuum *151*	
13.7	Infrared *152*	
13.7.1	Integrated Infrared Luminosity *152*	
13.7.2	Monochromatic Infrared Luminosities *153*	
13.8	Radio Continuum *153*	
13.9	Other Indicators *155*	
13.9.1	The X-ray Luminosity *155*	
13.9.2	Forbidden Lines *156*	
13.9.3	[CII] *157*	
13.9.4	Radio Recombination Lines *157*	
13.10	Population Synthesis Models *158*	
13.10.1	Dating a Star Formation Event *158*	
14	**Light Profiles and Structural Parameters** *161*	
14.1	The Surface Brightness Profile *161*	
14.1.1	Extended Radial Profiles *161*	
14.1.2	The Central Surface Brightness Profile of Early-Type Galaxies *162*	
14.1.3	The Vertical Light Profile of Late-Type Galaxies *166*	
14.2	Structural Parameters *166*	
14.2.1	Total Magnitudes, Effective Radii and Surface Brightnesses *166*	
14.2.2	Bulge to Disk Ratio *167*	
14.3	Morphological Parameters *168*	
14.3.1	Concentration Index *168*	
14.3.2	Asymmetry *168*	
14.3.3	Clumpiness *169*	
14.3.4	The Gini Coefficient G and the Second-Order Moment of the Brightest 20% of the Galaxy's Flux M_{20} *169*	
15	**Stellar and Dynamical Masses** *171*	
15.1	Stellar Mass Determination Using Population Synthesis Models *171*	
15.2	Dynamical Mass *175*	
15.2.1	Rotation Curves and the Dark Matter Distribution *177*	

15.2.2　The Total Mass of Elliptical Galaxies
　　　　from Kinematical Measurements　*184*
15.2.3　The Total Mass of Elliptical Galaxies from X-ray Measurements　*185*
15.2.4　The Mass of the Supermassive Black Hole　*187*

Part Three　Constraining Galaxy Evolution　*193*

16　Statistical Tools　*195*
16.1　Galaxy Number Counts　*195*
16.1.1　Observed Number Counts　*197*
16.2　Luminosity Function　*200*
16.2.1　Parametrization of the Luminosity Function　*203*
16.2.2　Luminosity Distributions and Bivariate Luminosity Functions　*204*
16.2.3　The Observed Luminosity Functions　*205*
16.3　Luminosity Density　*209*
16.3.1　The Cosmic Star Formation History
　　　　and Build Up of the Stellar Mass　*211*

17　Scaling Relations　*215*
17.1　Spectrophotometric Relations　*216*
17.1.1　The Color–Magnitude and Color–Color Relations　*216*
17.1.2　The Mass–Metallicity Relation　*218*
17.1.3　The Mass–Gas Relation　*220*
17.1.4　The Mass–Star Formation Rate Relation　*222*
17.2　Structural Relations　*223*
17.2.1　The Surface Brightness–Absolute Magnitude Relation　*223*
17.2.2　The Kormendy Relation　*224*
17.3　Kinematical Relations　*224*
17.3.1　The Tully–Fisher Relation　*225*
17.3.2　The Faber–Jackson Relation and the Fundamental Plane　*228*
17.3.3　The k-Space　*230*
17.4　Supermassive Black Hole Scaling Relations　*231*

18　Matter Cycle in Galaxies　*235*
18.1　The Star Formation Process　*236*
18.1.1　The Schmidt Law　*236*
18.2　Feedback　*239*
18.2.1　The Feedback of AGNs　*239*
18.2.2　The Feedback of Massive Stars　*242*

19　The Role of the Environment on Galaxy Evolution　*245*
19.1　Tracers of Different Environments　*245*
19.1.1　Detection of High-Density Regions　*246*
19.1.2　Other Quantitative Tracers of High-Density Environments　*249*
19.2　Measuring the Induced Perturbations　*250*
19.2.1　Other Tracers of Induced Perturbations　*253*

Appendix A Photometric Redshifts and *K*-Corrections *255*
A.1   The Photometric Redshifts   *255*
A.1.1   UV-Optical-Near-Infrared Photo-*z*   *255*
A.1.2   Far Infrared-Radio Continuum Photo-*z*   *258*
A.2   The *K*-Correction   *258*

Appendix B Broad Band Photometry   *263*
B.1   Photometric Systems   *263*

Appendix C Physical and Astronomical Constants and Unit Conversions   *267*

References   *269*

Index   *319*

Preface

In the late eighties, when I first started studying astronomy, the study of galaxies was mainly divided into well defined but quite disjoint research themes related to the spectral range within which the analyzed data were taken. Optical extragalactic astronomy, which was taking advantage of the arrival of the first charge-coupled devices (CCDs) on 4-m class telescopes, was principally devoted to studying stellar populations of intermediate age, while the first generation of near-infrared detectors provided information on the cold stellar component which is the dominant mass component of evolved stellar systems. Radio astronomy, boosted after the construction of the Westerbork radio telescope and the Very Large Array, was principally focused either on the study of the HI gas properties of nearby, late-type systems, or on the radio continuum emission of bright radio galaxies. In the X-ray domain, the Einstein satellite was producing the first images of nearby galaxies while the Infrared Astronomical Satellite (IRAS), thanks to its all-sky survey, was providing precious data for hundreds of thousands of galaxies still used today to study the properties of the dust frozen in the interstellar medium of all extragalactic sources.

Thanks to the pioneer work of Rob Kennicutt and of my first thesis advisor Giuseppe Gavazzi, it soon became obvious that combining data at different wavelengths was an extremely powerful way to study the physical processes at work in galaxies. I thus had the chance during my time at the University of Milano to be faced with both the technical problems related to the manipulation of sets of data at different wavelengths, in particular in the visible, near- and far-infrared, and centimeter radio, and with their physical interpretation in the framework of galaxy evolution. This expertise was further developed during my PhD at the Observatoire de Paris-Meudon, under the supervision of James Lequeux. There, I learned millimeter astronomy and I had privileged access to infrared astronomy as the person responsible for one of the guaranteed time key projects of the Infrared Space Observatory (ISO) mission. My admission at the Laboratoire d'Astrophysique de Marseille allowed me to be member of the Galaxy Evolution Explorer space mission (GALEX), and thus extend my expertise to the ultraviolet spectral domain. This multiwavelength approach in the study of galaxies is now widely developed and used by many, if not most, astronomers. At the same time, more and more refined models of galaxy evolution are able to reproduce different observables, shinning new light on the process that gave birth to these interesting objects.

This journey made me look at galaxies in a slightly different way than the one commonly described in most of the beautiful textbooks available in the literature, where these fascinating objects are generally divided according to their morphology and studied as separate entities. The multifrequency approach that I previously described forces us to look at all extragalactic sources as undefined objects whose physical properties can be determined and appreciated only after a combined analysis of their data collected at different wavelengths, despite their optical morphology.

To fulfill my work I had to learn how to handle the multifrequency data now available to the community, covering the whole electromagnetic spectrum, from X-rays to centimeter radio. I also had to understand how to derive the most important physical quantities necessary in the study of galaxies. Finding this information, though available in dedicated publications or targeted books, took me several years since it was never collected on a unique source. Furthermore, the experience acquired during these years allowed me to develop a critical view on the most widely used recipes for this exercise, to test the limits of the underlying assumptions, and to quantify their uncertainties. Thus, it seems timely to transfer my expertise to young students or to senior astronomers willing to approach the study of galaxies through a multifrequency analysis. The purpose of this book is to provide to the reader a working tool, useful as a starting point for such a study. I imagine this volume on the reader's desk with other useful manuals.

Obviously, if I have acquired considerable experience in multifrequency analyses, it has been possible to the detriment of other things. That is why this book should not be used as a reference for detailed and thorough studies of all the specific emitting processes, the nature of the emitting sources, the physical processes acting in all extragalactic systems, or the evolution of galaxies. I therefore warmly recommend the interested reader to refer to more specific publications, such as those I list in the text, for a more accurate use of their own data.

The conception and the first development of this book has been suggested to me by my friend and colleague Veronique Buat, who invited me on many occasions to share my expertise with students and young astronomers visiting our institute. I am thus particularly grateful to Veronique for her invaluable support in this beautiful project. The writing of the text and the widening of the subjects developed here have only been made possible thanks to the inestimable contribution of many of my expert friends that I prompted on several occasions to share stimulating discussions and constructive comments with me. I therefore want to warmly thank Philippe Amram, Samuel Boissier, Mederic Boquien, Albert Bosma, Emily Brageot, Jonathan Braine, Veronique Buat, Barbara Catinella, Andrea Cattaneo, Vassilis Charmandaris, Laure Ciesla, Monica Colpi, George Comte, Luca Cortese, Olga Cucciati, Emanuele Daddi, Daniel Dale, Jean Michel Deharveng, Lise Deharveng, Laura Ferrarese, Michele Fumagalli, Giuseppe Gavazzi, Sebastien Heinis, Olivier Ilbert, James Lequeux, Dario Maccagni, Henry McCracken, Paolo Padovani, Celine Peroux, Henri Plana, Bianca Poggianti, Dimitra Rigopoulou, Tsutomu Takeuchi, Daniel Thomas, Elisa Toloba, Marie Treyer, Ginevra Trinchieri, Bernd Vollmer and

Christine Wilson for their help. A special thanks is addressed to Yannick Roehlly for his inestimable help in the preparation of all the figures included in this volume, and to George Comte for a critical and complete reading of the manuscript. I also wish to thank Peppo Gavazzi and James Lequeux who gave me the opportunity to work on this fascinating subject for several years under their supervision, and with whom I still have fruitful collaborations. A special thanks to Peppo, who brought me to discover the world of astronomy in his office and during the huge number of nights that we spent together at the telescope.

I am also grateful to Anja Tschoertner and Ulrike Fuchs of Wiley for their help during the definition of this project and for finding the adapted solutions of the various technical problems that I encountered during the writing of the manuscript. Last, but not least, a special thank to my family, to my wife Beatrice and to my son Felice for their patience during the time I spent writing the book. I dedicate this work to my son Felice.

Marseille, September 2011 *Alessandro Boselli*

1
Introduction

1.1
Galaxies

Galaxies are gravitationally bound systems composed of a large number of stars, gas, dust, and other forms of matter. They span a large range in dimension, luminosity, and mass. Dwarf galaxies are composed of $\sim 10^7$ stars, while massive galaxies can have up to $\sim 10^{12}$. Massive galaxies have been classified according to their morphology into two main classes, ellipticals (E) and spirals (S) [1–8]. The first family is characterized by a smooth elliptical morphology, is dominated by an old stellar population, and generally contains little gas and dust (Figure 1.1). The motion of stars within ellipticals is mainly random. These objects are often called early-type galaxies. The second family is characterized by a flat disk with a spiral morphology, with the possible presence of a bar, a bulge, and a nucleus (S). These systems are undergoing episodes of star formation and thus include young stellar populations. They are rich in gas and can contain a significant amount of dust. Their kinematic is characterized by a more or less significant rotation. They are generally referred to as late-type systems. In addition to these two families, there exists an intermediate class of objects. The lenticular galaxies (S0) are characterized by a flat disk without any spiral structure, dominated by old stellar populations, and generally devoid of gas and dust. Besides these main families, astronomers defined the class of irregular galaxies (Irr or Pec) to include all those objects with a peculiar morphology that do not fit into the previously mentioned categories. This class includes, for instance, interacting and merging systems such as the Antennae (NGC 4038/4039) or the Mice (NGC 4676) shown in Figure 1.2. These are characterized by perturbed morphologies with clumpy regions of high surface brightness, and extended tidal tales with low surface brightness. Dwarfs are generally divided into gas-rich, star-forming Magellanic irregulars (Im) or blue compact dwarfs (BCD), and gas-poor, quiescent dwarf ellipticals (dE) and spheroidals (dS0) (Figure 1.3). Magellanic irregulars and blue compact dwarfs have irregular morphologies and are generally sustained by rotation. The former are extended and have moderate to very low stellar surface brightnesses, while the latter are very compact objects with high surface brightness and have aspects similar to that of bright HII regions. Dwarf ellipticals and spheroidals are objects with low surface brightness and are

A Panchromatic View of Galaxies, First Edition. Alessandro Boselli.
© 2012 WILEY-VCH Verlag GmbH & Co. KGaA. Published 2012 by WILEY-VCH Verlag GmbH & Co. KGaA.

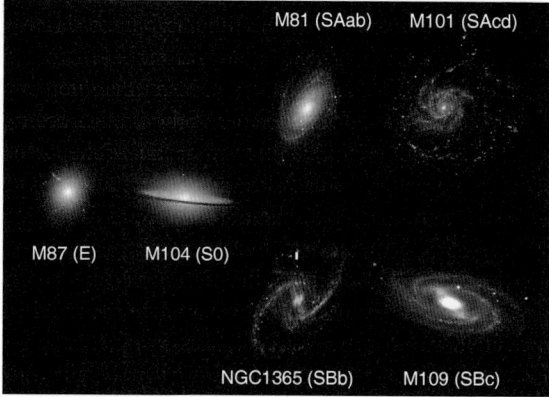

Figure 1.1 The Hubble diagram separates elliptical galaxies (E, left) from lenticulars (S0, middle left) and spirals (S, right). The spiral sequence is decomposed into normal (SA, upper) and barred (SB, lower) spirals.

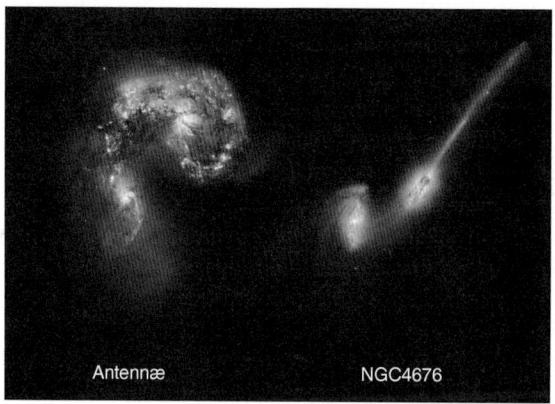

Figure 1.2 The interacting systems NGC 4038/4039 (the Antennae) and NGC 4676 (the Mice).

characterized by very smooth and regular morphologies. Galaxies can host an active galactic nucleus (AGN), where the energy necessary to sustain the powerful electromagnetic radiation is supplied by the accretion of mass onto the central supermassive black hole. Active galaxies can be divided into quasars or QSO (quasi stellar objects), which are compact sources with a stellar morphology, Seyfert galaxies, where the AGN dominates only the central region, and radio galaxies.

Galaxies are ordinary objects inhomogeneously distributed within the universe. If ρ_0 is the mean galaxy density of the local universe, whose value is ~ 0.006 objects Mpc^{-3} down to the absolute magnitude $M_i < -19.5$ [9], the density of galaxies spans from $\sim 0.2\rho_0$ in voids to $\sim 5\rho_0$ in superclusters and filaments, $\sim 100\rho_0$ in the core of rich clusters, and $\sim 1000\rho_0$ in compact groups [10, 11]. In the nearby universe, star-forming systems dominate the population by counting for roughly 70% of the extragalactic population, while ellipticals count for $\sim 10\%$

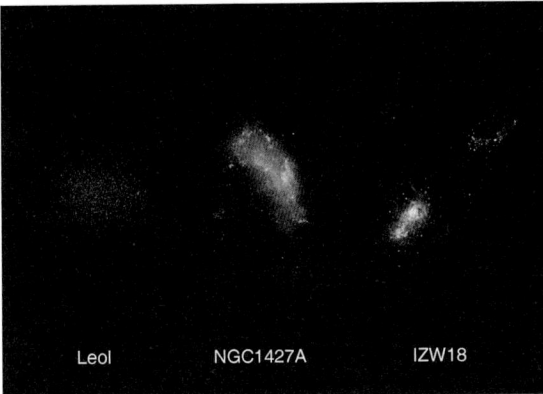

Figure 1.3 Dwarf galaxies can be divided into dwarf spheroidals (dS0, left, LeoI), Magellanic irregulars (Im, middle, NGC 1427A), and blue compact dwarfs (BCD, right, IZw18).

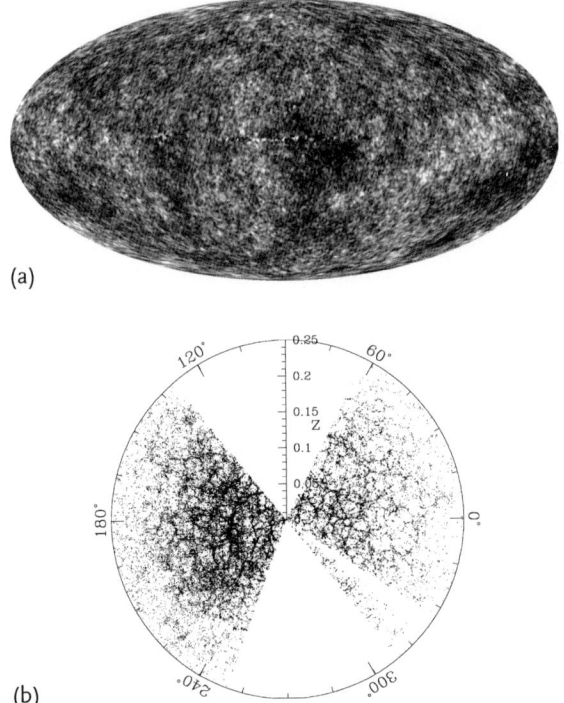

Figure 1.4 The WMAP background sky image (a) at 1.9 mm shows the temperature fluctuations of the universe when it was ~ 380 000 yr old (from [19]) (Reproduced by permission of the AAS; Credit: NASA/WMAP Science Team.). These temperature fluctuations, shown as color differences, correspond to the seeds that grew to form galaxies. The distribution of these temperature fluctuations resembles the distribution of galaxies observed by the Sloan Digital Sky Survey (SDSS) in the local universe (b) (from [20]) (Reproduced by permission of the AAS.).

and lenticulars for $\sim 20\%$ [12]. The fraction of late-type galaxies, however, strongly decreases in high-density environments such as in rich clusters [13, 14]. Dwarfs exceed in number massive objects. However, despite their relatively low number, ellipticals, lenticulars, and spiral bulges contain $\sim 75\%$ of the mass of stars within the universe [15, 16].

Owing to the new generation of ground-based and space facilities, galaxies are now observable at redshifts larger than 5, allowing us to picture the universe $\sim 1\,\mathrm{Gyr}$ after its formation. For this reason, galaxies can be used to probe the mass distribution at earlier and earlier epochs ($z \gtrsim 6$), providing a link between the primordial universe, now well represented by the cosmic microwave background (CMB; $z \sim 1100$), as traced by the Wilkinson Microwave Anisotropy Probe (WMAP) [17] and the Planck space mission [18], and the present epoch, as depicted in Figure 1.4. Indeed, galaxies are expected to develop within dark matter halos formed from primordial density fluctuations, whose distribution is well traced by the anisotropies observed in the microwave background image of the universe. The study of galaxy formation and evolution is thus tightly related to the evolution of the universe itself and is thus of paramount importance in cosmology.

1.2
A Multifrequency Approach

The understanding of galaxy evolution requires an accurate study of the different processes that gave birth and transformed them. Of fundamental importance is the study of the matter cycle, that is, understanding how the primordial gas component from the Big Bang has collapsed into giant molecular clouds to form stars of different mass. These stars later modified the physical and chemical properties of the interstellar medium, intervening in the process of star formation (feedback). Indeed, the dust formed by the aggregation of metals produced and injected into the interstellar medium by massive stars during the latest phases of their life plays a crucial role in galaxy evolution since it is a catalyst in the formation of molecular hydrogen and is a major coolant of the interstellar medium.

There is evidence showing that the matter cycle might be modulated by the environment to which galaxies belong. In high-density environments, for instance, the atomic gas, which is generally located in the outer disk of spiral galaxies, is easily removed in both gravitational interactions with nearby companions or in dynamical interactions with the hot and dense intracluster gas permeating rich clusters of galaxies, while the stellar component, tightly linked to the potential well of the galaxy, is perturbed only by gravitational interactions [10].

All the physical processes regulating the matter cycle and the perturbations induced by the environment must be considered and understood simultaneously. The typical properties of galaxies spanning a large range in luminosity and mass are generally traced using scaling relations, where structural, kinematical, chemical, and spectrophotometrical properties are combined and analyzed. These scaling relations are of paramount importance in the study of the physical processes at work in any kind of extragalactic source, and are thus widely used as constraints in models of galaxy formation and evolution. Because of their tightness, they are also powerful distance tracers generally used to study the large scale galaxy distribution within the universe.

Clearly, a coherent and complete understanding of galaxy evolution through cosmic time requires a simultaneous analysis of all the different galaxy components. This can be achieved only through a multifrequency analysis. Stars emit as black body in the ultraviolet (UV) to near-infrared (NIR) spectral domain, with a different contribution depending on their mean age. Massive young stars dominate the short wavelength range while evolved systems mainly emit in the near-infrared. The gaseous component, either atomic or molecular, can be observed using some specific emission lines; the former via the emission of the HI line at 21 cm while the latter via carbon monoxide lines emitted in the millimeter domain. The dust component is heated by the interstellar radiation field and emits in the mid- and far-infrared and submillimeter domain following different emission processes depending on whether or not it is in thermal equilibrium with the radiation. The millimeter radio and centimeter emission is due to the free–free emission of unbound electrons formed in HII regions by the ionizing radiation of the most young and massive stars, and by the synchrotron emission due to relativistic electrons accelerated in supernovae remnants spinning in week magnetic fields. Both emission processes are tightly related to recent star formation events and have a relative weight that changes with λ, where the free–free emissions is being important at short wavelengths while synchrotron dominating at long wavelengths. The accretion phenomenon on a compact source in binary systems, or the emission of the hot gas component permeating massive elliptical galaxies or forming the diffuse interstellar medium of gas-rich systems, can be observed in the X-rays.

In the last 25 years the advent of new technologies allowed astronomers to extend to large angular scales the observations of the universe with a continuous coverage, from X-rays to centimeter radio wavelengths, previously accessible only to the optical and centimeter radio bands. The nearby universe has been completely or partially mapped by all sky or shallow extended surveys such as ROSAT (Röntgensatellite; [21]) in X-rays, GALEX (Galaxy Evolution Explorer; [22]) in the ultraviolet (UV) band, SDSS (Sloan Digital Sky Survey, [23]) in the optical, 2MASS (Two Micron All Sky Survey, [24]) in the near-infrared, IRAS (Infrared Astronomical Satellite, [25]) and AKARI [26] in the infrared, and FIRST (Faint Images of the Radio Sky at Twenty cm, [27]) and NVSS (NRAO VLA Radio Survey, [28]) in the radio continuum. The nearby universe has also been observed in spectroscopic mode in its most important emission line, the HI hydrogen at 21 cm, by HIPASS (HI Parkes All Sky Survey, [29]), HIJASS [30], and ALFALFA (Arecibo Legacy Fast ALFA Survey, [31]), while the optical lines by the SDSS [23] and the 2dF Galaxy Redshift Survey (2-degree-Field Galaxy redshift Survey, [32]). Furthermore, imaging and spectroscopic multifrequency observations of galaxies at significant distances have been obtained using very deep exposures of relatively small sky regions accurately selected to be uncontaminated by the Galactic emission. Typical examples

Table 1.1 Emitting sources and emission processes at different wavelengths.

Domain λ	X-ray 0.1–10 keV (1–100 Å)	UV 912–3500 Å	Visible 3500–7500 Å	NIR 0.75–5 μm	MIR 5–20 μm	FIR-submm 20 μm–1 mm	Radio 1 mm–1 m
			Continuum				
Process	Black body; Thermal bremsstrahlung	Black body	Black body	Black body	Thermal emission	Modified black body	Synchrotron; Free-free
Emitting source	Accretion disk in binary systems; Hot gas	Young stars	Intermediate age stars	Old stars	PAH, hot dust grains	Cold dust grains	Relativistic electrons in weak magnetic fields; HII regions
			Main emission lines				
Emission lines		Atomic hydrogen, metals	Atomic hydrogen, metals	Atomic and molecular hydrogen, molecules	PAH	[CII], CO, molecules	HI(21 cm)
Origin		HII regions	HII regions	HII regions	PDR	Giant molecular clouds	Diffuse ISM
Absorption lines		Hydrogen, metals	Hydrogen, metals	Hydrogen			HI(21 cm)
Origin		Stellar atmosphere, ISM, IGM	Stellar atmosphere	Stellar atmosphere			Diffuse ISM

Notes: PAH: Polycyclic Aromatic Hydrocarbons; PDR: photodissociation region; ISM: interstellar medium; IGM: intergalactic medium.

are the GOODS (Great Observatories Origins Deep Survey, [33]) and COSMOS (Cosmic Evolution Survey, [34]) surveys.

The contribution of these multifrequency surveys to the study of galaxy evolution is invaluable, since it simultaneously provides us with information on the different components of any extragalactic object (see Table 1.1 for a summary). The modern instrumentation now in our hands enables us to measure the content and trace the distribution of the gaseous, stellar, and dust components in large samples of galaxies. A typical example is illustrated in Figure 1.5 which shows the image of the interacting system M51 observed at different wavelengths, from the X-ray (Figure 1.5a) to the 21 cm HI radio line (Figure 1.5l). The distribution of the old stellar population, as traced by the near-infrared image at 3.6 μm obtained by the Spitzer space mission, is relatively smooth. Old stars are the dominant stellar population in the companion NGC 5195, an early-type galaxy interacting with NGC 5194. The distribution of the young stellar component, as traced by the H_α narrow band (ground-based observation) and UV GALEX images, is very clumpy because of the presence of massive stars in compact HII regions primarily located along the spiral arms. The molecular gas component traced by the CO emission (IRAM) and, to a minor extent the atomic hydrogen (HI line at 21 cm; VLA) follows the distribution of the young stars. This is expected since the star formation process requires large amounts of gas to take place. A similar clumpy distribution is also evident in the 8 μm image which traces the distribution of the Polycyclic Aromatic Hydrocarbons (PAHs), in the hot dust component (24 μm) observed in the mid-IR by Spitzer, where the hot dust is primarily heated by the ultraviolet radiation emitted by the youngest stellar populations. To a lesser extent a similar morphology is present in the submillimeter domain (250 μm, Herschel space mission), where dust heating from the general interstellar field becomes more important. This clumpy structure can be also seen in the radio continuum (VLA) given that the relativistic electrons responsible for the synchrotron emission are accelerated in supernovae remnants produced after the explosion of the most massive stars, and in the X-ray where the emission is dominated by binary systems and supernovae remnants.

High quality morphological studies of the high-redshift universe are routinely undertaken with subarcsecond angular resolutions at either short wavelengths (UV, visible) or in the radio regime, thanks to millimeter and centimeter interferometers. In the infrared spectral domain, where the data are collected using space facilities of moderate size, the angular resolution is diffraction limited and thus sufficient for a detailed two-dimensional analysis only in the nearby universe, while high-redshift studies are soon limited by source confusion. The sensitivity and the spectral resolution of optical spectrometers is also sufficient to measure the kinematical properties of the brightest galaxies up to redshift $z \sim 1$.

Whereas multifrequency studies on galactic scales might still be prohibitive at high redshifts, the available data allows us to reconstruct the spectral energy distributions in the X-ray to centimeter radio regimes, such as those shown in Figure 1.6 for M51, for hundreds of thousands of galaxies up to high redshifts. Figure 1.6 clearly shows that the morphological differences between the late-type spiral NGC 5194 (Figure 1.6b) and the peculiar early-type NGC 5195 (Figure 1.6a), evident in

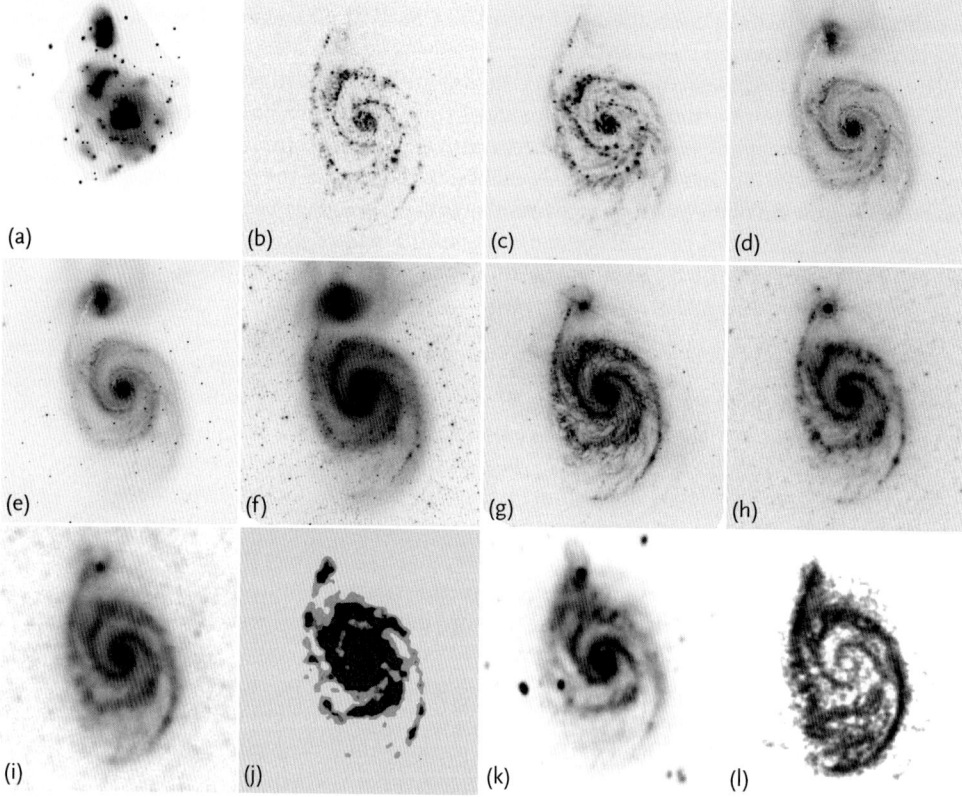

Figure 1.5 Multifrequency images of the interacting system M51 formed by the spiral galaxy NGC 5194 and the peculiar early-type NGC 5195: (a) X-ray image from Chandra ($\lambda \sim 10\,\text{Å}$), tracing the distribution of binary systems and supernova remnants [35]; (b) H_α+[NII] narrow band OHP image, ionizing radiation ($\lambda \leq 912\,\text{Å}$) produced by the youngest stellar populations [36]; (c) near-ultraviolet NUV ($\lambda = 2316\,\text{Å}$) GALEX image, distribution of the young stellar populations [37]; (d) optical g band ($\lambda = 4825\,\text{Å}$) SDSS image, young-intermediate age stellar populations; (e) optical i band ($\lambda = 7672\,\text{Å}$) SDSS image, intermediate–old age stellar populations; (f) near-infrared 3.6 μm Spitzer image, old stellar populations [38]; (g) mid-infrared 8.0 μm Spitzer image, Polycyclic Aromatic Hydrocarbons (PAHs; [38]); (h) infrared 24 μm Spitzer image, hot dust [38]; (i) far-infrared 250 μm Herschel image, cold dust; (j) 2.6 mm ^{12}CO(1–0) image, distribution of the molecular gas component (IRAM, [39]); (k) radio continuum 20 cm Westerbork image, relativistic electrons in week magnetic fields [40]; (l) 21 cm HI VLA image, distribution of the atomic hydrogen [41].

the multifrequency images shown in Figure 1.5, are also present in their spectral energy distributions. Indeed, NGC 5194 is characterized by a relatively intense UV emission due to the presence of recently formed stars and by an important mid-infrared, far-infrared, and submillimeter emission (dust heated by the young stellar population), while NGC 5195 is dominated by emission from an older stellar population, with an important contribution from the dust. These systematic differences

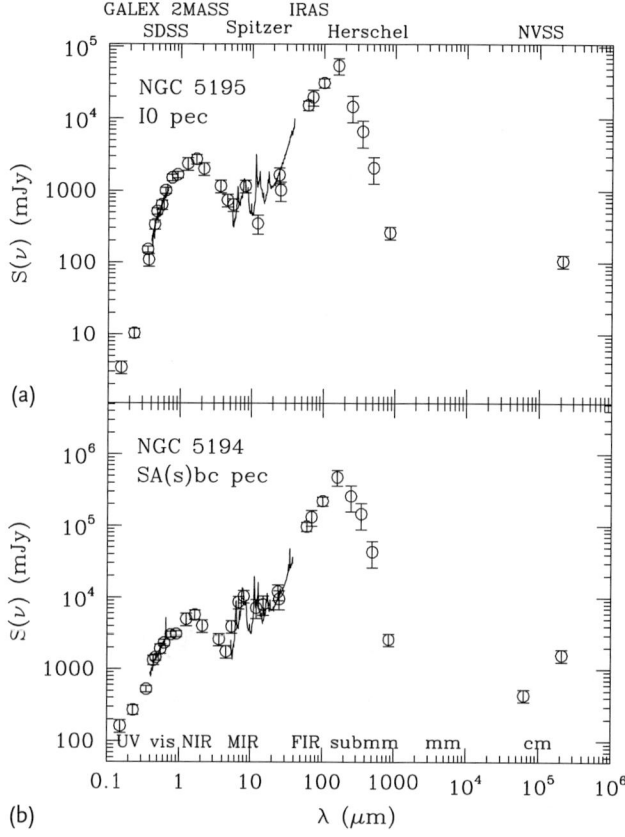

Figure 1.6 The UV to centimeter radio spectral energy distribution (SED) of the two galaxies NGC 5194 (M51) and NGC 5195. Data are taken from [42] or from NED.

are also evident both in the visible (Figure 1.7) and in the near-infrared (Figure 1.8) spectra when the two galaxies are observed at higher spectral resolution. The early-type NGC 5195 has a much redder optical spectrum than the late-type NGC 5194, indicating at the same time the presence of a dominating old stellar population and of an important dust attenuation. Its optical spectrum does not show strong emission lines, which are prominent in NGC 5194, due to the ionization of the gas in star-forming regions. The mid-infrared spectra obtained by Spitzer reveal the presence of PAHs emission in both galaxies (5 μm $< \lambda <$ 15 μm), with some atomic emission lines at longer wavelengths only present in the star-forming NGC 5194.

Spectral energy distributions and optical or mid-infrared spectra are useful tools since they can be reproduced by models and compared to observations in samples of objects at different redshift, thus providing us with one of the most popular and powerful tools for constraining galaxy evolution. The spectral resolution and wavelength coverage of both models and observations have significantly improved

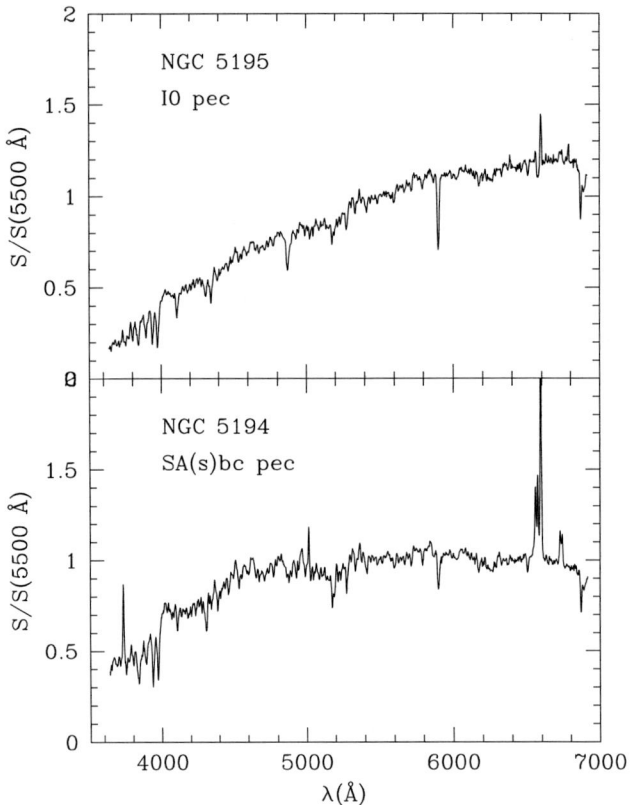

Figure 1.7 The 3600–6900 Å medium resolution ($R \sim 700$) optical spectrum of the two galaxies NGC 5194 (M51) and NGC 5195, from [43].

over the recent years, thus making it possible to remove some of the degeneracies affecting the reconstruction of the evolutionary history of the analyzed galaxies.

1.3
The Purpose of this Book

We are now entering a new era in the study of galaxy evolution: the surveys previously described are providing the community with hundreds of thousands of galaxies with imaging and spectroscopic data at various wavelengths. Astronomers had to slightly change the attitude of their approach to galaxies when dealing with these new huge datasets. In the past, the study of galaxies was generally carried out using heterogeneous samples of objects mainly selected according to their morphological type, separating ellipticals and lenticulars from spirals, irregulars and active galaxies. Today's extragalactic research is based on enormous datasets, where objects are selected independently of their morphological nature, but rather according to

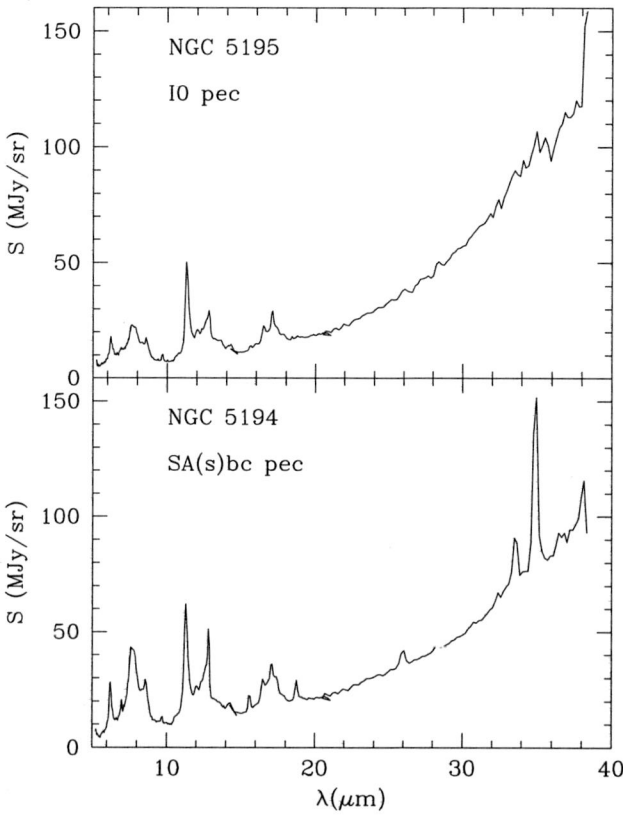

Figure 1.8 The Spitzer 5–40 μm low resolution mid-infrared spectrum of the two galaxies NGC 5194 (M51) and NGC 5195, from [44].

strict flux or luminosity criteria, in order to minimise selection biases. This new approach had several major consequences. First it contributed improving statistical tools which were developed on purpose for treating such large amounts of data. Among these, we can mention the determination of the galaxy counts, the luminosity or the mass function, and the density functions, which respectively give the number of galaxies per unit area, the number of galaxies per bin of luminosity or mass per unit volume, and the mean density of galaxies in the universe. Secondly, the spectral coverage combined with the angular and spectral resolution, the large field of view of the multifrequency surveys, and the depth of the data are allowing us to sample, at the same time, the whole parameter space necessary for a complete and coherent study of galaxy evolution. We can indeed see how the inter-relations between the different galaxy components on galactic scales change as a function of galaxy luminosity, redshift, and density. The early systematic studies of large, high-redshift samples were mainly limited to the determination of their statistical properties such as the luminosity function as well as to the reconstruction of the 3-D distribution of the detected objects. The availability of modern multifrequency datasets now allows us to undertake detailed studies of the physical processes at play in galaxies at different epochs, such as star formation and feedback, or determining the evolution of the scaling relations with time. Thus, works that were conceivable only for the nearby universe a few years ago can now be extended to high redshift. A third major change with respect to the past is that these deep, blind surveys naturally bring the discovery of unexpected objects. The most typical example are the Ultra Luminous Infra Red Galaxies (ULIRGs) discovered by IRAS [45]. These extragalactic objects are extremely luminous in the infrared and their origin is probably due to the merging of two gas-rich late-type galaxies [46].

The purpose of this book is to provide the community with a text that summarizes the information required to use the modern multifrequency datasets in a complete and coherent way, in order to gain insight on the structural and physical properties of galaxies selected according to various criteria, with the final aim of reconstructing their evolution. The book is thus structured as follows: in the first section I give a summary of the main radiative processes acting in galaxies, with an overview of the principal emitting sources dominating several classes of objects. The various emission processes are described for different wavelengths, starting from the X-ray and covering the UV, optical, near-infrared, the infrared and finally the millimeter and centimeter domain, separating the continuum emission from the line emission and absorption.

In Part Two, I show how the available multifrequency data can be used to derive physical quantities suitable for a complete study of the physical and structural properties of the selected galaxies. Given the practical purposes of this book, I list the relations most commonly used for determining those entities generally necessary in any study, such as the mass and the temperature of the gas and the dust, the mass of stellar components, and the dynamical mass of galaxies. At the same time, I report the most commonly used recipes necessary for correcting the data for dust extinction as well as those used to quantify the star formation activity in galaxies. I describe how the contribution of the different components can be separated in radio data and explain how photometric and spectroscopic data can be combined to reconstruct the spectral energy distribution of any extragalactic source. I also give the general recipes for classifying galaxies using different spectral features or for measuring gas metallicities.

In Part Three, I give some representative examples on how these derived quantities can be used to study galaxy evolution. I first summarize the statistical tools most frequently used (galaxy counts, luminosity functions). I then explain how the most important scaling relations can be determined. I also describe how the multifrequency data can be used to study the matter cycle in galaxies and to reconstruct the star formation history of the universe. This third section, whose purpose is that of presenting the most recent observational results, is certainly the one most lacking completeness given the huge amount of publications that are continuously made available on the web, which cannot all be mentioned due to lack of space. The book also includes some practical appendices useful for determining photometric redshifts and K-corrections and for homogenizing data obtained in different photometric systems.

I naturally focus my attention on the nearby universe, where angular and spectral resolution, combined with sensitivity, permit an accurate study of the different physical components (gas, dust, stars...) on galactic scales for almost all kinds of extragalactic sources. This choice is also justified by the fact that both the emitting sources and the emission processes and, as a consequence, all the relations used to quantify the physical entities described in Part Two, do not change with redshift. The relationships described in Part Three are standard and can be determined without distinction in samples of nearby and far galaxies, provided that appropriate K-corrections are applied to the data, as indeed explained in the text. The only difference resides in the fact that generally all these relationships are better determined in the nearby than in the far universe, not only because of the quality of the data (higher sensitivity, angular and spectral resolution, wavelength coverage), but also because the dwarf galaxy population can be investigated only locally. It is therefore only in the nearby universe that the whole parameter space of galaxy properties is fully sampled.

The novelty of this book resides in the fact that it describes galaxies as seen at different wavelengths and explains how the multifrequency data can be used in studying their properties. This approach is quite different from that proposed by classical text books, where galaxies are generally first divided according to their morphological type (ellipticals and lenticulars, spirals and irregulars, dwarfs, AGNs) and then described separately according to their properties (structure, kinematic, stellar populations, interstellar medium).

Given the extent of the subject of this book, a panchromatic view of galaxies, all the different topics are treated in a relatively basic way. The idea is to provide the reader with several practical instruments useful for a direct interpretation of the data. I want to convince the reader that the multifrequency analysis is a very powerful tool for studying the properties of any extragalactic source and for constraining models of galaxy formation and evolution. At the same time, I hope to stimulate new ideas in the use of multifrequency datasets in new and original studies of any extragalactic source. There are certainly more specific and complete books or review articles focused on single topics to which the reader should refer if he/she wants to develop some specific analyses. For this reason, I give extensive references to works that I consider important for a deeper investigation. Given the vastness of the topic and my limited knowledge, I certainly missed several important works and i apologize. The book might also suffer from another weakness related to my personal lack of experience in some of the covered fields. Despite this weakness, my intention is to give an overview of the tools that astronomers apply to multifrequency analyses of galaxies, as complete as possible. I hope that this text can be used by PhD students and senior scientists working on multifrequency data to easily find the information they need to deal with unfamiliar data.

Part One Emitting Sources and Radiative Processes in Galaxies

2
X-ray

The first study of the X-ray emission of galaxies has been made possible by the launch of the Einstein Observatory in 1978 [47]. Thanks to the data acquired by following space facilities such as ROSAT [48], ASCA [49], and in particular to the subarcsecond angular resolution of Chandra [50] and the spectral resolution of XMM [51], we now have a relatively clear vision of the X-ray properties of galaxies. Their X-ray emission is either due to compact sources of stellar origin such as X-ray binaries, to resolved sources such as supernovae remnants, or to the extended hot plasma with a relative weight which depends on galaxy type [52–54]. In active galaxies (AGNs, Seyferts) the X-ray emission is dominated by the accretion disk around the central supermassive black hole associated with the AGN [52].

The study of the X-ray binary population is critical for constraining the evolution of the stellar component of galaxies: while low-mass binary systems are associated with the evolved stellar population, massive binaries trace the star formation activity of galaxies. The observation of AGNs is fundamental for understanding the evolution of massive black holes and the feedback of nuclear activity on the parent galaxy, as well as the accretion process, while the study of the metal abundances via X-ray emission lines in the hot plasma and the associated outflows is strictly related to the chemical evolution of the universe [55].

2.1
Continuum

Spiral galaxies have X-ray luminosities ranging between $\sim 10^{38}$ and $\sim 10^{41}$ erg s^{-1} in the 0.2–3.5 keV spectral range [52]. Their emission is dominated by X-ray binaries in relatively massive objects and by supernovae remnants in star-forming dwarf systems, such as the Magellanic Clouds [52]. Spirals with a very active star formation rate have a gaseous component, whose emission is limited to the soft X-ray band, mostly below 1.5 keV, as observed in a few nearby objects [56]. Bulges can also contain a small amount of hot gas, while no emitting gas has been found in the halo of normal spirals [56].

In starburst galaxies, the X-ray emission is associated with ultra-luminous X-ray sources (intermediate mass black holes), generally located in star-forming regions,

A Panchromatic View of Galaxies, First Edition. Alessandro Boselli.
© 2012 WILEY-VCH Verlag GmbH & Co. KGaA. Published 2012 by WILEY-VCH Verlag GmbH & Co. KGaA.

and to the hot gaseous component which might be the dominant component in starburst galaxies such as M82 [57]. The X-ray emission of the hot gas might come from the central regions, the disk and the halo. Prominent nuclear outflows are frequent. X-ray sources, whose variability testify their compact nature, can have X-ray luminosities exceeding L_X (0.5–10 keV) $\geq 10^{41}$ erg s^{-1} [56].

The X-ray luminosity of ellipticals ranges from $10^{40} \leq L_X \leq 10^{43}$ erg s^{-1}. In these objects the emission is dominated by the halo of hot gas ($T \sim 10^7$ K) surrounding the galaxy. Most of this gas originates from the stellar mass loss of evolved stars and planetary nebulae [53]. In massive ellipticals inhabiting the core of rich clusters and groups, some of the gas can be accreted from the surrounding intergalactic medium through cooling flows. Here the distribution of hot gas is significantly different than that of stars, and the hot gas extends well outside the optical body and is thus difficult to separate from the continuum emission due to the hot intergalactic medium. Discrete sources, mainly low-mass X-ray binaries, become important at low X-ray luminosities. Models predict that the hot gas should be retained in the massive objects while partly or fully expelled by stellar winds in low-mass systems where the X-ray emission can be dominated by discrete sources [53, 56].

2.1.1
Discrete Sources

The high sensitivity and the subarcsecond angular resolution of Chandra made it possible to spatially and spectrally separate the emission of discrete sources similar to the Galactic X-ray binaries from the diffuse emission of the hot interstellar medium in galaxies up to distances on the order of ~ 20 Mpc. A detailed review on the populations of X-ray sources in galaxies can be found in [54]. To summarise, these sources can be divided into compact sources:

1. Low-Mass X-ray Binaries (LMXB) formed by neutron stars or black holes associated with stars of type later than A, whose X-ray emission is time-variable due to orbital periods, flares, and bursts. On average, they have soft spectra ($kT \sim 5$–10 keV). These sources are associated with the old stellar population and are found in elliptical galaxies, spiral bulges, and globular clusters. They are expected to have relatively long lifetimes ($\sim 10^8$–10^9 yr) and are thus tightly related to the old stellar population. Their typical luminosity is $L_X < 2 \times 10^{39}$ erg s^{-1}.
2. High-Mass X-ray Binaries (HMXB) formed by neutron stars or black holes associated with young OB stars, with time-variable luminosities and spectra. They have, on average, harder spectra than LMXBs (maybe except for perhaps black hole binaries, which might have soft spectra). These sources are generally associated with the young stellar populations and are thus present in spiral arms and are the dominant emitting source in star-forming galaxies in the luminosity range $L_X < 10^{39}$ erg s^{-1}. Their lifetime is relatively short ($\sim 10^6$–10^7 yr), thus their presence witnesses ongoing star formation activity.

3. Ultra-luminous X-ray sources (ULX), whose nature is still poorly known, with luminosities $L_X > 10^{39}$ erg s^{-1}, thus larger than the Eddington luminosity of neutron stars or black holes of mass $\sim 5\,M_\odot$. As rare sources, they are probably associated with intermediate mass black holes (IMBH of mass 10^2–$10^4\,M_\odot$; [58]) or extreme high-mass X-ray binaries and are generally found in active star-forming, low metallicity environments off-centered with respect to the nucleus of the galaxy [59]. This evidence suggests that, contrary to nuclear massive black holes, they have a stellar origin.
4. Super Soft Sources (SSS), characterized by a black body spectrum of temperature $kT \sim 15$–80 eV, probably associated with nuclear burning white dwarf binaries.
5. Quasi Soft Sources (QSS), with a black body spectrum of temperature $kT \sim 100$–300 eV, whose nature is still unknown.

Among the compact sources, ULX, SSS and QSS are quite rare in galaxies.
In binary systems and in massive black holes the X-ray emission is related to the transfer of matter form the companion star or from the accreting disk to the compact source. Here the emission is due to the infalling material which is spiralling around the compact object heating up via viscous friction between the different matter layers (those closer to the compact source spin more rapidly than the outer ones). The potential energy provided by the infalling material is transformed into heat and re-radiated as a black body.

A fraction of the X-ray emission of galaxies comes from resolved, extended sources:
6. Supernovae remnants (SNR), among the brightest X-ray emitting systems, with X-ray luminosities exceeding 10^{37} erg s^{-1}. They are the dominant emitting sources in dwarf star-forming systems such as the Magellanic Clouds. The X-ray emission of old supernovae remnants is due to thermal bremsstrahlung ([60, 61], see below) while it is dominated by the synchrotron emission of relativistic electrons accelerated by the shock in young systems [62, 63].

2.1.1.1 Emission Process in Binary Systems

The emission process of binary systems related to the accretion of infalling matter spinning around a compact object has been extensively studied by [64–66]. Following the description of [67], if δM is the mass accreted by a star of mass M and radius R, the gravitational energy acquired by the star is given by the relation

$$\delta E_{\mathrm{grav}} = \frac{GM\delta M}{R} \quad (2.1)$$

and the rate of gravitational energy released per unit time δt is

$$\dot{E}_{\mathrm{grav}} = \frac{\delta E_{\mathrm{grav}}}{\delta t} = \frac{GM\dot{M}}{R} \quad (2.2)$$

where

$$\dot{M} = \frac{\delta M}{\delta t} \qquad (2.3)$$

is the accretion rate. If the star keeps its thermal equilibrium, the acquired gravitational energy is radiated, producing an accretion luminosity

$$L_{acc} = \dot{E}_{grav} = \frac{GM\dot{M}}{R} \qquad (2.4)$$

Equation (2.4) shows that the luminosity accounted by the accreted material of a star of a given mass M and accretion rate $\dot{M}$ is more and more important in compact sources. The accretion rate is however limited by the fact that the resulting luminosity does not exceed the Eddington luminosity[1], otherwise the radiation pressure would push back the infalling material preventing its accretion. Thus $L_{acc} < L_{Edd}$, defined as

$$L_{Edd}[L_\odot] = \frac{4\pi c G M}{k} = 3.2 \times 10^4 \left(\frac{M}{M_\odot}\right)\left(\frac{k_{es}}{k}\right) \qquad (2.5)$$

where k is the opacity and k_{es} the electron-scattering opacity. This leads to

$$\dot{M} < \frac{4\pi c R}{k} < 10^{-3} \frac{R}{R_\odot} [M_\odot \ \text{yr}^{-1}], \qquad (2.6)$$

indicating that the upper limit of the accretion rate depends only on the stellar radius.

To keep its thermal equilibrium, the potential energy acquired by the accreting star is absorbed by the surface boundary layers, increasing their temperature, and then remitted as blackbody radiation. Since the size of compact objects is not modified by the accreting material, the temperature of the boundary layers rises up to

$$T = \left(\frac{L_{acc}}{4\pi R^2 \sigma}\right)^{1/4} = \left(\frac{GM\dot{M}}{4\pi R^3 \sigma}\right)^{1/4} \qquad (2.7)$$

where σ is the Stefan–Boltzmann constant, with a limiting temperature obtained by substituting $\dot{M}$ with the limiting accretion rate given in Eq. (2.6). For white dwarfs the temperature of the boundary layers is on the order of $T \sim 10^6$ K, for a neutron star it is $T \sim 1.5 \times 10^7$ K, and for a black hole it is $T > 3 \times 10^7$ K.

2.1.2
X-ray Emission in Active Galaxies

The X-ray emission of galaxies hosting an AGN is dominated by the circumnuclear region surrounding the central black hole. The viscous friction between the different layers of the infalling material located on the accreting disk heats up the matter

---

[1] The Eddington luminosity in a star is defined as the limiting luminosity where radiation pressure equals gravitational forces.

which re-emits as a black body at UV and X-ray wavelengths [68]. The accreted material can be provided either from the galaxy itself (gas expelled from ordinary stars via stellar winds and supernovae explosions) or from nearby objects in interacting systems. It can also come from the debris of stars tidally disrupted by the central black hole. In active galaxies matter is generally accreted along an accretion disk: as it loses angular momentum, the matter falls into the black hole, providing gravitational energy to the system. The accretion through a thin disk is the most simple representation of this phenomenon. However, if the internal pressure increases because the radiation pressure becomes competitive with gravity or because the material is unable to radiate the energy dissipated by viscous friction, the disk becomes thick. For a detailed description of the accretion phenomenon in black holes hosted by active galaxies and their relative emission processes we refer the reader to [68–70].

2.1.3
Hot Gas

The hot phase of the interstellar medium emits in X-rays for thermal bremsstrahlung (free–free) and free–bound emission. Recombination lines dominate for energies less than 1 keV, when the temperature is still sufficiently low for the capture of electrons in some elements such as Iron, while at higher temperatures (> 1 keV), bremsstrahlung dominates [71]. The X-ray emission of a hot plasma with characteristics similar to those of the interstellar medium of galaxies has been calculated by [72] and [73].

2.1.3.1 Thermal Bremsstrahlung (Free–Free)

The bremsstrahlung emission process is due to the energy loss of electrons in the electric field of ions, without capture. When the velocity distribution of the electrons is Maxwellian (electrons are in thermal equilibrium), as in the case of the hot plasma in the interstellar medium (ISM) of galaxies or of the intergalactic medium in clusters of galaxies, the bremsstrahlung volume emissivity $\epsilon_{\mathrm{ff}}(\nu)$ (defined as the total power radiated per unit volume, per unit solid angle, per unit frequency interval ν to $\nu + d\nu$) from a plasma of temperature T is given by the relation [74]

$$\epsilon_{\mathrm{ff}}(\nu)d\nu = \frac{8}{3}\left(\frac{2\pi}{3}\right)^{1/2}\frac{n_\nu Z^2 e^6}{m^2 c^3}\left(\frac{m}{kT}\right)^{1/2} N_i N_e g(\nu, T) e^{-\frac{h\nu}{kT}} d\nu \qquad (2.8)$$

$$\approx 5.4 \times 10^{-39} n_\nu Z^2 \frac{N_i N_e}{T^{1/2}} g(\nu, T) e^{-\frac{h\nu}{kT}} d\nu \quad [\mathrm{erg\,s^{-1}\,cm^{-3}\,Hz^{-1}\,rad^{-2}}] \quad (2.9)$$

where n_ν is the index of refraction of the plasma (whose value is close to 1), e and m are the charge and the mass of the electron respectively, N_i and N_e are the ions and electron density, and Z is the charge of the ion (in practice $Z \sim 1$ and $N_i \sim N_e$). The Gaunt factor $g(\nu, T)$ corrects for the quantum mechanical effects and for the effects of distant collisions, and is a slowly varying function of frequency and temperature (references for appropriate Gaunt factors in various domains can be found in [74]).

The Gaunt factor is given by the relation [74]

$$g(\nu, T) = \frac{\sqrt{3}}{\pi} \ln \Lambda \approx 0.54 \ln \Lambda \tag{2.10}$$

In the X domain,

$$\Lambda \approx 4.7 \times 10^{10} \left(\frac{T}{\nu}\right) \tag{2.11}$$

In the case of the interstellar medium where the gas is optically thin, the resulting volume emissivity $\epsilon_{ff}(\nu)$ per unit frequency is proportional to [75, 76]

$$\epsilon_{ff}(\nu) \propto N_e^2 T^{-1/2} \exp\left(-\frac{h\nu}{kT}\right) \tag{2.12}$$

while the emissivity ϵ_{ff} per unit volume is

$$\epsilon_{ff} \propto N_e^2 T^{1/2} \tag{2.13}$$

Thus the gas temperature can be determined by fitting the X-ray spectrum. The total bremsstrahlung luminosity, L_{ff}, of a thermal plasma is then given by the relation [74]

$$L_{ff} = 4\pi V \int \epsilon_{ff}(\nu) d\nu = \frac{32\pi}{3} \left(\frac{2\pi}{3}\right)^{1/2} \frac{e^6}{mc^3 h} \left(\frac{kT}{m}\right)^{1/2} N_i N_e V Z^2 g(\nu, T) \tag{2.14}$$

$$\approx 1.43 \times 10^{-27} T^{1/2} N_i N_e V Z^2 g(\nu, T) \quad [\text{erg s}^{-1}] \tag{2.15}$$

2.1.3.2 Recombination Emission (Free–Bound)

The free–bound emission due to the recombination of an electron with a ion is dominant whenever the temperature of the gas is not too high to totally ionize even the heavy elements of the interstellar medium (≤ 1 keV, [71]). A lower limit of the recombination luminosity $L_{fb}^{\lim}$ can be determined as indicated in [74]:

$$L_{fb}^{\lim} \approx 10^{-21} N_e N_i T^{-1/2} V Z^4 \quad [\text{erg s}^{-1}] \tag{2.16}$$

Comparing Eqs. (2.14) and (2.16) indicates that the total luminosity due to free–bound radiation dominates with respect to the total luminosity due to free–free for temperatures less than 10^6 K, as depicted in Figure 2.1. Indeed, for temperatures $kT \leq 1$ keV, the 0.1–5 keV spectra of a gas with characteristics similar to those of the typical interstellar medium of galaxies is dominated by line emission, while the continuum is dominant only at higher temperatures (5 keV).

▶ **Figure 2.1** The model X-ray spectral energy distribution in the range 0.1–5 keV for a source characterized by an interstellar medium of temperature (a) $kT = 0.1$ keV; (b) 1 keV, and (c) 5 keV. Recombination lines dominate the emission for temperatures $kT \leq 1$ keV, while the continuum emission (bremsstrahlung) dominates for higher temperatures (courtesy of G. Trinchieri).

(a) model with lines, kT=0.5 keV, abun 100% solar

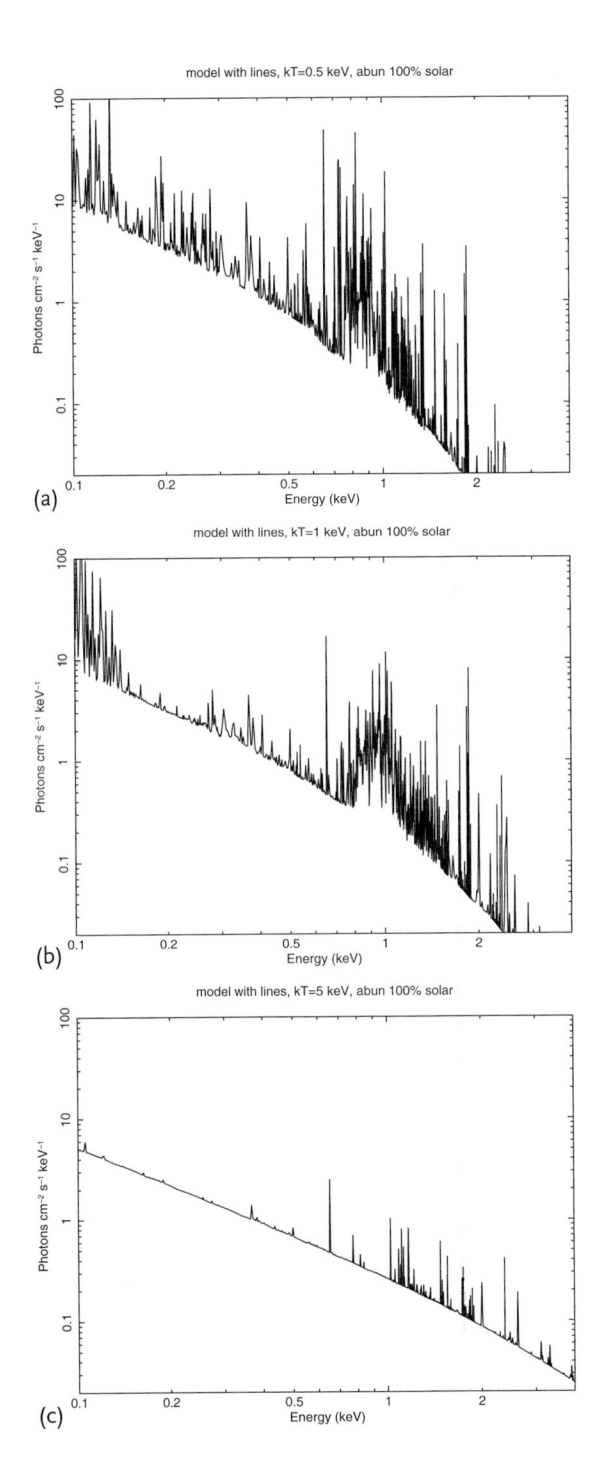

(b) model with lines, kT=1 keV, abun 100% solar

(c) model with lines, kT=5 keV, abun 100% solar

3
UV-Optical-NIR

The emission of galaxies from the far-ultraviolet (≥ 912 Å) to the near-infrared (≤ 3 μm) spectral domain is due to the emission of the different underlying stellar populations whose relative weight changes with galaxy type. At these wavelengths the emission is modulated by the interstellar medium, which can absorb and scatter the stellar radiation. While at long wavelengths ($\lambda \gtrsim 3$ μm) the contribution of dust becomes important, shortward of 912 Å the UV radiation (Lyman continuum) is almost completely absorbed by the interstellar medium of the emitting source. In star-forming systems, young stars (age $\leq 10^8$ yr) dominate the ultraviolet domain, those of intermediate age ($\sim 10^9$ yr) dominate the visible range while the oldest one ($\geq 10^{10}$) the near-infrared. However intermediate age stellar populations from the thermal pulsing asymptotic giant branch, TP-AGB, might contribute to the near-IR emission [77]. The color of these star-forming galaxies, defined as the ratio of the fluxes observed in two filter bands (see Chapter 9), is generally blue and their spectrum is characterized by several emission lines.

Elliptical and lenticular galaxies have red colors and several strong absorption features in their spectra. The UV emission of early-type galaxies might be dominated by an old stellar component (the UV upturn, [78]). In the wavelength range 3 μm $\leq \lambda \leq 10$ μm, the contribution of the Rayleigh–Jeans tail of the cold stellar component is important, in particular in dust-poor, quiescent ellipticals.

Besides the stellar component, other sources can contribute to the emission at these frequencies. These include the thermal emission of the accretion disk in AGNs and the jets associated with powerful radio sources, as in the case of M87, whose emission is due to synchrotron radiation.

The study of stellar populations is fundamental for reconstructing the star formation history of galaxies. Star formation is indeed the process that transformed most of the baryonic mass issued from the primordial nucleosynthesis into stars, giving birth to galaxies. At the same time, this process is particularly important since, within stars, the primordial gas has been processed to form the heavy elements, generally called metals, now present in the universe.

3.1
Continuum: Stellar Emission

Stars are bodies bound by self-gravity which radiate energy supplied by an internal engine [67]. They can be represented by a black body with effective temperature T_{eff}. This approximates the temperature of the star's outer layer, the photosphere, where the bulk of the emitted radiation originates. If L_{star} is the luminosity of a star, its effective temperature is given by the relation

$$T_{eff} = \left(\frac{L_{star}}{4\pi R^2 \sigma}\right)^{1/4} \tag{3.1}$$

where R is the radius of the star and σ is the Stefan–Boltzmann constant. The peak emission from a star of effective temperature T_{eff} is given by the Wien's law:

$$\lambda_{max} T_{eff} = 2.9 \times 10^7 \quad [\text{Å K}] \tag{3.2}$$

Thus the color of a star mainly depends on its effective temperature. The surface gravity g of a star is a parameter which fixes the pressure gradient in the atmosphere, determining the densities at which spectral lines are formed

$$g = \frac{GM}{R^2} \tag{3.3}$$

where G is the gravitational constant. The third of parameter characterising stellar atmospheres is their chemical composition (metallicity).

Stars have been classified according to their spectral type. The main sequence stars, those burning hydrogen to helium in their cores, have been coded as O, B, A, F, G, K, M and L (and have subclasses ranging from 0 to 9) according to the presence and the intensity of different absorption line features. As defined, these codes follow a sequence in mass and temperature where O represents the most massive and hottest objects (with stellar masses and temperature up to $100 M_\odot$ and 45 000 K) while L the coldest (< 3000 K) and least massive ($\sim 0.1 M_\odot$) objects. The red giants, supergiants and white dwarfs are stars that already left the main sequence (see Figure 3.1).

Luminosity and effective temperature are the two most fundamental stellar parameters that can be easily measured observationally. The Hertzsprung–Russell (H–R) diagram shows the presence of a well defined relationship between the effective temperature and luminosity of the main sequence stars. Within this sequence, the luminosity increases with effective temperature. Outside the main sequence, red giants are luminous stars with low effective temperatures, and white dwarfs are faint objects with blue colors.

In the main sequence, the luminosity of a star is related to its mass as [67]

$$L_{MS} [L_\odot] \propto (M [M_\odot])^\nu \tag{3.4}$$

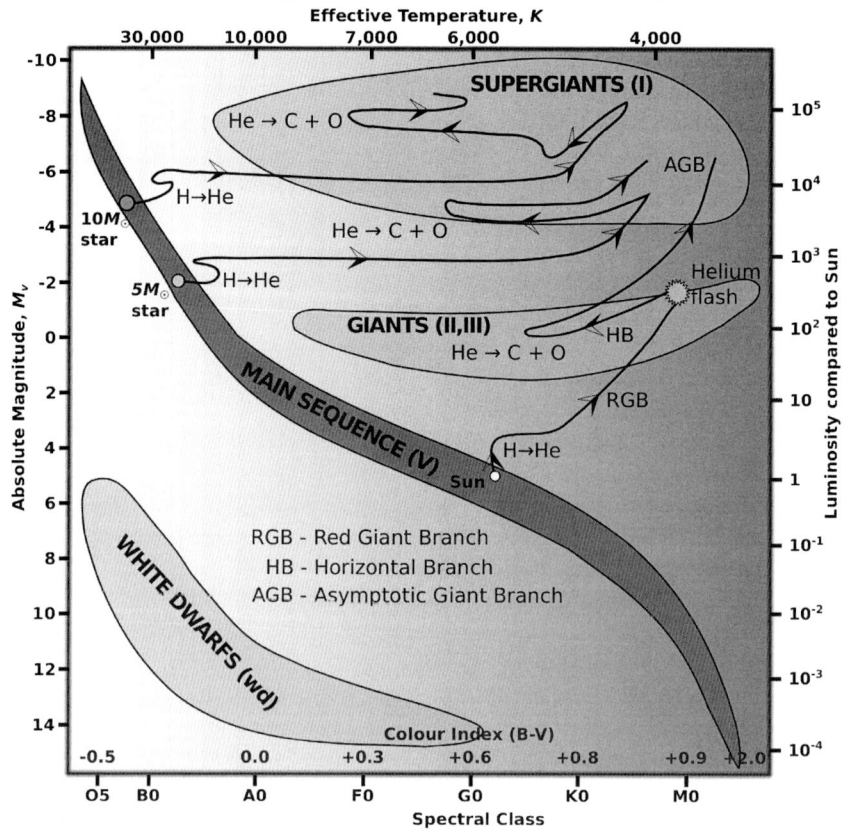

Figure 3.1 The Hertzsprung–Russell (H–R) diagram.

with ν ranging between 2 and 5. The models of [79] give the following results:

$$L_{MS}\,[L_\odot] = 81(M\,[M_\odot])^{2.14} \quad M > 20\,[M_\odot] \tag{3.5}$$

$$L_{MS}\,[L_\odot] = 1.78(M\,[M_\odot])^{3.5} \quad 2\,[M_\odot] < M \leq 20\,[M_\odot] \tag{3.6}$$

$$L_{MS}\,[L_\odot] = 0.75(M\,[M_\odot])^{4.8} \quad M \leq 2\,[M_\odot] \tag{3.7}$$

Massive stars are thus much more efficient in producing light than low-mass objects. For the same reason, their mean lifetime on the main sequence, τ_{MS}, which is proportional to the mass of available gas divided by the radiated energy, is much shorter than in small objects:

$$\tau_{MS} \propto \frac{M}{L} \propto M^{1-\nu} \tag{3.8}$$

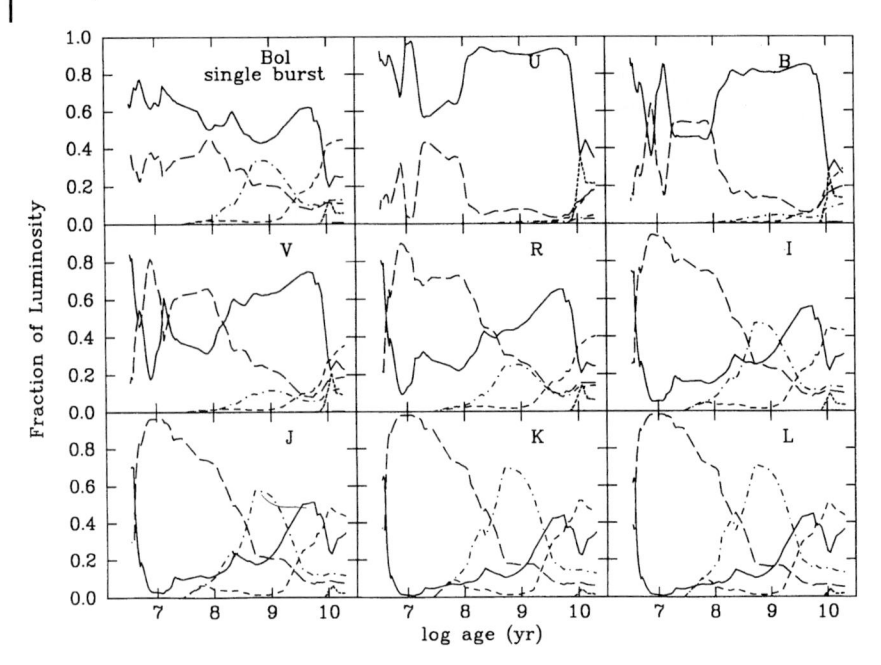

Figure 3.2 The contribution of the different stellar populations as a function of time to the emission of a synthetic galaxy formed after a single burst of star formation, determined using the stellar population synthesis models of [81]. The solid line indicates main sequence stars, the short-dashed line the sub-giant branch (SGB), the dashed line indicates the red giant branch (RGB), the long-dashed line the core-He burning (CHeB), the dot and short-dashed line the asymptotic giant branch (AGB), and the dot and long-dashed line the post-asymptotic giant branch (PAGB) stars. Reproduced by permission of the AAS.

The huge progress in the study of physical processes acting in stars has allowed astronomers to build stellar evolution models that accurately reproduce synthetic stellar spectra which span the entire range of parameter space (effective temperature, gravity, and metallicity) [80]. Population synthesis models combine libraries of different observed or synthetic stellar spectra with the evolutionary tracks for different initial mass functions (see Section 13.1) and star formation histories to reproduce the global UV to near-IR emission of galaxies. Figures 3.2–3.4 show the contribution, in various photometric bands, of the different stellar populations to the total emission of a synthetic galaxy, formed after a single burst of star formation or undergoing constant activity as determined from the population synthesis models of [77, 81].

3.2
Emission Lines

The UV-optical-near-infrared spectrum of galaxies is characterized by a number of important emission and absorption features, whose presence and intensity are

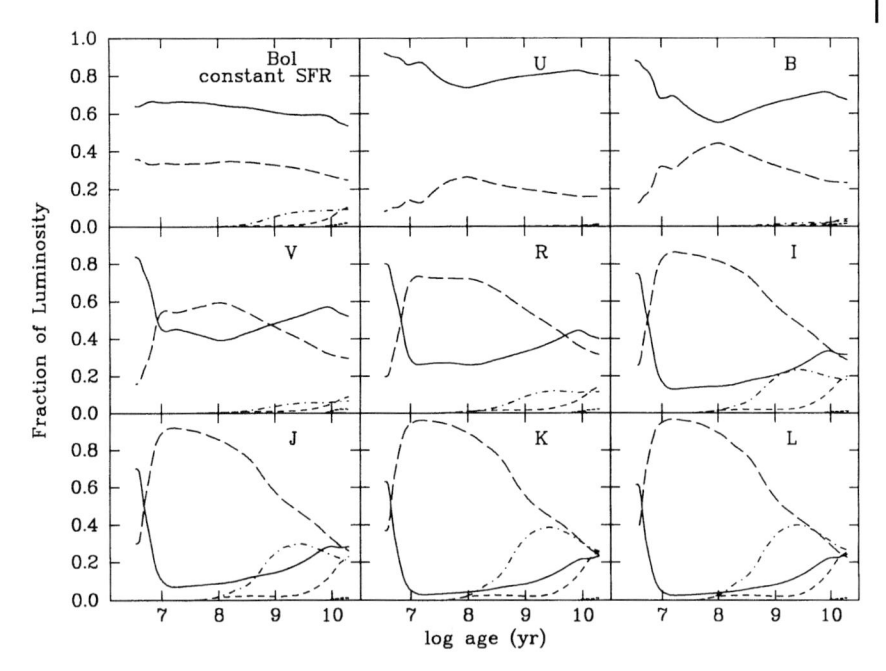

Figure 3.3 The contribution of the different stellar populations as a function of time to the emission of a synthetic galaxy formed through constant star formation activity determined using the stellar population synthesis models of [81]. The solid line indicates main sequence stars, the short-dashed line the sub giant branch (SGB), the dashed line the red giant branch (RGB), the long-dashed line the core-He burning (CHeB), the dot and short-dashed line the asymptotic giant branch (AGB) and the dot and long-dashed line the post-asymptotic giant branch (PAGB) stars. Reproduced by permission of the AAS.

strongly related to the nature of the underlying stellar population and to the physical conditions of the interstellar medium. For an accurate description of the physical processes responsible for the different line emissions, we refer the reader to the books of [82, 83]. To summarize, the line emission of galaxies is due to the emission of diffuse nebulae or HII regions, to planetary nebulae, to supernovae remnants, and to active galactic nuclei. These emission lines are present in star-forming, gas dominated systems (see Figure 3.5) but are rare in quiescent ellipticals and lenticulars (see Figure 3.10).

The overall line emission of normal galaxies is dominated by HII regions. In these regions the gas, which generally has densities on the order of $10-10^4$ cm^{-3}, is photoionized by the UV radiation emitted by hot (effective temperatures 30 000 K $\leq T_e \leq$ 45 000 K) and massive O and early B type stars recently formed within the nebulae. The spectra of these nebulae are characterized by strong HI hydrogen recombination lines and [NII], [OII], [OIII], [NeIII], [SII], [ArIII] collisionally excited lines, which arise from energy levels within a few volts of the ground state level and can therefore be excited by collisions with thermal electrons. In the optical region, all these lines are forbidden because in the abundant ions all the excited levels within a few volts of the ground level arise from the same electron configuration

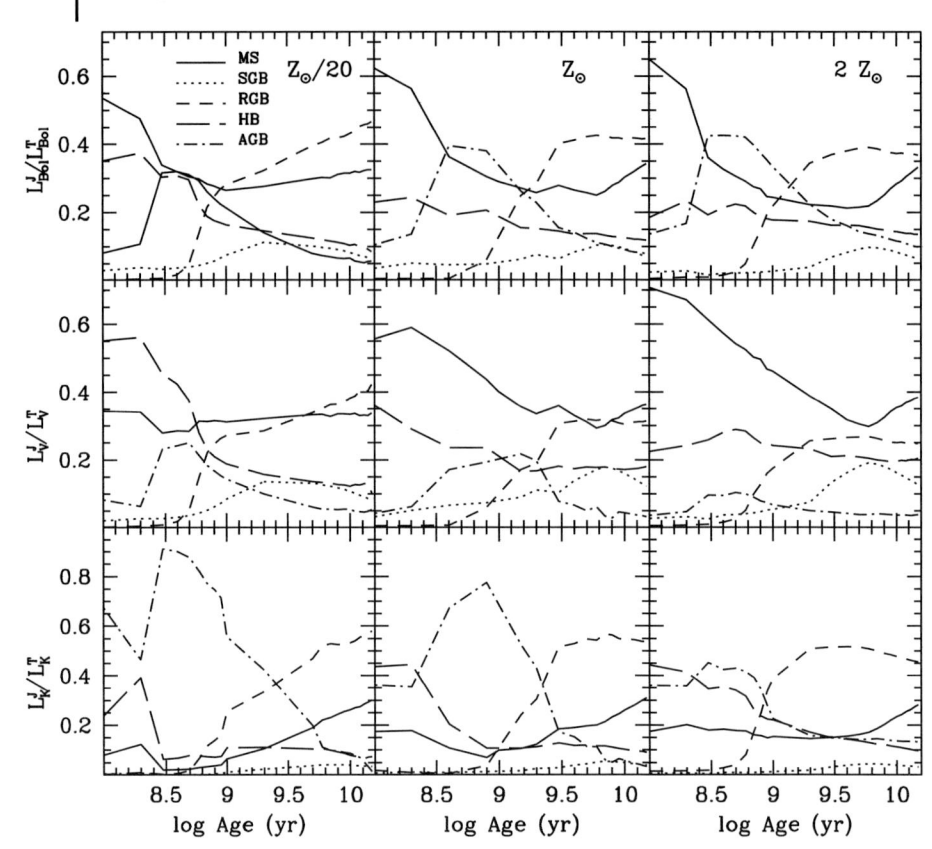

Figure 3.4 The contribution of the different stellar populations as a function of time to the emission of a synthetic galaxy formed after a single burst of star formation determined using the stellar population synthesis models of [77] for different metallicities. The solid line indicates main sequence stars (MS), the dotted line the sub giant branch (SGB), the short dashed line red giant branch (RGB), the long-dashed line the horizontal branch (HB), and the dot-dashed line the asymptotic giant branch (AGB). Reproduced with the permission of John Wiley & Sons Ltd.

as the ground level itself, and thus radiative transitions are forbidden by the parity selection rule [82].

Line emission can also be generated within planetary nebulae, isolated symmetric objects formed by the gas expelled by an old star previous to its transformation into a white dwarf. In these objects, the gas is photoionized at a rate higher than in HII regions by the hot star whose effective temperature is on the order of 45 000 K, producing HI, HeI and HeII recombination lines, with intense collisionally excited [OIII] and [NeIII] lines. Due to the relatively high velocity of the expanding shells (~ 25 km s^{-1}), planetary nebulae have a typical lifetime on the order of 10^4 yr.

In supernova remnants, the excitation of the gas is due to the transfer of kinetic energy from the expanding remnant (shock excitation) to the surrounding gas (recombination and collisionally-excited emission).

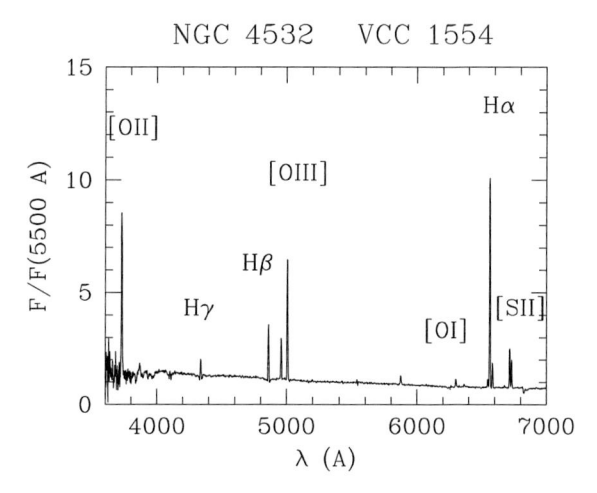

Figure 3.5 The integrated optical spectrum of the star-forming late-type galaxy NGC 4532 (adapted from [84]). The strong hydrogen emission lines are due to the presence of young O and early B stars which ionize the gas within HII regions, while the oxygen and sulfur lines are due to metals produced and injected into the interstellar medium by massive stars.

Some galaxies host an Active Galactic Nucleus (AGN) whose emission is characterized by several strong emission lines. The observed range in velocity of these nuclei (≥ 100 km s^{-1}) is generally significantly wider than that observed in normal HII regions (10 km s^{-1}). Here the emitted gas is photoionized by the continuum emission due to the accretion disk around the black hole or by relativistic particles. Broad emission lines are generally permitted recombination lines, while nebular forbidden emission lines are observed in the regions surrounding the accretion disk, where the density is low enough to prevent collisional de-exitation. Figure 3.6 shows the high resolution (~ 1 Å) composite spectrum in the range $800 \text{ Å} \leq \lambda \leq 8500 \text{ Å}$ of a quasar (Quasi Stellar Object, an AGN with a compact, quasi stellar morphology) as determined by combining the spectra of more than 2200 quasars observed by the SDSS [85]. Seyfert galaxies, where the emission of the central AGN is contaminated by that of the surrounding stars inhabiting the galaxy, have spectra with intermediate properties between those of quasars and normal galaxies. They are generally divided into Seyfert 1, with broad permitted lines and narrow forbidden lines, and Seyfert 2, with narrow permitted and forbidden lines of similar width. Seyfert 1 and 2 galaxies are believed to be similar objects seen with different viewing angles. Low-ionization nuclear emission-line galaxies, LINERs, have spectral characteristics in between those of Seyfert 2 galaxies and HII regions, with [OI]λ6300 and [SII]$\lambda\lambda$6716, 6731 emission lines that are stronger than those of starbursts and HII regions since they are not photoionized by massive OB stars [82].

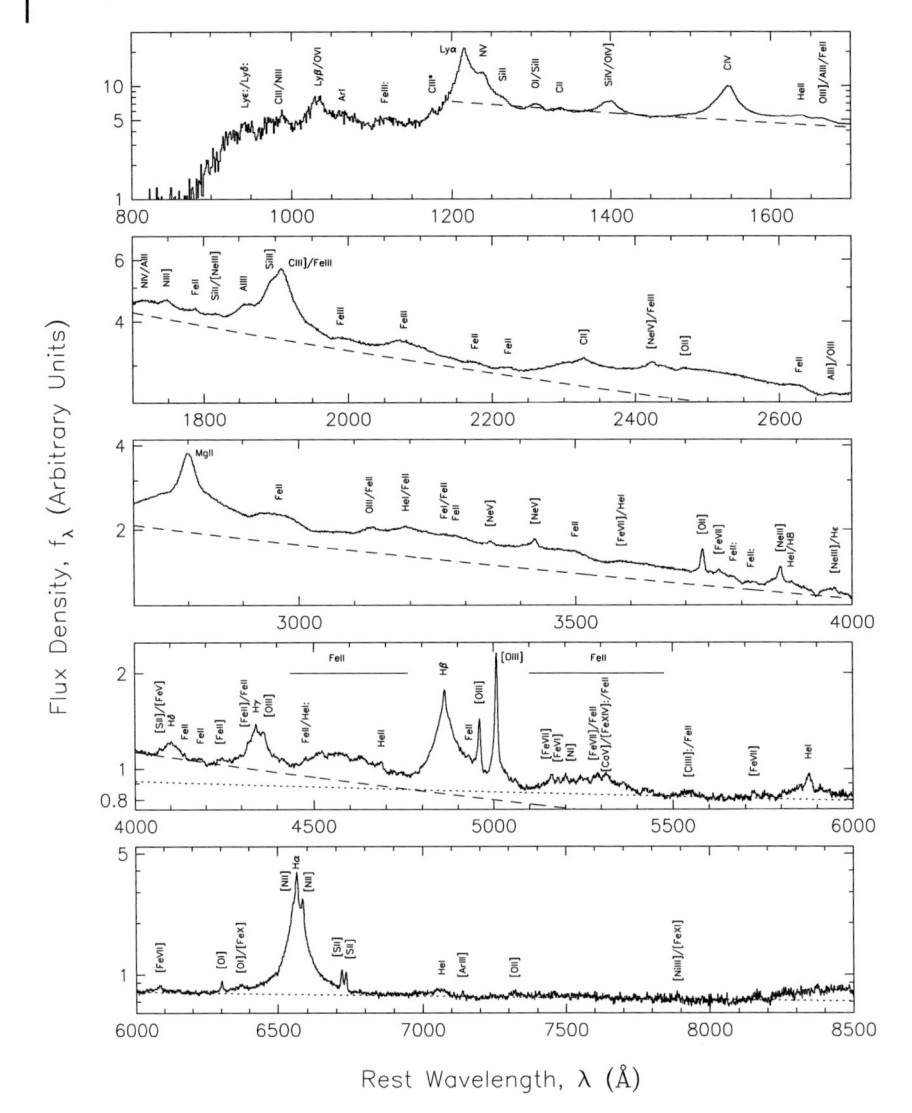

Figure 3.6 The high resolution rest frame composite quasar spectrum obtained by combining the spectra of more than 2200 quasars in the SDSS, from [85]. The spectrum of quasars is dominated by broad emission lines emitted by the gas photoionized by the continuum emission due to the accretion disk around the supermassive black hole. Reproduced by permission of the AAS.

3.2.1
Hydrogen Lines

Hydrogen recombination lines are easily observed in the spectra of galaxies. UV photons of wavelength shortward than 912 Å have sufficient energy (13.6 eV) to ful-

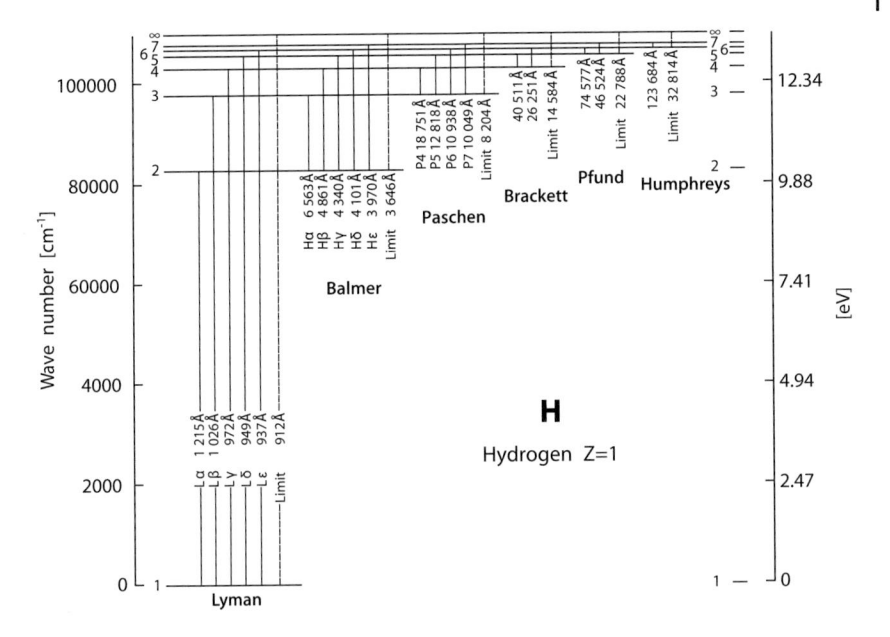

Figure 3.7 The Grotrian diagram for hydrogen.

ly ionize the diffuse atomic hydrogen (HI) of the interstellar medium within a given volume called the Strömgren sphere, thus creating an HII region.[1] Recombination due to a proton capturing an electron, causing a cascade to the ground level, produces a series of lines in the ultraviolet (Lyman series), optical (Balmer), and near-infrared (Paschen, Brackett, Pfund, Humphreys) spectral domains (see Figure 3.7). The interstellar medium is generally optically thick to the Lyman resonance lines of HI. Thus Lyman line photons are converted to a lower-series photon and a lower member of the Lyman series, producing the aforementioned cascade [82]. Whenever these conditions are satisfied, that is, all the Lyman line photons are absorbed by the diffuse gas located around the emitting star (Case B conditions), which is generally the case in a typical interstellar medium of density $n_e = 10^4 \text{ cm}^{-3}$ and $T_e = 10^4$ K, then the expected ratios of the different hydrogen recombination lines are as indicated in Table 3.1 (from [86]).

The intensity of the recombination lines is proportional to the ionization rate, which in turns depends on the luminosity of the exciting source. The relative intensity of hydrogen recombination lines depends only weakly on the temperature and only marginally on density for densities such as those measured in the interstellar medium. To measure the dust extinction of an emission line, the observed ratio of different hydrogen recombination lines can be compared to the expected value (see Chapter 12). Their expected ratio is scaled in a medium of density $n_e = 10^4 \text{ cm}^{-3}$ and $T_e = 10^4$ K when the emitted photons are absorbed by the diffuse gas located around the star (case B) as indicated in Table 3.1 (from [86]).

1) HII indicates fully ionized atomic hydrogen.

Table 3.1 Hydrogen recombination line, wavelength (in μm) and intensity relative to H_β for a density $n_e = 10^4$ cm^{-3} and $T_e = 10^4$ K (case B), from [86].

Line	Balmer λ μm	$I(\lambda)/I(H_\beta)$	Line	Brackett λ μm	$I(\lambda)/I(H_\beta)$
H_α	0.6563	2.85	Br_α	4.05	7.77×10^{-2}
H_β	0.4861	1	Br_β	2.63	4.47×10^{-2}
H_γ	0.4340	0.469	Br_γ	2.17	2.75×10^{-2}
H_δ	0.4101	0.260	Br_δ	1.94	1.81×10^{-2}
H_ϵ	0.3970	0.159	Br_ϵ	1.82	1.26×10^{-2}

Line	Paschen λ μm	$I(\lambda)/I(H_\beta)$	Line	Pfund λ μm	$I(\lambda)/I(H_\beta)$
P_α	1.875	0.332	Pf_α	7.46	2.45×10^{-2}
P_β	1.282	0.162	Pf_β	4.65	1.58×10^{-2}
P_γ	1.094	9.01×10^{-2}	Pf_γ	3.74	1.04×10^{-2}
P_δ	1.005	5.53×10^{-2}	Pf_δ	3.30	7.25×10^{-3}
P_ϵ	0.9546	3.65×10^{-2}	Pf_ϵ	3.04	5.24×10^{-3}

By making some assumptions on the initial mass function (IMF, see Section 13.1) and on dust extinction, any hydrogen recombination line can be combined with population synthesis models to quantify the number of ionizing sources – that is, to measure the star formation activity of galaxies (see Chapter 13).

3.2.2
Metals

The intensity I_l of an emission line l is given by the relation [82]

$$I_l = \int n_i n_e \epsilon_l(T) ds \tag{3.9}$$

where n_i and n_e are the density of the ions responsible for the emission and the electron density respectively, ds is the line of sight through the emitting nebula, and $\epsilon_l(T)$ is the emission coefficient. In the case of recombination lines, the recombination emission coefficients are not particularly sensitive to the temperature, and thus their intensity depends mostly on the abundances of the emitting ions. For this reason, the emission of recombination lines due to elements heavier than He are particularly difficult to observe.

However, many of the forbidden lines that are associated with the emission of heavy elements – most frequently O, N, and S – can be easily seen in the UV, optical, and near-IR spectra of galaxies. The excess energy of the ionizing photons, $h\nu - 13.6$ eV, is transferred as kinetic energy to the ionized gas which cools down mainly through the emission of collisionally excited lines of different ions such as O^{2+}, O^+ and N^+. Within the ionized gas, energy equilibrium is reached with temperatures on the order of $T_e \sim 10^4$ K, with T slightly changing with chemical composition (the electron temperature T decreases with increasing abundance because of the presence of more coolants) and stellar temperature. The energy excess $h\nu$, however, is generally relatively low, on the order of a few eV, thus the only excited states of ions that will be significantly populated are those with energies close to the relevant ground state energy. Transitions to their ground states are forbidden, but given the low density of the interstellar medium (the electron density is sufficiently low to prevent collisional de-excitation), the collisionally excited ions have the time to decay radiatively emitting a photon before being excited by another collision. Forbidden lines are thus very intense in the low-density interstellar medium of galaxies [75, 82, 87].

The electron density of the ionized gas can be measured with the intensity ratio of two lines from the same ion, emitted by different levels with similar excitation energy kT. If the two levels have different collisional de-excitation rates, the ratio of the intensity of the two lines depends only on the density of the gas. Electron densities are generally measured using the [OII]$\lambda 3726/\lambda 3729$ and [SII]$\lambda 6716/\lambda 6731$ doublets, as shown in Figure 3.8 [82]. Among these two lines, the [SII] doublet is often preferred since it needs a lower spectral resolution to be resolved and since it suffers from less extinction. Other forbidden line ratios vs. electron density diagrams for UV and infrared lines can be found in [82].

Electron temperatures inside a nebula can be determined from the measurement of the intensity ratio of pairs of emission lines relative to the same ion but with significantly different excitation energies. Among these, the one generally used is [OIII]$\lambda 4959 + \lambda 5007$/[OIII]$\lambda 4363$, measurable only in metal-poor extragalactic sources ([OIII]$\lambda 4363$ is undetectable in metal-rich galaxies). Figure 3.9 shows the expected variation of different line ratios as a function of the electron temperature as calculated for a density of $n_e = 1 \text{ cm}^{-3}$ (from [82]).

Alternative techniques for determining the density and temperature of the gas within HII regions using UV, infrared, optical, and radio data can be found in [82].

3.3
Absorption Lines

Stellar spectra are characterized by several absorbtion features whose intensity and shape depends on the physical conditions (temperature, density, velocity, magnetic field, and element abundance) of the stellar atmosphere. These parameters change according to the spectral type of the emitting star. For a simple description of the classification of stellar spectra the reader should see [88].

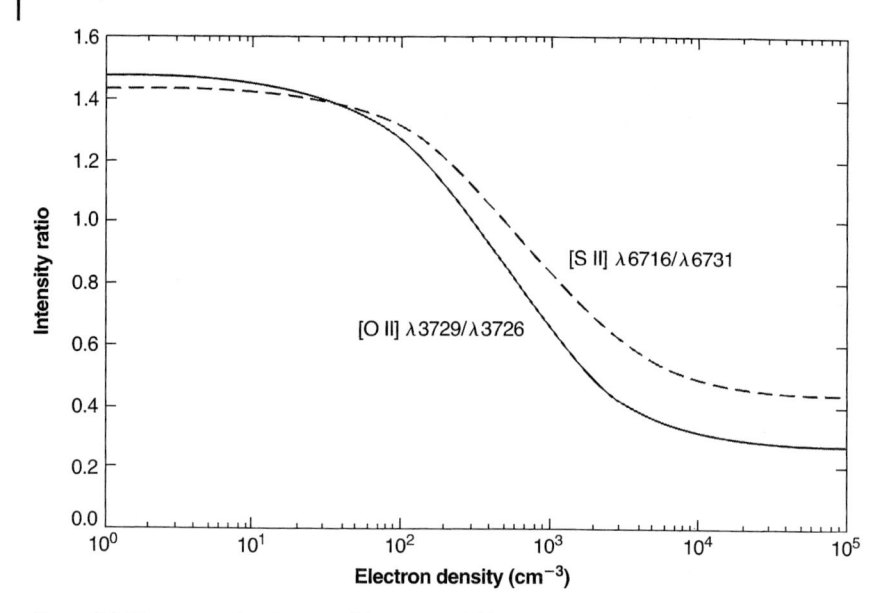

Figure 3.8 The expected variations of the [OII] (solid line) and [SII] (dashed line) intensity ratio as a function of the electron density n_e calculated for $T_e = 10\,000$ K, from [82].

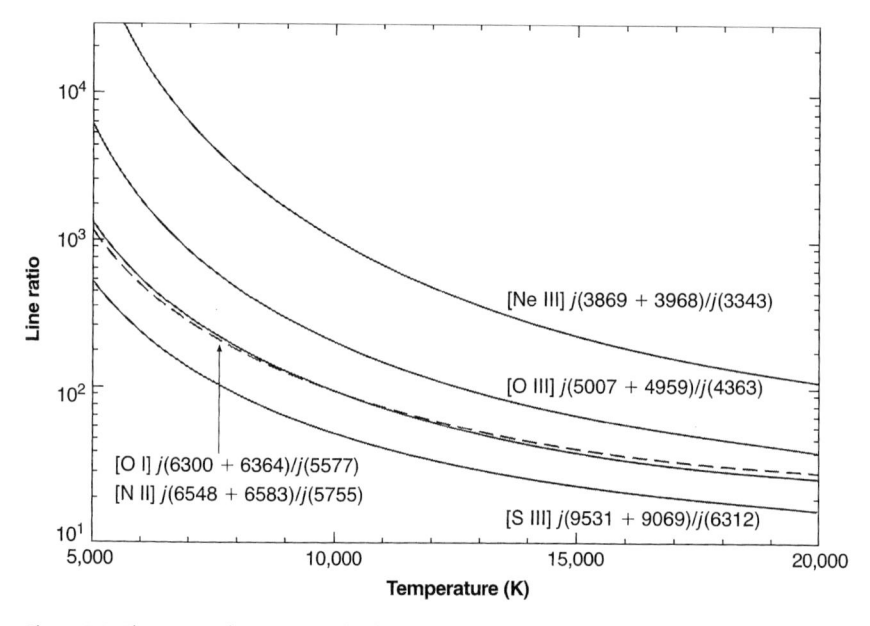

Figure 3.9 The expected variations of different line intensity ratios as a function of the electron temperature calculated for an electron density of $n_e = 1\,\text{cm}^{-3}$, from [82].

In external galaxies, stars are resolved only within the very local universe. Spectra of galaxies are thus generally used to study their underlying stellar population. This can be done by simultaneously measuring several absorption line features in the spectra: they can be used to date the age of the stellar population, to estimate its average metallicity and individual chemical element abundances and abundance ratios, or, when the spectral resolution is sufficiently high, to measure dynamical masses [89]. Absorption lines also have the advantage of being relatively insensitive to dust extinction.

As for stellar spectra, the properties of the underlying stellar populations of galaxies can be derived by fitting their observed spectra (continuum, absorption, and emission lines) with synthetic spectra produced by population synthesis models.

3.3.1
Hydrogen Lines

3.3.1.1 Balmer Absorption Lines

Balmer absorption lines are easily observable in the continuum spectra of galaxies, particularly in gas-poor systems such as ellipticals where no emission is present. When an electron in the first excited state absorbs a sufficiently energetic photon ($\lambda < 3646$ Å) that was produced in a stellar photosphere, the electron jumps to a higher energy state, producing a Balmer absorption line. The absorption of these energetic photons in the stellar atmosphere produces a sharp cut-off in the continuum spectrum below 3600 Å (Balmer limit). In O and B stars, Balmer absorption lines are weak since their dominating UV radiation is able to ionize the surrounding hydrogen whenever the emitted photons have $\lambda < 912$ Å. Balmer absorption lines are at their maximum in stars with effective temperatures $8000 \text{ K} \leq T_{\text{eff}} \leq 12\,000 \text{ K}$ (A stars), and decrease for cooler spectral types. The presence of dominant Balmer absorbtion features in the spectra of galaxies, generally measured in terms of equivalent width, indicates recent star formation activity that rapidly ceased [90]. In general, this spectral index can be used to estimate the age of the underlying stellar population, although it should be used with some caution because of the possible degeneracy due to the contribution of the blue horizontal branch population [91]. Balmer absorption features are evident in the spectra of elliptical galaxies such as M49 (see Figure 3.10).

3.3.1.2 HI UV Absorption Lines

Atomic hydrogen can be observed through UV absorption lines. Neutral hydrogen in the line of sight of a QSO causes absorption of the QSO UV continuum by the redshifted Ly$_\alpha$ (1215.67 Å) line. Given the large cross-section for the Ly$_\alpha$ transition, absorption line measurements are the most sensitive method for detecting baryons at any redshift [92]. The high sensitivity combined with the high ($\sim 25 \text{ km s}^{-1}$) spectral resolution of the 10 m class telescopes enable the detection of HI clouds with column densities $N(\text{HI}) \geq 10^{12} \text{ cm}^{-2}$. The HI column density $N(\text{HI})$ is defined as the product of the density of the gas and the path length along the line of sight through the gas. Each absorbing gas system produces an absorption feature in the QSO spectrum at a wavelength $\lambda_{obs} = \lambda_{Ly_\alpha}(1 + z_{abs})$, where z_{abs} is the redshift of the absorbing cloud.

A sharp cut-off in the UV continuum is also present shortward 912 Å. This cut-off is due to the highly efficient absorption of ionizing photons produced by O and early-type B stars by the ISM gas.

3.3.2
Other Elements

The spectra of galaxies, in particular spectra of quiescent systems such as ellipticals and lenticulars, are characterized by other absorption features on the UV to near-IR spectral domain (Figure 3.10). In G stars such as the Sun, the most important absorption features are the magnesium triplet MgI at 5167, 5173 and 5184 Å (generally indicated with Mgb), the two sodium NaI lines at 5890 and 5896 Å, the calcium doublet CaII (H and K) at 3969 and 3934 Å and the calcium triplet at 8498, 8542 and 8662 Å. A prominent feature, the so called G band, is present at ~ 4300 Å. This feature results from a combined effect due to a series of lines caused by the absorption of CH molecules and of iron. Other absorption features such as TiO can be detected when late-type stars are present. Other important lines are those of the iron Fe 5270 and 5335 Å. The optical spectral indices most commonly used in the analysis of the underlying stellar population of galaxies are the Lick/IDS indices ([93, 94]) described in [89, 95], although new higher resolution indices are now available in the literature.

Of particular importance is the $D_n(4000)$ break, the strongest discontinuity in the optical spectra of galaxies. This feature is due to the accumulation of a large number of spectral lines in a narrow wavelength region whose intensity is related to the presence of ionized metals. In young and massive hot stars, the UV radiation

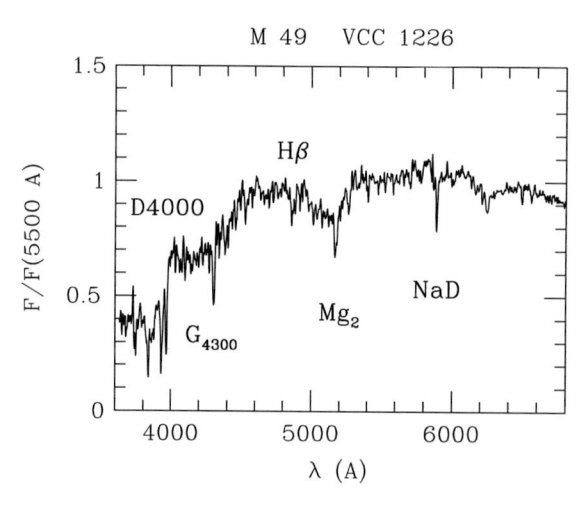

Figure 3.10 The integrated optical spectrum of the elliptical galaxy M49 (adapted from [84]).

multiply ionizes the gas decreasing the opacity, and thus reducing the $D_n(4000)$ break. This discontinuity is therefore moderate in galaxies dominated by young stellar populations (spirals and irregulars) while it is dominant in early-type objects (ellipticals, lenticulars) [96].

3.4
Molecular Lines

Molecules have three different types of transitions: electronic, vibrational, and rotational. Electronic transitions are equivalent to atomic transitions. For the typical molecules in the interstellar medium, the energy of these transitions is several eV, and they lie in the far-UV. Vibrational transitions of molecules are due to the quantization of the possible modes of vibration, and can be stretching modes corresponding to the variation of the distance between the different atoms composing the molecule, the only possiblity in diatomic molecules such as H_2 or CO, or bending and deformation modes in more complex structures. The energy of these transitions is on the order of a fraction of eV, with corresponding wavelengths in the near-infrared. Rotational transitions correspond to a quantization of the rotation of molecules, which can be a global rotation around the principal axes of inertia, or some internal rotation for complex molecules [75]. These lines have energies on the order of meV, and thus the transitions are in the millimeter and centimeter domain.

3.4.1
H_2 Near-Infrared Emission Lines

In symmetric molecules such as H_2, vibrational transitions are forbidden. Their emission is weak but observable in the ISM of galaxies because of the high abundance of molecular hydrogen. In shocks and in photodissociation regions, defined as those regions where the UV radiation is strong enough to photodissociate molecules, and having physical properties in between those of fully ionized HII regions and giant molecular clouds (not reached by the UV radiation), the excited vibrational levels can be populated by collisions if the kinetic energy of the gas is larger than 2000 K. They can also be populated by fluorescence, that is, radiative cascades from electronic levels populated by the absorption of far-UV photons. The roto-vibrational H_2 line $S(1)(v = 1 - 0, J_u - J_l = 3 - 1)$ at 2.122 µm, easily observable in the near-IR domain, traces the molecular gas currently participating in star formation shocked by newly formed stars.

3.4.2
H_2 UV Absorption Lines

The electric dipole moment of symmetric molecules is zero at rest, and thus in molecular hydrogen only electronic transitions are permitted. These electronic transitions of H_2 are in the far-UV ($\lambda \leq 1150$ Å). When observed at high resolution, each electronic transition is resolved into a series of different lines, each one coming from a transition between the vibrational levels of the two electronic states. At higher resolution, each one of these vibrational lines can be further decomposed into several lines corresponding to different rotational sub-levels. When observed at low resolution, each electronic transition shows up as a broad absorption feature resulting from the superposition of all the roto-vibrational lines [75, 97].

4
The Infrared

During their evolution, stars produce and inject metals into the interstellar medium via stellar winds and supernova explosions. These metals aggregate to form dust grains of different sizes and compositions. Dust is generally associated with the gaseous phase of the ISM, and is distributed inhomogeneously in the disk of spirals, inside ellipticals, or in starburst galaxies.

Although the dust mass is just a small fraction of the gaseous component ($\sim 1\%$), it plays a major role in the physics of the ISM and in the process of star formation, and is thus a key parameter in the study of galaxy evolution. Dust is an efficient filter of the stellar radiation, in particular at UV and optical wavelengths. Dust attenuation, resulting from a combination of absorption (which induces a reddening of the stellar spectrum) and scattering (bluening), must be considered for determining the underlying stellar population of all extragalactic objects. In normal galaxies, a significant fraction of the energy emitted by stars ($\sim 30\%$) is absorbed by dust and re-radiated in the far-IR, in the $\sim 3\,\mu m$ to $1\,mm$ spectral range. This fraction is much larger in starburst systems such as those dominating the early universe. As a major coolant of the ISM, dust prevents gas heating up from the general interstellar radiation field, thus playing an important role in the process of star formation. It also contributes to the formation of molecular hydrogen from HI, acting as catalyst in a reaction otherwise forbidden in the gas phase. It can also transfer kinetic energy to the gas through gas-dust collisions if the density of the ISM is sufficiently high, or through the release of electrons in a photoelectric effect whenever dust is irradiated by UV photons. For a detailed discussion on all these effects we refer the reader to [75, 97].

Dust dominates the emission of galaxies, in particular of gas-rich, star-forming systems, in the infrared domain, from $\simeq 5\,\mu m$ to $\simeq 1\,mm$. In this spectral range, the emission of galaxies is characterized by a continuum with prominent emission lines, the most important associated with Polycyclic Aromatic Hydrocarbons (PAHs). These atomic and molecular lines are an important coolant of the ISM.

A Panchromatic View of Galaxies, First Edition. Alessandro Boselli.
© 2012 WILEY-VCH Verlag GmbH & Co. KGaA. Published 2012 by WILEY-VCH Verlag GmbH & Co. KGaA.

4.1
Continuum: Dust Emission

Dust is composed of a mixture of carbonaceous and amorphous silicate grains of different sizes and compositions (for a review see [98]). Big grains, of size $a > 200$ Å, composed of both graphite and silicate, are in thermal equilibrium with the UV and the optical radiation field and dominate the IR emission for $\lambda > 60$ μm. In the range 10 μm $\leq \lambda \leq$ 60 μm, the emission is usually dominated by very small, three-dimensional grains (VSG) mostly composed of graphite (10 Å $\leq a \leq$ 200 Å), absorbing mainly in the UV, whose emission is however not negligible at shorter wavelengths [99, 100]. In the range 3 μm $\leq \lambda \leq$ 15 μm, the emission of the ISM is generally dominated by the Unidentified Infrared Emission Bands (UIB), also called the Infrared Emission Features (IEF) or Aromatic Infrared Bands (AIB). The carriers of the AIBs and associated continuum, often called Polycyclic Aromatic Hydrocarbons (PAHs), are probably planar molecules heated stochastically by the absorption of single photons that temporarily reach very high temperatures, at which most of the emission occurs [99, 100]. The size distribution of dust grains is generally represented by the often called MRN distribution [101]

$$dn_i = A_i n_H a^{-3.5} da, \quad a_{\min} < a < a_{\max} \tag{4.1}$$

where a is the grain radius, dn_i is the number of grains of species i with radii in the interval $[a, a + da]$, A_i is a normalization constant, and n_H is the abundance of hydrogen nuclei taken as reference, with $a_{\min} \sim 0.005$ μm and $a_{\max} \sim 0.25$ μm (a more recent and accurate description of the dust grain size distribution can be found in [102]).

The largest dust grains are generally in thermal equilibrium with the radiation: they are heated by UV and optical photons and cool by the thermal emission of infrared photons, although other heating and cooling mechanisms are possible [75]. The energetic balance between the absorbed and emitted radiation can be described by the following relations: in the simple assumption that dust grains are spherical (the polarization of the stellar light scattered by dust indicates the presence of aligned nonspherical grains, but geometrical effects should be minor in the determination of the statistical dust emission properties [103]), the total energy E_{abs} absorbed by a dust grain of radius a hit by a radiation field of density $u(\nu)$ is

$$E_{abs} = \int_0^\infty 4\pi a^2 Q_a(\nu) \pi \frac{cu(\nu)}{4\pi} d\nu \tag{4.2}$$

where $Q_a(\nu)$ is the absorption efficiency, $4\pi a^2$ is the area of the grain. Thus $Q_a(\nu)\pi I(\nu)$ is the energy absorbed per unit area by a grain, and $I(\nu) = cu(\nu)/4\pi$ is the interstellar radiation flux per steradian. The absorption efficiency is higher in the UV than in the optical or near-IR, and can be approximately represented by the relation

$$Q_a(\nu) = Q_0 \left(\frac{\nu}{\nu_0}\right)^\beta \frac{a}{a_0} \tag{4.3}$$

with the grain emissivity index $\beta \approx 2$ although this value might change with the spectral range as shown by [103], which gives $\beta = 1$ for 50 μm $\leq \lambda \leq$ 100 μm, $\beta = 1-2$ for 100 μm $\leq \lambda \leq$ 250 μm, and $\beta = 2$ for $\lambda > 250$ μm. The same grain, at a temperature T, emits a total energy

$$E_{em} = \int_0^\infty 4\pi a^2 Q_a(\nu) \pi B(\nu, T) d\nu \quad (4.4)$$

where $B(\nu, T)$ is the Planck function [76],

$$B(\nu, T) = \frac{2h\nu^3}{c^2} \frac{1}{e^{\frac{h\nu}{kT}} - 1} \quad (4.5)$$

or equivalently:

$$B(\lambda, T) = \frac{2hc^2}{\lambda^5} \frac{1}{e^{\frac{hc}{\lambda kT}} - 1} \quad (4.6)$$

If the grain is in thermal equilibrium, the flux density has the form

$$S_{dust}(\nu, T) \propto \nu^\beta B(\nu, T) \quad (4.7)$$

which, integrated, gives the energy flux $W(T)$:

$$W(T) = \int_0^\infty Q_a(\nu) \frac{cu(\nu)}{4\pi} d\nu = \int_0^\infty Q_a(\nu) B(\nu, T) d\nu \quad (4.8)$$

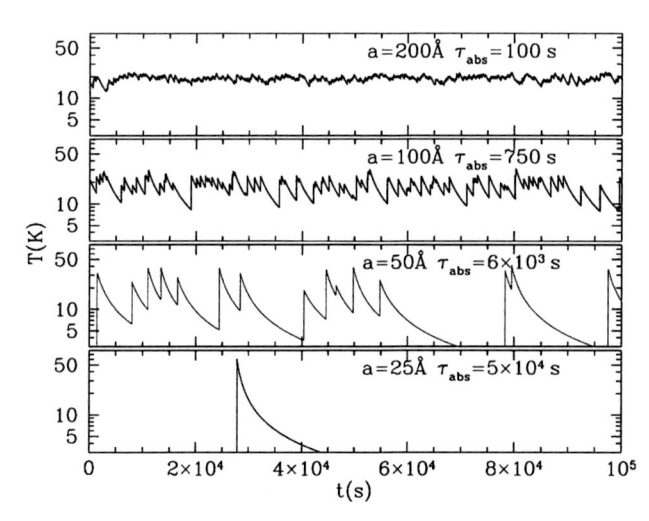

Figure 4.1 The temperature of dust grains of different sizes as a function of the mean time between absorption of photons from the general interstellar radiation field, from [104]. Only the largest grains ($a = 200$ Å) are in thermal equilibrium with the radiation and thus emit as a modified black body.

from which we can derive, for $\beta = 2$, the relation

$$W(T)\,[\text{erg cm}^{-2}\,\text{s}^{-1}] = 4.6 \times 10^{-11} \left(\frac{a}{0.1\,\mu\text{m}}\right) T\,[\text{K}]^6 \qquad (4.9)$$

The behavior of dust grains of smaller size is quite different. Very small grains and PAHs have a small heat capacity: they instantaneously increase their temperature when hit by a single UV or visible photon, and they rapidly cool afterwards, keeping a low temperature until another photon is absorbed, as graphically shown in Figure 4.1, where the temperature of dust grains of different sizes is plotted as a function of the mean time between absorption of successive photons from the general interstellar radiation field. Very small grains and PAHs are thus not in thermal equilibrium with the interstellar radiation field and cannot be represented with a simple modified black body.

4.2
Emission Lines

The infrared spectral domain of extragalactic sources is characterized by the presence of several atomic and molecular emission lines that are fundamental to the study of the physics of the ISM and the process of star formation. The mid-IR domain is dominated by the presence of PAHs, while fine-structure atomic lines responsible for the cooling of the ISM are present in the far-IR. The emission of rotational transitions of dipolar molecules, among which the most important is probably carbon monoxide (CO), are observable in the far-IR and submillimeter domain (the ground level ^{12}CO($J = 1-0$) transition is at 2.6 mm), while the weak emission of the most abundant molecule in the universe, H_2, is observable in the mid-IR domain.

4.2.1
PAHs

The Aromatic Infrared Bands have an important contribution to the total IR emission of galaxies and generally dominate their mid-IR emission in the $3\,\mu\text{m} \leq \lambda \leq 15\,\mu\text{m}$ wavelength range as depicted in Figure 4.2. Polycyclic Aromatic Hydrocarbons (PAH), as very small grains, have a small heat capacity and are thus not in thermal equilibrium with the interstellar radiation field. They are probably responsible for the UV bump at 2175 Å observed in the Galactic extinction curve [105, 106] (see Section 12.1.1). The most intense emitting PAH lines are at 3.3, 6.2, 7.7, 8.6, 11.3, and 12.7 μm, and are probably due to optically active vibrational modes typical of molecules formed by an aromatic ring skeleton with H atoms attached at the edge. For these major features, the vibrational modes are the C–H (3.3 μm) and C–C (6.2 and 7.7 μm) stretching modes, C–H in-plane bending modes (8.6 μm), and C-H out-plane bending modes (11.3, 12.7 μm) [98].

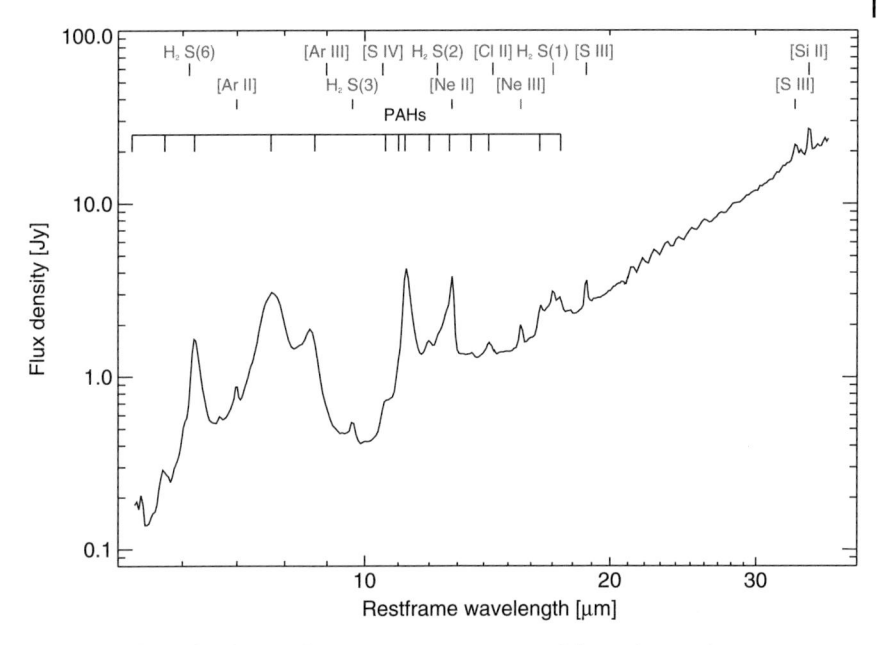

Figure 4.2 The mid-IR, low resolution IRS Spitzer spectrum of the starburst galaxy M82 [111]. The spectrum shows the emission of PAHs and fine structure lines superposed to the continuum, whose emission is due to very small dust grains. Reproduced by permission of the AAS.

PAHs have been found in almost all of the ISM environments where they can be excited by UV or optical photons, in particular in the photodissociation regions between HII regions and giant molecular clouds, while they are rare in those regions with a very intense UV radiation field which is able to destroy them. Indeed they are the dominant emission features in the mid-IR spectrum of massive spirals, contributing to $\lesssim 10\%$ of their total 4–1000 μm far-IR luminosity, while they become less important in low-metallicity systems where a combined effect of the intense UV radiation field and the low dust content favor their dissociation [44, 107]. They are present in ultra luminous infrared galaxies (ULIRGs) dominated by starburst activity, in particular those with cold dust temperatures, while their emission is very weak in systems dominated by a central AGN. The PAH emission of LINERs is intermediate, between that of AGNs and starburst galaxies [108–110].

4.2.2
Cooling Lines in PDR

The ISM provides the gas reservoir to feed star formation in galaxies. Atomic hydrogen, originating from the primordial nucleosynthesis, has to first cool down to collapse in molecular clouds to form stars. The newly formed stars inject metals and energy via UV radiation into the ISM, determining its physical conditions. The heating and cooling of the gas is predominantly regulated in neutral regions of the

ISM, generally called photodissociation regions (PDR), from which comes most of the CO emission of galaxies [112].

The heating and cooling of the gaseous phase of the ISM occurs via kinetic energy transfer from and to atoms, molecules, and ions of the interstellar gas. As extensively described in [75], the principal heating process starts with the removal of an electron from an atom or a dust grain of the ISM. This electron generally has a kinetic temperature higher than the average temperature of the gas phase, so that it can transfer its energy through elastic collisions to the gas electrons and ions (elastic Coulomb interactions). For energy equipartition, the heated ions transfer their energy to atoms and molecules. The time scale for thermalization in the diffuse neutral gas is rapid, on the order of one year, much shorter than the time scale for evolution of the interstellar medium (10^6 yr).

The removal of the electrons can originate from different physical processes: in the densest regions of the ISM, such as the core of giant molecular clouds, the gas can be heated by low energy cosmic rays that can easily penetrate the dense molecular clouds that are opaque to UV radiation. In the cold, diffuse neutral ISM, in photodissociation regions, and in the warm phase, the gas is principally heated by electrons which are removed from dust grains by the photoelectric effect. In HII regions, the dominant heating process is the photoelectric heating by photoionization of atoms and molecules due to the ionizing UV radiation. The hot gas that emits in X-ray can penetrate and ionize the ISM whenever the density of the gas is very low, as in the lowest density regions of the diffuse gas or in the warm gas (see Figure 4.3).

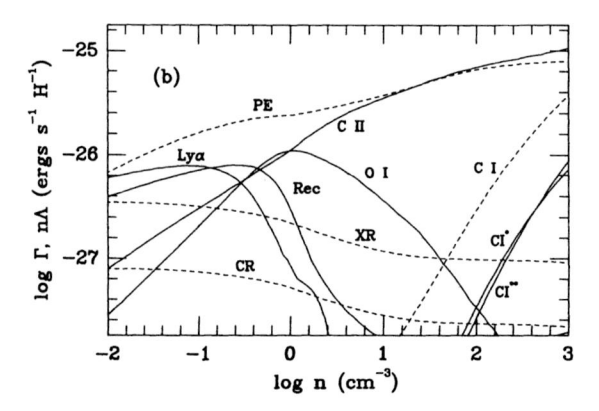

Figure 4.3 Heating and cooling rates per hydrogen nucleus (Γ, $n\Lambda$, in erg s^{-1} H^{-1}) as a function of the density of the gas of the ISM (n, in cm^{-3}) for a given pressure function. Heating rates, indicated by dashed lines, are due to: photoelectric heating from small grains and PAHs (PE); X-ray (XR); cosmic rays (CR); photoionization of C (CI). Cooling rates, indicated by solid lines, are due to: CII fine structure, [CII] at λ 158 µm (CII); OI fine structure, [OI] at λ 63 µm (OI); CI fine structures at λ 609 µm (CI*) and λ 370 µm (CI**); recombination onto small grains and PAHs (Rec); cooling by excitation of Ly$_\alpha$ and other transitions (from [113]). Reproduced by permission of the AAS.

Other processes can contribute to the heating of the ISM. Chemical reactions such as the formation of H_2 on dust grains yields 4.48 eV of energy that is partly transferred as kinetic energy to the newly formed H_2 molecule. This energy can be later transferred to other molecules by elastic collisions. This process is efficient in shocks and dense photodissociation regions, or in high-density regions where collisions between molecules are more frequent. In the core of molecular clouds, gas can also be heated by the thermal exchange with dust grains: indeed in these regions the gas is not reached by the interstellar radiation field, and is colder than the dust which is heated by the infrared radiation that is not absorbed by the gas. Supernova explosion, stellar winds, expansion of HII regions, and gravitational interactions or gas collapse can transfer kinetic energy to the gas of the ISM (hydrodynamic and magneto-hydrodynamic heating).

At the same time, gas cools thanks to two different families of processes. The atoms, molecules and ions excited by collisions transform part of their kinetic energy into radiation, generally in the infrared. Among these radiative processes, fine-structure line cooling is the most efficient in the ISM, except in hot gas or in the densest regions inside molecular clouds. The atoms and ions of the ISM have their fundamental energy level split by the fine-structure interaction between the total orbital momentum of the electron and their total spin. Collisions with electrons or ions populate fine-structure levels, then later de-excite, emitting a line generally in the far-IR. The most efficient elements in the neutral medium are CII, which has an excitation temperature of 92 K, and OI (228 K). Their emission is at 157.7 and 63.2 μm, respectively. In HII regions, the most important cooling elements are OII and OIII, NII and NIII, NeII, and NeIII. Gas can also be cooled by the collisional excitation of permitted lines as is the case of the hydrogen atom that is excited from the ground level $n = 1$ to $n = 2$, whose de-excitation produces a Lyman-α photon. This cooling process is particularly efficient at high gas temperatures ($T > 8000$ K). In shocks and photodissociation regions, the collisional excitation and de-excitation of various molecules (H_2, H_2O, CO) and their isotopes is important. CO rotational lines and CI (which has a fine structure line excitation temperature of ~ 23 K) are the dominant cooling lines within molecular clouds. In warm and hot ionized gas, cooling is dominated by the collisional excitation due to electrons of permitted and forbidden lines. The recombination of electrons with dust grains is important only at high temperatures, while gas cooling through its thermal exchange with dust grains is negligible in the diffuse ISM [75, 113]. Table 4.1 gives the most important emission lines responsible for gas cooling in PDR [114]. Kaufman *et al.* [114] also give the variation of these cooling lines as a function of the UV radiation field, metallicity, and gas volume density predicted by PDR models.

4.2.3
H_2 Lines

Given its symmetric structure, H_2 rotational dipole transitions, which corresponds to the emission of a rotating electric dipole, are forbidden. Only quadrupole transitions are allowed. These are extremely weak, but given the high density of the

Table 4.1 Emission lines responsible for the cooling of the ISM.

Species	Transition	Wavelength (μm)	E_{upper}/k (K)	n_{cr} [a] (cm^{-3})
[CII]	$^2P_{3/2}-^2P_{1/2}$	157.74	92	3×10^3
[OI]	$^3P_1-^3P_2$	63.18	228	5×10^5
[OI]	$^3P_0-^3P_1$	145.53	326	1×10^5
[CI]	$^3P_1-^3P_0$	609.14	23.6	5×10^2
[CI]	$^3P_2-^3P_1$	369.87	62.5	3×10^3
CO	$J = 1-0$	2600.78	5.53	3×10^3
CO	$J = 2-1$	1300.39	16.59	1×10^4
CO	$J = 3-2$	867.00	33.19	5×10^4
CO	$J = 6-5$	433.56	116.16	4×10^5
CO	$J = 15-14$	173.62	663.36	8×10^6

[a] Critical density for collision with hydrogen.

molecular hydrogen in the interstellar medium, they can be observed in the 5–30 μm spectral domain. The intensity of their emission depends on both the kinetic temperature and the column density of the gas.

Molecular hydrogen can be excited by colliding with other molecules of the ISM or radiatively excited absorbing UV photons. If the density of the gas is low enough ($< 10^4$ cm^{-3}) then the molecule de-excites radiatively through the ground electronic roto-vibrational states, while if the density of the gas is high ($> 10^4$ cm^{-3}), then the populations of rotational and vibrational states are redistributed by collision before the de-excitation to the ground level. The collision rate of the H$_2$ molecules with other particles (molecules, atoms) increases with density and temperature. Collisional de-excitation is thus more important than radiative de-excitation in high-density and high temperature regimes. The critical density, defined as the density of the gas at which the collisional de-excitation rate equals the radiative de-excitation rate, is a weak function of the gas temperature. When collisional excitation and de-excitation dominates with respect to radiative process, thus whenever the density of the gas is above the critical density, the gas is in local thermodynamic equilibrium (LTE) and the molecular hydrogen has an ortho-to-para ratio of 3 [115]. The lowest levels of the purely rotational quadrupole transitions of the ortho (J-odd, corresponding to parallel spins of the hydrogen atoms) and para (J-even, anti-parallel spins) series are at $S(0)(v = 0-0, J_u - J_l = 2-0)$ at 28.2188 μm and $S(1)(v = 0-0, J_u - J_l = 3-1)$ at 17.0348 μm, respectively. Their emission is due to relatively warm molecular gas, with temperatures on the order of ∼ 150–250 K.

4.2.4
Dust Absorption of Ly$_\alpha$ Scattered Photons

In galaxies, Ly$_\alpha$ photons are produced by the recombination in hydrogen of the interstellar medium ionized by the radiation of OB stars, by shocks in supernova remnants, or by the continuum flux of AGNs. These photons suffer a large number of resonant scattering in the ambient neutral atomic hydrogen. That is, each Ly$_\alpha$ photon absorbed by an HI atom is remitted as a new Ly$_\alpha$ photon more isotopically than the continuum. This effect significantly increases the mean free path of Ly$_\alpha$ photons within galaxies, thus enormously increasing the chances of dust absorption with respect to continuum photons in dust-rich systems [116, 117]. For this reason, in dust-rich galaxies such as starbursts, Ly$_\alpha$ photons do not satisfy the hydrogen recombination line case B condition [118] described in Section 3.2.1 required for quantifying the effects of dust attenuation (see Section 12.2.1). As a consequence, the Ly$_\alpha$ luminosity is not a good tracer of star formation (see Chapter 13).

5
Millimeter and Centimeter Radio

The millimeter and centimeter radio domain is of fundamental importance in the study of galaxies. Here the continuum emission is due to a thermal component originating from HII regions (free–free) and to a nonthermal component due to relativistic electrons spinning around weak magnetic fields (synchrotron), with a contribution to the thermal emission by the coldest dust at short wavelengths. In normal galaxies, the electrons responsible for both the free–free and synchrotron emission are tightly related to the star formation activity of galaxies: in the first case these electrons are produced by the photoionization of the diffuse gas surrounding young and massive stars inside HII regions, while those emitting synchrotron radiation are accelerated to relativistic velocities in supernova remnants. In AGNs, the radio emission is synchrotron radiation associated with the radio jet. Radio continuum polarization observations are also a powerful tool for measuring magnetic fields.

The two most important emission lines to the study of galaxies are in the radio domain: HI gas, the principal baryonic component of the universe, is observable through the emission of 21 cm line (1.4 GHz), while the strongest molecular emission line, ^{12}CO(1–0), generally coupled with molecular hydrogen, is at 2.6 mm (115 GHz). The high spectral resolution (~ 1 km s^{-1}) achieved at millimeter frequencies, combined with interferometric observations, enables accurate studies of the kinematical properties of galaxies down to arcsecond angular scales. In the next years the Square Kilometer Array (SKA) and the Atacama Large Millimeter Array (ALMA) telescope projects will allow for detailed studies of the kinematical and physical properties of the gaseous component of the ISM of galaxies at extremely high distances.

5.1
Continuum

The continuum emission of normal galaxies is dominated by the thermal emission of the coldest dust component up to ~ 1.5 mm, with a contribution from the nonthermal synchrotron and the thermal free–free emission, as shown in Figure 5.1.

A Panchromatic View of Galaxies, First Edition. Alessandro Boselli.
© 2012 WILEY-VCH Verlag GmbH & Co. KGaA. Published 2012 by WILEY-VCH Verlag GmbH & Co. KGaA.

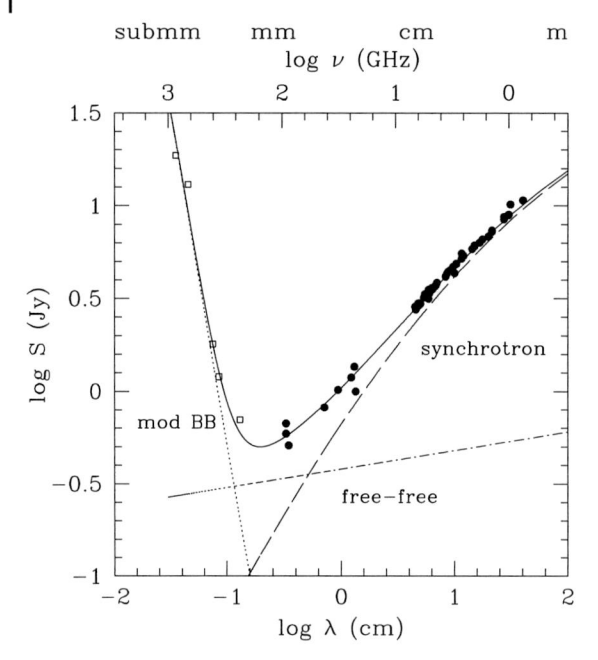

Figure 5.1 The submillimeter–radio continuum spectral energy distribution of the starburst galaxy M82, adapted from [119]. The submillimeter domain is dominated by the Rayleigh–Jeans tail of the modified black body emission due to the cold dust (dotted line), the radio continuum at $\lambda \geq 1\,\mathrm{cm}$ is due to synchrotron emission (long dashed line), while free–free dominates the 1 mm–1 cm spectral range (dotted dashed line). The solid line gives the total emission. The curvature in the synchrotron emission is due to self-absorption.

At longer wavelengths, the galactic radio emission is due to both free–free and synchrotron radiation, the later dominating at high wavelengths ($\gtrsim 1\,\mathrm{cm}$).

5.1.1
Free–Free Emission

Thermal free–free emission, also called thermal Bremsstrahlung, is due to the energy loss of free electrons (with a thermal velocity distribution) deflected in the electric field of an ion inside HII regions. The flux density for a free–free radio continuum emission has the form

$$S_{\mathrm{ff}}(\nu) \propto \nu^{-\alpha_{\mathrm{ff}}} \tag{5.1}$$

In the millimeter spectral domain and at high frequencies, where the medium is optically thin, $\alpha_{\mathrm{ff}} = 0.1$, while at low frequencies $\alpha_{\mathrm{ff}} = -2$ since the medium becomes optically thick to its own radiation (and behaves like a black body) [75].

5.1.2 Synchrotron Emission

Synchrotron emission from galaxies is characterized by a power spectrum whose flux density is given by [120]

$$S_{\text{syn}}(\nu) \propto \nu^{-\alpha_{\text{syn}}} \tag{5.2}$$

whose spectral index α_{syn} is directly connected to the spectral index of the energy distribution of radiating ultra-relativistic electrons. If the energy distribution of cosmic ray electrons in the energy interval between E and $E + dE$ is given by

$$N(E)dE = KE^{-\gamma}dE \tag{5.3}$$

where K is their volume density, then the nonthermal synchrotron spectral index is

$$\alpha_{\text{syn}} = \frac{\gamma - 1}{2} \tag{5.4}$$

We note that the emissivity $\epsilon_{\text{syn}}(\nu)$ of relativistic electrons also depends on the strength of the magnetic field B to the power $\alpha_{\text{syn}} + 1$ [121]

$$\epsilon_{\text{syn}}(\nu) \propto B^{\alpha_{\text{syn}}+1} \nu^{-\alpha_{\text{syn}}} \tag{5.5}$$

Self-absorption becomes important at long wavelengths. In the case where the absorbing gas is well mixed with the emitting sources, as is generally the case in starburst galaxies, the synchrotron emission corrected for self-absorption is given by [119]:

$$S_{\text{syn-sab}}(\nu) = S_{\text{syn}}(\nu) \frac{1 - e^{-\tau}}{\tau} \tag{5.6}$$

where the optical depth depends on the frequency as:

$$\tau = \tau_l \nu_g^{-2.1} \tag{5.7}$$

where τ_l is the optical thickness at 1 GHz and ν_g is the frequency in GHz. Typical values of τ_l for starburst galaxies can be found in [119]: they range from $\tau_l = 0.04$ in M82 to $\tau_l = 0.27$ in NGC 253.

5.1.3 Dust Emission

Thanks to the advent of new generation bolometers and to the transparency of the atmosphere at given frequencies, dust emission can be observed from the far-IR to the sub-mm and millimeter spectral domain by combining observations from ground-based (450 and 850 µm, 1.3 mm) and space facilities (e.g., 250, 350, 500 µm for SPIRE on Herschel). As in the far-IR, dust emission follows a modified black body, and in this spectral domain can be represented by the Rayleigh–Jeans approximation whenever $h\nu \ll kT$ [76]:

$$B_{\mathrm{RJ}}(\nu, T) = \frac{2\nu^2 kT}{c^2} \tag{5.8}$$

Here, the Rayleigh–Jeans spectrum is modified by the grain emissivity with the same ν^β factor given in Eq. (4.7). The contribution of dust to the emission around 1 mm can be determined only after subtracting the contribution of synchrotron and free–free emission from the continuum, and whenever the observed spectral range includes important molecular lines, their contamination, as described in Section 8.1.2.

5.2
Emission Lines

The radio domain includes several of the most important lines necessary for the study of galaxy evolution. The HI emission line at 21 cm is used to determine the atomic gas reservoir of galaxies, while the ^{12}CO(1–0) line at 2.6 mm is used to determine the molecular hydrogen content. If the atomic hydrogen emission line has been detected only up to $z \sim 0.25$ [122], the higher level CO transition lines such as ^{12}CO(3–2) fall into the millimeter bands for high-redshift objects, and the ^{12}CO(1–0) line becomes observable at high frequencies with centimeter radio telescopes such as the VLA. The importance of these lines resides in the fact that they trace the gaseous component which feeds star formation. The spectral and angular resolution obtained through interferometer observations make also these lines powerful tools for studying the two-dimensional kinematical properties of galaxies at different redshifts.

5.2.1
Molecular Lines

Because of its strong dipolar moment, carbon monoxide (CO) is the most observed molecule in the ISM. CO rotational lines, which are generally optically thick, are in the millimeter and submillimeter domain (see Table 5.1). For the ^{12}CO($J = 1$–0) rotation line, the most intense molecular line in the ISM of galaxies, whose emission is at 2.6 mm (115 GHz), the excitation temperature, T_{ex}, can be measured through the relation [75]

$$T_{\mathrm{ex}}\,[\mathrm{K}] = 5.53 \left[\ln\left(\frac{1}{T_{\mathrm{B}}^*/5.53 + 0.151} + 1\right)\right]^{-1} \tag{5.9}$$

where T_{B}^* is the brightness temperature, and $T_{\mathrm{B}}^* = T_{\mathrm{A}}^*/\eta_{\mathrm{mb}}$ (T_{A}^* = antenna temperature; η_{mb} = main beam efficiency). The column density of the CO gas, and

Table 5.1 Molecular emission lines in the ISM (data taken from [126]).

Species	Transition	Frequency GHz	Wavelength mm
CS	$J = 1-0$	48.99	6.119
HCN	$J = 1-0$	88.63	3.383
HCO$^+$	$J = 1-0$	89.19	3.361
HNC	$J = 1-0$	90.66	3.307
CS	$J = 2-1$	97.98	3.060
C^{18}O	$J = 1-0$	109.78	2.731
^{13}CO	$J = 1-0$	110.20	2.720
CO	$J = 1-0$	115.27	2.601
CS	$J = 3-2$	146.97	2.040
C^{18}O	$J = 2-1$	219.56	1.365
^{13}CO	$J = 2-1$	220.40	1.360
CO	$J = 2-1$	230.54	1.300
CS	$J = 5-4$	243.94	1.229
HCN	$J = 3-2$	265.89	1.128
HCO$^+$	$J = 3-2$	267.56	1.120
HNC	$J = 3-2$	271.98	1.102
CO	$J = 3-2$	345.79	0.867

thus indirectly that of H$_2$ ($N(H_2) \simeq 10^5 N(CO)$), can be determined by combining the observations of the optically thick ^{12}CO lines with those of other less abundant, optically thin lines [75]. This is generally done using other isotopic varieties such as ^{13}CO or with the optically thin C^{18}O line. Adopting a CO to H$_2$ conversion factor (see Section 11.1.2.1), the ^{12}CO($J = 1-0$) line is indeed the molecular gas line most frequently used to estimate the bulk of the molecular hydrogen content of external galaxies. Other ^{12}CO rotational transitions [123], or other lines such as the HCN($J = 1-0$) [124, 125], which is a good tracer of the molecular gas with densities $n(H_2) \geq \sim 3 \times 10^4$ cm^{-3} and whose emission is at 88.63 GHz (3.38 mm), have been preferred for tracing the content and distribution of the densest gas in the core of molecular clouds.

5.2.2
HI

Atomic hydrogen constitutes 75% of the mass and 90% of the number of atoms of the baryonic matter in the universe. At the same time, HI is the principal constituent of the ISM of galaxies. Most of the hydrogen atoms are in the fundamental level (the energy of the level immediately above this is ~ 10 eV, thus relatively high), and are distributed into two hyperfine sub-levels of different statistical weight. The energy of the hydrogen atom is larger by 6×10^{-6} eV when the spin of the proton and the electron are parallel compared to when they are anti-parallel [75]. The statistical equilibrium between the populations $n(F=1)$ and $n(F=0)$ of the two sub-levels can be expressed by the Boltzmann equation, characterized by a spin temperature T_S,

$$\frac{n(F=1)}{n(F=0)} = \frac{g(F=1)}{g(F=0)} e^{-\frac{h\nu}{kT_S}} = 3e^{-\frac{h\nu}{kT_S}} \tag{5.10}$$

where 3 is the ratio of the statistical weights for the parallel (3) and anti-parallel (1) sub-levels. The transition between these two sub-levels is responsible for the emission of photons at a frequency of 1420 MHz (21.106 cm). This transition is highly forbidden, and has a probability for spontaneous transition $A_{10} = 2.868 \times 10^{-15}\,\mathrm{s}^{-1}$, and hence the radiative lifetime of the upper sub-level is $1/A_{10} = 1.1 \times 10^7$ yr [75]. During this long time scale, however, other processes can populate the upper sub-level, as summarized by [127]: the absorption and emission stimulated by photons of any existing radiation field (including the cosmic background radiation), collisions of hydrogen atoms with other particles (especially with other hydrogen atoms and electrons), and "pumping" by Lyman-α photons (an atom of hydrogen in its ground state level ($n=1$) can be excited to the $n=2$ state by a Ly$_\alpha$ photon, then de-excite to come back to the $n=1$ level, and can populate to some extent the upper parallel spin sub-level). Collisional equilibrium becomes dominant with respect to radiation whenever the density of the gas is higher than the critical density $n_{\mathrm{crit}} = 10^{-2}\,T_K^{-1/2}\,\mathrm{cm}^{-2}$, where T_K is the kinetic temperature of the gas. In this case $T_S \simeq T_K$. This condition is generally satisfied in normal galaxies, with a possible exception in very low-density regions associated with tidal debris in interacting systems or in ram-pressure stripped tails in cluster galaxies [127].

The brightness temperature of an HI gas cloud characterized by an optical thickness τ and a spin (or kinetic) temperature T_S, bathed in a radiation field T_R is given by the relation [127]

$$T_B(V) = T_R e^{-\tau(V)} + T_S[1 - e^{-\tau(V)}] \tag{5.11}$$

where the optical depth $\tau(V)$ varies with velocity (because of the Doppler effect) across the line profile. Since the optical thickness in the line of sight $\tau(V)$ is proportional to the HI column density $N(\mathrm{HI})$ (the number of hydrogen atoms per unit surface area along the line of sight) divided by the spin temperature T_S, for small optical thicknesses ($\tau(V) \ll 1$) the HI column density can be derived from the relation

$$N(\mathrm{HI})\,[\mathrm{cm}^{-2}] = 1.822 \times 10^{18} \int \left(\frac{T_S}{T_S - T_R}\right) T_{\mathrm{bl}}\,[\mathrm{K}](V)\,dV\,[\mathrm{km\,s}^{-1}] \tag{5.12}$$

where T_{bl} is the brightness temperature of the line profile at the velocity V (in km s^{-1}). In the ISM of normal galaxies, the spin temperature T_S (or the kinetic temperature T_K) is on the order of ~ 100 K. This is significantly larger than $T_R = 2.7$ K, and thus the ratio ($T_S/(T_S - T_R)$) is generally ~ 1. In the outskirts of HI disks, in tidal debris or in any other low-density region, the density of the gas can be too low to be collisionally excited. T_S could thus drop to low values, making the fraction in brackets larger than 1. However it has been shown that in these low-density regions the upper hyperfine sub-level of the hydrogen atoms can be efficiently populated by UV pumping, inducing an increase of T_S, thus reducing Eq. (5.12) to a simple integral of the brightness temperature of the line profile over the velocity space.

All this derivation, however, is valid in the assumption that the HI gas is optically thin and that the UV radiation is pervasive. This is probably not the case in the densest regions of the ISM, where the HI gas is optically thick and hardly excited by penetrating UV photons [128], leading to an underestimate of the total HI column density. In these high density regions, HI fluxes should be corrected for self-absorption using standard recipes generally based on the line of sight inclination of the galaxy, as described in Section 11.1.1.

5.3
Absorption Lines

5.3.1
HI

As for the UV continuum, high-redshift radio-loud QSOs can be used to observe the HI 21 cm absorption line in the intergalactic gas of damped Lyman-α systems with very high gas column densities. Combined with UV-optical spectra, HI 21 cm absorption lines can be used to derive the spin temperature of the gas (the excitation temperature of the line), as shown in [129] and [130].

Part Two Derived Quantities

6
Properties of the Hot X-ray Emitting Gas

X-ray data can be used to determine the temperature of the hot gas surrounding elliptical galaxies and, under some assumptions, their total dynamical mass (see Section 15.2.3).

6.1
X-ray Luminosity

The X-ray luminosity of a galaxy in a given band (low $< \lambda <$ up, generally expressed in keV), can be determined by the relation

$$L_X \,[\mathrm{erg\,s^{-1}}] = 4\pi D^2 \int_{\mathrm{low}}^{\mathrm{up}} S(\nu)\, d\nu \qquad (6.1)$$

where D is the distance of the galaxy and $S(\nu)$ is the flux density at the frequency ν. For the Einstein observatory the typical band is 0.2–3.5 keV, for the ROSAT Position Sensitive Proportional Counter (PSPC) and High Resolution Imager (HRI) it is 0.1–2.4 keV. The nominal band for XMM European Photon Imaging Camera (EPIC) is 0.15–15 keV, although its sensitivity drops at ~ 10 keV, while for Chandra it is ~ 0.1–10 keV, but limited to the range 0.3–7 keV for most scientific purposes.

6.2
Gas Temperature

In elliptical galaxies, the hot gas is principally produced by the mass loss of evolved stars, which eject gas at a rate of [131]

$$\dot{M}\,[M_\odot\,\mathrm{yr}^{-1}] \approx 1.5 \times 10^{-11} \left(\frac{L_B}{L_{B,\odot}}\right) t_{15}^{1.3} \qquad (6.2)$$

where t_{15} is the stellar age in units of 15 Gyr. This gas is heated by shocks due to stellar motion within the galaxy and is raised to the stellar kinematical temperature

A Panchromatic View of Galaxies, First Edition. Alessandro Boselli.
© 2012 WILEY-VCH Verlag GmbH & Co. KGaA. Published 2012 by WILEY-VCH Verlag GmbH & Co. KGaA.

of [53]

$$T_{gas} \approx T_{star} \approx T_{virial} \approx \frac{\mu m_p \sigma^2}{k} \sim 10^7 \text{ K} \sim 1 \text{ keV} \tag{6.3}$$

where μ is the mean atomic weight of the gas, m_p is the proton mass, k is the Boltzmann constant, and σ is the galaxy velocity dispersion. Equation (2.12) shows that the temperature of the X-ray emitting gas can be determined by fitting the X-ray spectrum of the galaxy. Gas densities and temperature gradients should be accurately measured for determining the total dynamical mass of the galaxy using X-ray data (see Section 15.2.3). This has been done in a sample of nearby early-type galaxies by [132]. Determining the gas temperature of the galactic interstellar medium is only possible in compact objects, where the emission is not contaminated by the emission of the diffuse hot plasma associated with the cluster or the group to which the galaxy might belong. The temperature radial profile of compact objects increases towards the center of the galaxy. An increase of the gas temperature in the outskirts of galaxies is a clear sign of contamination from the hot gas of the intracluster medium [132].

Whenever X-ray spectra are not available, gas temperatures (and thus total dynamical masses) can be determined by making some assumptions on the properties of the emitting gas. In this case, the temperature radial profile of an elliptical galaxy is generally parameterized by the relation [74]

$$T(r) = T_0 \left(1 + \frac{r}{r_T}\right)^{-\alpha} \tag{6.4}$$

where r_T is a characteristic radius and T_0 the temperature at the center, while the gas density distribution by an isothermal sphere

$$\rho(r) = \rho_0 \left[1 + \left(\frac{r}{r_c}\right)^2\right]^{-3\beta/2} \tag{6.5}$$

and the electron density by

$$n_e(r) \propto r^{-3\beta} \tag{6.6}$$

Within these hypothesis (assuming an isothermal sphere), the dynamical mass of the elliptical galaxy can be easily determined (see Section 15.2.3).

7
Dust Properties

Formed by the aggregation of metals injected into the interstellar medium principally by massive stars in the latest phases of their evolution, dust plays a major role in the physical equilibrium of the interstellar medium. By absorbing the most energetic photons of the interstellar radiation field, dust participates in the cooling of the atomic and molecular gas inside molecular clouds and is thus a key parameter in the study of the process of star formation.

Infrared and submillimeter (5 µm–1 mm) data are generally used to determine the mass and temperature of the interstellar dust. While the longest wavelengths are crucial for an accurate estimate of the total dust mass, the shortest wavelengths are necessary for measuring the temperature of the warm dust component associated with the process of star formation.

7.1
The Far-IR Luminosity

The monochromatic far-infrared (far-IR, or FIR) luminosity of an extragalactic source at a given IR frequency can be determined using the relation

$$L_{IR}(\nu) = 4\pi D^2 S(\nu) \tag{7.1}$$

where D is the distance to the galaxy and $S(\nu)$ is the flux density at the frequency ν. Since the advent of IRAS, astronomers tried to estimate the total infrared luminosity of extragalactic sources by integrating the flux density over the whole infrared spectral domain:

$$F_{IR} = \int_{low}^{up} S(\nu) d\nu \tag{7.2}$$

A first attempt in determining the infrared luminosity of galaxies using data in the four different IRAS bands centered at 12, 25, 60 and 100 µm has been made by the IRAS team. The definition firstly proposed by [133] is still widely used [46];

A Panchromatic View of Galaxies, First Edition. Alessandro Boselli.
© 2012 WILEY-VCH Verlag GmbH & Co. KGaA. Published 2012 by WILEY-VCH Verlag GmbH & Co. KGaA.

whenever flux densities, $S(\nu)$, are expressed in Jy:

$$F_{\text{FIR}}\,[\text{W m}^{-2}] = \int_{42}^{122} S(\nu)d\nu = 1.26 \times 10^{-14}(2.58\,S(60) + S(100)) \tag{7.3}$$

and

$$L_{\text{FIR}}\,[L_\odot] = 4\pi D^2 F_{\text{FIR}} \tag{7.4}$$

or

$$F_{\text{IR}}\,[\text{W m}^{-2}] = \int_{8}^{1000} S(\nu)d\nu = 1.8 \times 10^{-14}$$
$$\times [13.48\,S(12) + 5.16\,S(24) + 2.58\,S(60) + S(100)] \tag{7.5}$$

and

$$L_{\text{IR}}\,[L_\odot] = 4\pi D^2 F_{\text{IR}} \tag{7.6}$$

with infrared luminosities generally expressed in solar bolometric luminosities ($L_\odot = 3.83 \times 10^{33}$ erg s^{-1}). Referring to their infrared luminosity definition, [46] defined luminous infrared galaxies (LIRGs) as those objects with $L_{\text{IR}} > 10^{11} L_\odot$, ultra-luminous infrared galaxies (ULIRGs) as those objects with $L_{\text{IR}} > 10^{12} L_\odot$ and hyper-luminous infrared galaxies (HyLIRGs) as those objects with $L_{\text{IR}} > 10^{13} L_\odot$. The advent of ISO and more recently Spitzer and Herschel, which gave access to a larger frequency window with a more complete sampling of the infrared spectral energy distribution of extragalactic sources, enabled to re-calibrate the total infrared flux of galaxies, F_{TIR} [134],

$$F_{\text{TIR}} = \int_{3}^{1100} S(\nu)d\nu \tag{7.7}$$

with the empirical relation

$$\log\left(\frac{F_{\text{TIR}}}{F_{\text{FIR}}}\right) = a_0 + a_1 x + a_2 x^2 + a_3 x^3 + a_4 x^4 \tag{7.8}$$

with

$$L_{\text{TIR}} = 4\pi D^2 F_{\text{TIR}} \tag{7.9}$$

where $x = \log[S(60)/S(100)]$ and a_0, a_1, a_2, a_3, and a_4 are 0.2738, -0.0282, 0.7281, 0.6208, and 0.9118, respectively, at $z = 0$. An updated version of the L_{TIR} calibration is given by [135]:

$$L_{\text{TIR}}(\text{IRAS}) = 2.403\nu L(25) - 0.2454\nu L(60) + 1.6381\nu L(100) \tag{7.10}$$

for the IRAS bands, or

$$L_{TIR}(\text{Spitzer}) = 1.559\nu L(24) + 0.7686\nu L(70) + 1.347\nu L(160) \tag{7.11}$$

for the Spitzer bands. An alternative calibration based on Spitzer data is given by [136]:

$$L_{TIR}(\text{Spitzer}) = 0.95\nu L(8) + 1.15\nu L(24) + \nu L(70) + \nu L(160) \tag{7.12}$$

The error in the total infrared luminosity determined using this alternative calibration is generally < 10% (the uncertainty changes as a function of the interstellar radiation field, and it can go up to ~ 30% in the worst cases) [136], and is slightly higher (~ 25%) using the calibration of [135]. The far-IR luminosity of normal galaxies ranges in between 10^{42} erg s^{-1} ≤ L_{FIR} ≤ $10^{44.5}$ erg s^{-1} (~ $10^8 L_\odot$ ≤ L_{FIR} ≤ $10^{11} L_\odot$) [137], and is slightly larger (in between 0.3 and 0.45 dex) when expressed as L_{TIR} [134]. The infrared luminosity of normal late-type galaxies is generally ≤ 25% of the stellar luminosity, with slightly higher values in AGNs [137–139]. The contribution of the infrared luminosity to the total bolometric luminosity of galaxies increases at higher infrared luminosities and can reach, in ultra luminous infrared galaxies, up to 95% of the total emitted energy [46].

As noticed by [134], the F_{TIR}/F_{FIR} ratio strongly changes with redshift because of a strong IR K-correction. The above calibrations are thus valid only at a redshift of zero. Various authors have thus tried to determine some empirical relationships between the infrared monochromatic luminosity at a given wavelength and the total IR or far-IR luminosity of galaxies with the idea of using different measurements of rest frame flux densities to estimate the total infrared luminosity of galaxies at different redshifts. The most widely used are the calibrations of [140, 141] for IRAS (12, 25 μm) and ISO (6.7, 15 μm) bands, or those of [142–144] for Spitzer (8, 24, 70, 160 μm) bands. The first calibrations done using Herschel data are presented in [145]. The uncertainty on the determination of L_{IR} using these simple relations is relatively high, up to 50%, thus significantly higher than that reached using several bands at the same time. In particular, total infrared luminosities determined using rest frame mid-IR bands (≤ 15 μm) are the most uncertain because of the contribution from PAHs, whose emission depends on the different physical properties of the ISM (metallicity, UV radiation field) in normal galaxies [107, 146] and starbursts [147].

7.2 Dust Mass and Temperature

The dust mass of an ideal cloud formed of spherical grains of uniform size, composition, and temperature, can be easily determined by following the calculations of [103] once the distance, D, the flux density, $S(\nu)$, and the temperature, T, of the cloud are known. Considering a cloud formed by N particles of radius a, the flux density of the cloud is

$$S(\nu) = N\left(\frac{\pi a^2}{D^2}\right) Q_a(\nu) B(\nu, T) \propto \nu^\beta B(\nu, T) \tag{7.13}$$

with $1 \leq \beta \leq 2$, from which we can derive N:

$$N = \frac{S(\nu) D^2}{\pi a^2 Q_a(\nu) B(\nu, T)} \tag{7.14}$$

The volume of the cloud is

$$V_{\text{cloud}} = N\frac{4}{3}\pi a^3 = \frac{4}{3} a \frac{S(\nu) D^2}{Q_a(\nu) B(\nu, T)} \tag{7.15}$$

Assuming a dust density ρ, the dust mass is then given by

$$M_{\text{dust}} = \rho N \frac{4}{3}\pi a^3 = \rho \frac{4}{3} a \frac{S(\nu) D^2}{Q_a(\nu) B(\nu, T)} \tag{7.16}$$

As discussed in [103], this relation for an ideal cloud can be generalized to realistic clouds and thus used to estimate dust masses of external galaxies once the typical, mean size and emission efficiency of the dust grains are known. This has been done by [148] on a sample of nearby, normal late-type galaxies using the IRAS 60 and 100 μm flux densities. The dust mass of their sample galaxies can thus be obtained from the relation

$$M_{\text{dust}}[M_\odot] = C\, S(100\,\mu\text{m}\,[\text{Jy}])\, D^2\,[\text{Mpc}](e^{(144/T)} - 1) \tag{7.17}$$

where C is a constant that depends on grain opacity $K_{100\,\mu\text{m}} = 3Q_{100\,\mu\text{m}}/4a\rho$, taken by [148] to be $C = 4.58$ for $K_{100\,\mu\text{m}} = 25\,\text{cm}^2\,\text{g}^{-1}$ [103], although other values of C can be adopted ($C = 1.81$ for a graphite grain opacity $K_{100\,\mu\text{m}} = 63\,\text{cm}^2\,\text{g}^{-1}$ and $C = 3.57$ for a silicate grain opacity $K_{100\,\mu\text{m}} = 32\,\text{cm}^2\,\text{g}^{-1}$ [149]). As suggested by [103], the dust temperature can be inferred from the observed $S(100\,\mu\text{m})/S(60\,\mu\text{m})$ flux density ratio assuming an emissivity law for the dust grains $Q_a(\lambda) \propto \lambda^{-\beta}$, with $\beta = 1$ which is appropriate for the $50\,\mu\text{m} \leq \lambda \leq 250\,\mu\text{m}$ spectral range, and that for typical temperatures on the ISM on the order of $\sim 20\text{--}30\,\text{K}$ scales as

$$T\,[\text{K}] \sim \frac{95.9}{\ln\left[7.69\frac{S(100\,\mu\text{m})}{S(60\,\mu\text{m})}\right]} \tag{7.18}$$

Recent observations of nearby galaxies with ISO, Spitzer and SCUBA allowed us to sample the far-IR sub-mm spectral domain, where the emission is due to the cold dust dominating in mass. These observations have consistently proved that, while in starburst galaxies such as Arp 220 the peak of the dust emission is in the 60–100 μm spectral range, with corresponding dust temperatures on the order of $T \sim 50\,\text{K}$ [150, 151], in normal galaxies it is shifted to longer wavelengths (100–200 μm) because of a significantly colder dust ($T \sim 20\,\text{K}$; [152, 153]). These

Table 7.1 The Milky Way grain opacity K_ν at different frequencies, from [98].

λ μm	K_ν cm² g⁻¹	Instrument
100	27.15	IRAS
160	10.20	Spitzer, AKARI
250	4.00	Herschel
350	1.90	Herschel
450	1.15	SCUBA
500	0.95	Herschel
850	0.38	SCUBA
1300	0.19	IRAM

Data taken from http://www.astro.princeton.edu/~draine/dust/dustmix.html (accessed August 2011).

observations have also shown that galaxies might have different dust components: one cold, associated with the diffuse ISM, and one warmer, associated with the star-forming HII regions. The determination of the total dust mass of galaxies using a simple relation based only on IRAS 60 and 100 μm data is thus highly uncertain. As the bulk of the dust is cold, in particular in normal galaxies, dust masses can only be accurately determined whenever the sub-mm fluxes at 450, 850 μm or 1.3 mm are available. The Herschel satellite significantly helps to overcome this problem, providing the community with sub-mm data for large samples of nearby and high-redshift galaxies. Equation (7.17) can be generalized as

$$M_{\text{dust}} = \frac{S(\nu) D^2}{K_\nu B(\nu, T)} \qquad (7.19)$$

where K_ν is the grain opacity at a given frequency (also called absorption cross-section per mass of dust, measured in cm² g⁻¹), whose value can be approximatively estimated with ~ 10% uncertainty by the relations

$$K_\nu \,[\text{cm}^2\,\text{g}^{-1}] \sim 2.42 \times 10^5 \left(\frac{\lambda}{\mu m}\right)^{-2} \qquad (7.20)$$

for $20 < \lambda < 700$ μm, and

$$K_\nu \,[\text{cm}^2\,\text{g}^{-1}] \sim 3.15 \times 10^4 \left(\frac{\lambda}{\mu m}\right)^{-1.68} \qquad (7.21)$$

for 700 μm $< \lambda < 10^4$ μm (from the relations given by [154], modified as suggested by [98]).[1] The updated values of K_ν for the most important photometric infrared, submillimeter and millimeter bands are given in Table 7.1.

1) Tabulated values of K_ν can be found in the B. Draine personal web-page, at the following address: http://www.astro.princeton.edu/~draine/dust/dustmix.html (accessed August 2011).

Vlahakis et al. [155] have shown how SCUBA 450 and 850 μm can be used to estimate the temperature and mass of the different dust components. Assuming that the flux density of a given galaxy is due to a warm and a cold dust component, the flux density $S(\nu)$ at a given frequency is given by the relation

$$S(\nu) = N_W \times \nu^\beta B(\nu, T_W) + N_C \times \nu^\beta B(\nu, T_C) \tag{7.22}$$

where N_W and N_C represent the relative dust masses in the warm and cold components, T_W and T_C the dust temperatures, $B(\nu, T)$ the Planck function, and β the emissivity index. Assuming a value for β, ($\beta = 2$ for [155]), N_W, N_C, T_W, and T_C are determined by fitting the infrared to sub-mm galaxy spectral energy distribution. The dust mass is then given by

$$M_{dust} = \frac{S(850\,\mu m)\,D^2}{K_d(850\,\mu m)} \times \left[\frac{N_C}{B(850\,\mu m, T_C)} + \frac{N_W}{B(850\,\mu m, T_W)} \right] \tag{7.23}$$

where $B(850\,\mu m, T)$ is the Planck function at 850 μm and $K_d(850\,\mu m)$ is the dust mass opacity coefficient at 850 μm, defined as $K_{850\,\mu m} = 3Q_{850\,\mu m}/4a\rho$. Vlahakis et al. [155] assume $K_{850\,\mu m} = 0.077\,m^2\,kg^{-1}$, which is an intermediate value for graphite and silicates, as given by [156, 157].

An alternative but quite indirect method is that of considering a metallicity-dependent gas to dust ratio and assuming the relation [158]

$$M_{dust} = M_{gas}\,Z\,\epsilon\,f \tag{7.24}$$

where M_{gas} is the gas mass, Z the metallicity relative to the solar metallicity, ϵ is the ratio of the mass of metals in the dust to the total mass of metals, and f is the ratio of the mass of metals to the mass of gas for gas with solar metallicity, that [158] assume to be $f = 0.019$. This method is quite uncertain given the difficulty in determining the molecular gas mass, the later often determined by inverting this equation (see Section 11.1.2). This difficulty should be added to the large uncertainty in the absolute metallicity calibration.

In the last years several models have been developed to reproduce the dust emission of galaxies. These models are generally accessible on the web, and can thus be used to study different kind of galaxies. Some of them are mostly empirical and are constrained by simultaneously fitting the observed stellar and dust spectral energy distributions with population synthesis models by adopting a realistic dust attenuation law, and a star formation history to reproduce the stellar population of the galaxy. They also assume that the absorbed radiation is re-emitted in the far-IR by different dust components associated with PAHs, to the cold, diffuse ISM and to the hot star-forming regions [159]. Considering different relative geometries of the dust and stars within the galaxies might also be required [160, 161]. Other models also take into account the physical and chemical properties of the different dust components such as PAHs, amorphous silicate, and carbonaceous grains [136]. Dust masses and temperatures are some of the outputs of these models. By comparing dust masses of a subset of the SINGS sample determined using flux densities limited to $\lambda \leq 160$ μm to those measured including 450 and 850 μm data, Draine et al. [162] have shown that they randomly differ by $\sim 50\%$. Muñoz-Mateos et al. [163] applied the models of [136] to the SINGS galaxy sample and later derived some empirical relations to estimate the total dust mass of galaxies directly using far-IR Spitzer data:

$$M_{\text{dust}}[M_\odot] = \frac{4\pi D^2 [\text{Mpc}]}{1.616 \times 10^{-13}} \left(\frac{\langle \nu S(\nu)\rangle_{70}}{\langle \nu S(\nu)\rangle_{160}}\right)^{-1.801}$$
$$\times \frac{0.95\langle \nu S^{ns}(\nu)\rangle_8 + 1.150\langle \nu S^{ns}(\nu)\rangle_{24} + \langle \nu S(\nu)\rangle_{70} + \langle \nu S(\nu)\rangle_{160}}{[\text{erg s}^{-1}\text{ cm}^{-2}]} \quad (7.25)$$

where $S^{ns}(\nu)$ are the flux densities with a subtracted stellar contribution, determined by extrapolating the stellar continuum from the 3.6 μm flux density, whose mean values are $\langle S(\nu)_\lambda\rangle/\langle S(\nu)_{3.6}\rangle = 0.660, 0.453, 0.269$, and 0.032 for $\lambda = 4.5$, 5.8, 8.0 and 24 μm, respectively. In the absence of IRAC data, dust masses can be measured using the following relation

$$M_{\text{dust}}[M_\odot] = \frac{4\pi D^2 [\text{Mpc}]}{1.616 \times 10^{-13}} \left(\frac{\langle \nu S(\nu)\rangle_{70}}{\langle \nu S(\nu)\rangle_{160}}\right)^{-1.801}$$
$$\times \frac{1.559\langle \nu S^{ns}(\nu)\rangle_{24} + 0.7686\langle \nu S(\nu)\rangle_{70} + 1.347\langle \nu S(\nu)\rangle_{160}}{\text{erg s}^{-1}\text{cm}^{-2}}$$
$$(7.26)$$

Normal galaxies have dust masses on the order of $M_{\text{dust}} = 10^6-10^8 M_\odot$, with $\sim 3\%$ in PAHs, this fraction being slightly smaller ($\sim 1\%$) in low-metallicity systems. The dust mass represents $\sim 0.5\%$ of the gaseous phase [162]. Normal galaxies have a dust temperature of $\sim 15-25$ K in the diffuse ISM and $\sim 40-60$ in star-forming HII regions [159, 164]. Starburst and ultra-luminous infrared galaxies (ULIRGs) have a warm dust component associated with the starburst, with temperatures of $\sim 30-60$ K, with a peak of the emission at 60 μm. In Seyfert galaxies there is an extra component with a temperature on the order of $\sim 150-250$ K peaking at 25 μm, probably heated by the central AGN [46].

8
Radio Properties

Radio data are fundamental for deriving several important properties of the emitting sources such as the presence of an active nucleus, measuring the activity of star formation, or determining the intensity of the magnetic field. The high angular resolution of interferometric observations are also crucial for measuring the size of the nuclear emitting regions down to parsec scales. The study of the radio properties of galaxies can be done once the possible nuclear contamination is determined and subtracted, and the synchrotron and free–free contribution are separated. In normal galaxies, the central radio source is brighter than the diffuse, disk emission, but its contribution to the total radio emission is generally relatively small ($\lesssim 20\%$; [165]). The most luminous radio sources in normal galaxies are generally compact and confined in the nuclei of interacting systems [166].

8.1
Determining the Contribution of the Different Radio Components

8.1.1
Synchrotron vs. Free–Free Radio Emission in the Centimeter Domain

The contribution of the thermal and nonthermal components to the radio emission of galaxies can be obtained by fitting the centimeter radio continuum spectral energy distribution (in the range between ~ 1 and $50\,\mathrm{cm}$) with a thermal (free–free) and nonthermal (synchrotron) component [120, 167]:

$$S(\nu) = C_{\mathrm{ff}} \nu^{-0.1} + C_{\mathrm{syn}} \nu^{-\alpha_{\mathrm{syn}}} \tag{8.1}$$

The fit also provides the spectral slope of the synchrotron emission α_{syn}, which in normal galaxies is generally $\alpha_{\mathrm{syn}} \simeq 0.8$. The contribution of the thermal emission to the total radio continuum emission of galaxies can also be determined using hydrogen recombination lines [168, 169], assuming a thermal electron temperature T_e. Niklas et al. [120] give the following relation for estimating the thermal radio continuum flux density using H_α data, assuming a thermal electron temperature

A Panchromatic View of Galaxies, First Edition. Alessandro Boselli.
© 2012 WILEY-VCH Verlag GmbH & Co. KGaA. Published 2012 by WILEY-VCH Verlag GmbH & Co. KGaA.

of $T_e = 10^4$ K:

$$S_{ff}(\nu)\,[\text{mJy}] = 2.238 \times 10^9\, F(H_\alpha)\,[\text{erg cm}^{-2}\,\text{s}^{-1}](T_e\,[\text{K}])^{0.42}$$
$$\times \left[\ln\left(\frac{0.04995}{\nu\,[\text{GHz}]}\right) + 1.5\ln(T_e\,[\text{K}])\right] \quad (8.2)$$

An alternative relation using H_β is given by [168] and [166],

$$F(H_\beta)\,[\text{erg cm}^{-2}\,\text{s}^{-1}] \sim 0.28 \times 10^{-12} \left(\frac{T_e\,[\text{K}]}{10^4}\right)^{-0.52} (\nu\,[\text{GHz}])^{0.1}\, S_{ff}(\nu)\,[\text{mJy}] \quad (8.3)$$

where the ratios of other frequently observed optical or near-IR recombination line fluxes to H_β, $F(H_\beta)$, are:

$$\frac{F(H_\alpha)}{F(H_\beta)} = 2.86 \left(\frac{T_e}{10^4\,\text{K}}\right)^{-0.07} \quad (8.4)$$

$$\frac{F(Br_\alpha)}{F(H_\beta)} = 0.079 \left(\frac{T_e}{10^4\,\text{K}}\right)^{-0.36} \quad (8.5)$$

$$\frac{F(Br_\gamma)}{F(H_\beta)} = 0.028 \left(\frac{T_e}{10^4\,\text{K}}\right)^{-0.24} \quad (8.6)$$

8.1.2
The Emission of the Cold Dust Component at $\lambda \leq 1.5$ mm

At wavelengths shorter than ~ 1.5 mm, the emission is dominated by the cold dust component. The contribution of dust to the emission around 1 mm can be determined only after subtracting the contribution of synchrotron and free–free emission from the continuum, and whenever the observed spectral range includes important molecular lines, their contamination. This is indeed the case for nearby galaxies at 1.3 mm, where the continuum emission can be contaminated by the ^{12}CO(2−1) line at 230 GHz and, to a minor extent, by the ^{13}CO(2−1) line at 220 GHz, CS(5−4) at 244 GHz, HCN(3−2) at 266 GHz and C^{18}O(2−1) at 219 GHz. The synchrotron and free–free contributions can be easily determined following the prescriptions that we just described in the previous section, and extrapolating the fit to the appropriate millimeter wavelength and integrating it within the bandpass for the continuum observations. This contribution is generally small, on the order of a few percent, as shown in Figure 8.1 [39, 170].

The contribution of any emission line to the 1.3 mm dust continuum can be estimated using the prescriptions of [39, 170, 171]. If

$$I_{\text{line}}\,[\text{K km s}^{-1}] = \int_{\text{line}} T_{\text{mb}}(\text{line})\,d\nu \quad (8.7)$$

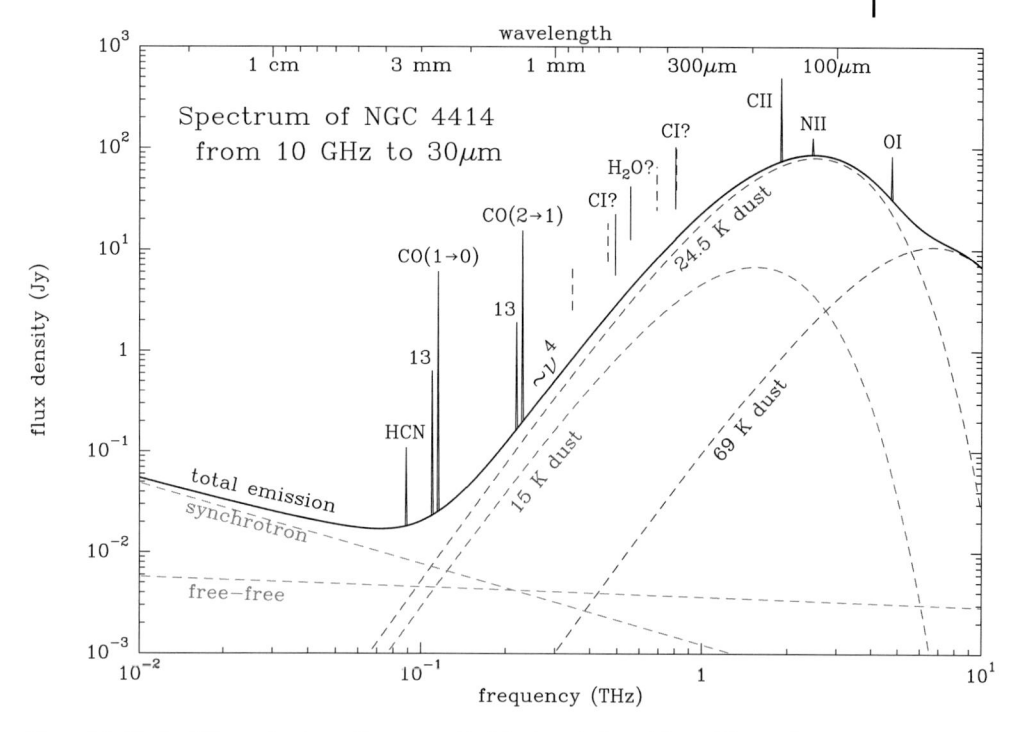

8.1 Determining the Contribution of the Different Radio Components

Figure 8.1 The far-IR to centimeter radio spectral energy distribution of the nearby late-type spiral galaxy NGC 4414 [172]. The spectrum shows the continuum and the major emission lines. Dashed lines indicate the modified black body emission of three different dust components at a temperature of 69, 24.5, and 15 K respectively, and the radio continuum emission due to synchrotron and free–free. Reproduced with permission (c) ESO.

is the specific intensity of the line (expressed in terms of surface brightness temperature) and the Gaussian beam Ω_{beam} is defined as

$$\Omega_{\text{beam}} = \frac{2\pi \theta_{\text{FWHM}}^2}{8 \ln 2} \tag{8.8}$$

where θ_{FWHM} is the FWHM of the beam, then the contribution of the emitting line to the continuum emission measured with the bolometer can be determined using the relation

$$S_{\text{line}} = \frac{2k\nu^3}{c^3 \Delta \nu_{\text{bol}}} \Omega_{\text{beam}} I_{\text{line}} \tag{8.9}$$

where $\Delta \nu_{\text{bol}}$ is the bolometric bandwidth. I_{line} can either be the surface brightness of the line within a single beam, which should be compared to the continuum flux density within the beam, or the total line emission of a galaxy, to be compared to the total galaxy continuum flux density. For the ^{12}CO(2−1), Eq. (8.9) reduces to

$$S_{\text{line}} \, [\text{mJy}] = \frac{3.31 \times 10^{-2}}{\Delta \nu_{\text{bol}} \, [\text{GHz}]} (\theta_{\text{FWHM}} \, [\text{arcsec}])^2 \, I_{\text{CO}(2-1)} \, [\text{K km s}^{-1}] \tag{8.10}$$

In normal galaxies, the ^{12}CO(2–1) line emission contributes $\sim$ 10% of the continuum emission within a $\sim$ 50–70 GHz wide band. An extra $\sim$ 1% should be added to account for the contribution of other emitting lines. The contribution of the dust emission to the total observed continuum is then given by

$$S_{\text{dust}} = S_{\text{total}} - S_{\text{ff}} - S_{\text{syn}} - S_{\text{line}} \tag{8.11}$$

8.2 The Radio Luminosity

The radio luminosity of a galaxy at a given frequency is given by the relation

$$L_{\text{radio}}(\nu) = 4\pi D^2 S(\nu) \tag{8.12}$$

where $S(\nu)$ is the flux density at a given frequency. At 1.4 GHz, normal galaxies have luminosities in the range $\sim 10^{19}$ W Hz^{-1} $\leq L_{\text{radio}} \leq 10^{23}$ W Hz^{-1} [166], while radio-loud AGNs lie in the luminosity range 10^{20} W Hz^{-1} $\leq L_{\text{radio}} \leq 10^{25}$ W Hz^{-1} [173]. The radio continuum luminosity of a normal galaxy only accounts for $< 10^{-4}$ of its bolometric luminosity [166]. Diffuse radio emission is associated with most of the star-forming and starburst galaxies but also to some ellipticals and lenticulars, in particular those with some residual star formation, although these early-type galaxies are generally characterized by a compact radio source. Some nuclear emission can also be present in all kind of galaxies, and is dominant in active radio sources. Ultra-luminous infrared galaxies also have compact radio emission [174].

The radio luminosity corresponding to the synchrotron emission is given by the relation [74, 121, 175]

$$L_{\text{syn}}(\nu) \, [\text{erg s}^{-1} \, \text{Hz}^{-1}] = 4\pi \left(\frac{D}{[\text{cm}]}\right)^2 \frac{I(\nu)}{[\text{erg s}^{-1} \, \text{cm}^{-2} \, \text{Hz}^{-1} \, \text{rad}^{-2}]} \frac{\Omega}{[\text{rad}^2]} \tag{8.13}$$

where Ω is the surface of the emitting source (in rad^{-2}). $I(\nu)$ is given by

$$I(\nu) \, [\text{erg s}^{-1} \, \text{cm}^{-2} \, \text{Hz}^{-1} \, \text{rad}^{-2}] = 13.5 \times 10^{-23} b(\alpha_{\text{syn}}) l K$$

$$\times \left(\frac{B}{[\text{G}]}\right)^{1+\alpha_{\text{syn}}} \left(\frac{\nu \, [\text{Hz}]}{6.26 \times 10^{18}}\right)^{-\alpha_{\text{syn}}} \tag{8.14}$$

for a random magnetic field B, where $b(\alpha_{\text{syn}})$ is a parameter which only weakly depends on the spectral index α_{syn} (and is $b(\alpha_{\text{syn}}) = 0.0806$ for $\alpha_{\text{syn}} = 0.8$ [121]), l is the linear size of the emitting source along the line of sight, and K the volume density of ultra-relativistic electrons (see Eq. (5.3)). This relation is also valid for the synchrotron emission of galaxies outside the radio domain, i.e., in the optical and X-ray regimes.

The radio continuum emission of galaxies at 20 cm, dominated by the nonthermal component, is tightly correlated with the recent activity of star formation [176–178]. This is due to the fact that the synchrotron emission is due to ultra-relativistic electrons accelerated in supernovae remnants losing their energy in weak magnetic fields. Equation (8.14) shows that the synchrotron component of the radio luminosity is proportional to the density of electrons, K, whose value is proportional to the number of supernovae and is thus directly related to the youngest stellar population. The contribution of the total and regular magnetic field to the synchrotron emission of galaxies requires polarization measurements and the assumption of an energy-density equipartition between cosmic rays and magnetic fields [179]. Observations have shown that the random magnetic field is dominant along spiral arms, while the regular component has an increasing importance in the interarm regions [179, 180].

In the Milky Way, where the magnetic field can be measured, the synchrotron luminosity can be directly related to the supernova rate, ν_{SN}, by the relation [166] (see Section 13.8)

$$L_{syn}\,[\mathrm{W\,Hz^{-1}}] \sim 13 \times 10^{22} \left(\frac{\nu}{\mathrm{GHz}}\right)^{-\alpha_{syn}} \left(\frac{\nu_{SN}}{[\mathrm{yr^{-1}}]}\right) \tag{8.15}$$

Given the small scatter and large dynamic range in the far-infrared–radio correlation, it is expected that this relation is valid not only for normal spiral galaxies but also in starbursts such as M82 [166].

The free–free component is also related to the recent star formation activity since it is typically produced in the ionized gas of HII regions (Section 13.8). Indeed, the thermal emission can be used to estimate the production rate of Lyman continuum photons, N_{UV} [166]:

$$N_{UV}\,[\mathrm{s^{-1}}] \geq 6.3 \times 10^{52} \left(\frac{T_e}{10^4\,[\mathrm{K}]}\right)^{-0.45} \left(\frac{\nu}{[\mathrm{GHz}]}\right)^{0.1} \left(\frac{L_{ff}\,[\mathrm{W\,Hz^{-1}}]}{10^{20}}\right) \tag{8.16}$$

9
The Spectral Energy Distribution

The energetic output of any extragalactic source can be determined by constructing its spectral energy distribution (SED). The stellar component emits in the UV to near-IR domain, with young and massive stars dominating the UV[1] while old stars are the main contributors in the near-IR. Dust, produced by the aggregation of metals injected into the interstellar medium by massive stars through stellar winds, efficiently absorbs the stellar light, particularly at short wavelengths, and re-emits it in the infrared domain (5 µm–1 mm). At longer wavelengths, the emission of normal galaxies is generally dominated by the loss of energy of relativistic electrons accelerated in supernovae remnants [176, 177] spinning in weak magnetic fields (synchrotron emission). Reconstructing SEDs is thus of fundamental importance for quantifying the relative contribution of the different emitting sources to the bolometric emission of galaxies and to study the physical relations between the various galaxy components (interstellar radiation field, metallicity, dust and gas content, magnetic fields, etc.). At the same time, SEDs are necessary for measuring K-corrections and photometric redshifts (see Appendix A.1).

Several works have been devoted to the reconstruction, the study, and the modelization of the SEDs of different kind of galaxies. Some of them were sampling the whole electromagnetic spectrum, from UV to centimeter radio wavelengths, of different extragalactic sources [140, 152, 182–185], including AGNs [186], and others were focused on the stellar continua [85, 187–189] or on the infrared emission of dust [134, 135, 153, 190–192]. Physical models have been proposed to reproduce the observed UV to radio continuum SEDs of galaxies (e.g., [160]).

The observed spectral energy distribution of any extragalactic source can be determined by plotting the variation of the energy, defined as $\nu S(\nu)$ or $\lambda S(\lambda)$, as a function of frequency ν or wavelength λ, where $S(\nu)$ is the flux density measured in Jansky or $W\,m^{-2}\,Hz^{-1}$ and $S(\lambda)$ in $W\,m^{-2}\,Å^{-1}$. Today it is common for astronomers to identify spectral energy distributions with the spectral distribution of flux densities by tracing the variation of $S(\nu)$ or $S(\lambda)$ with ν or λ. Whenever constructed using photometric data obtained only in broad band filters, spectral energy distributions reduce to very low resolution spectra and their mean properties can

---

[1] The UV emission of early-type galaxies is however due to an old stellar population responsible for the UV upturn [78, 181].

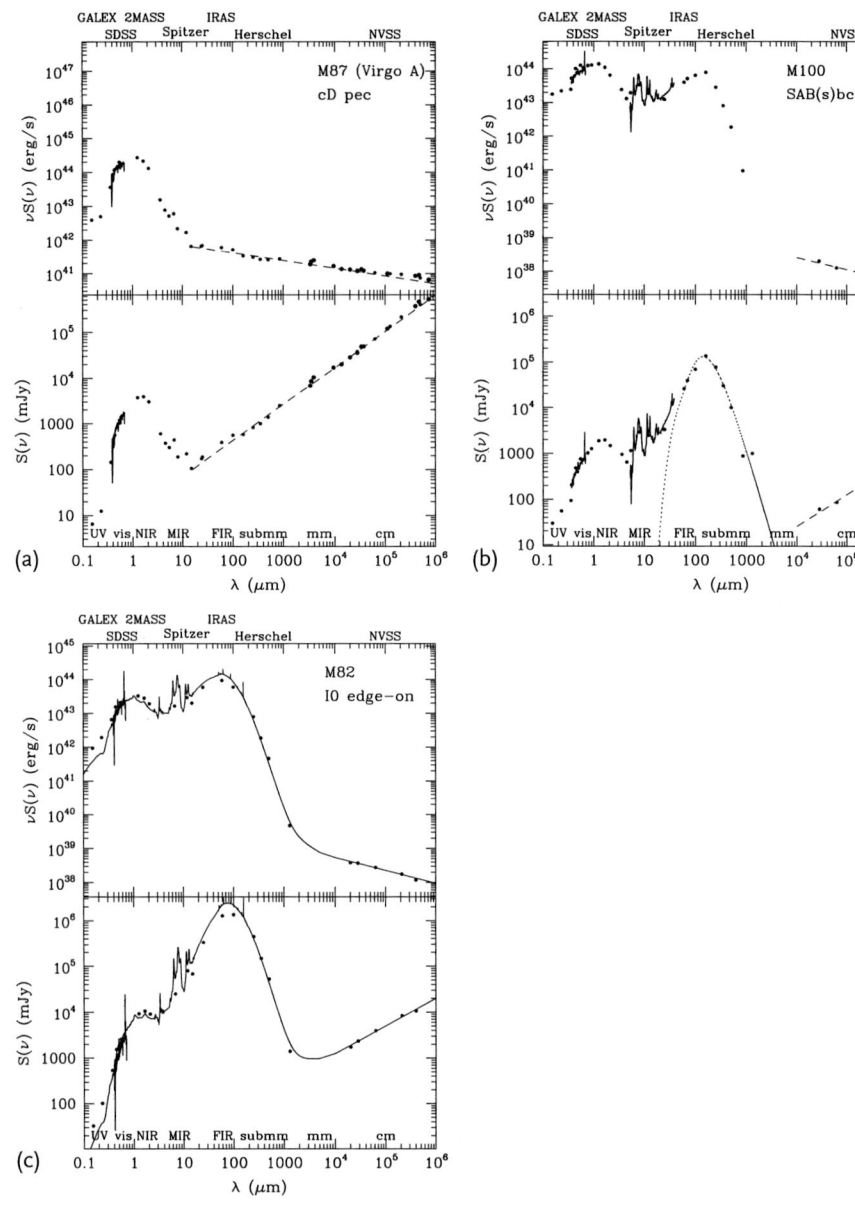

Figure 9.1 The observed UV to centimeter radio spectral energy distribution of (a) the elliptical galaxy M87; (b) the spiral M100, and (c) the starburst M82, adapted from [194]. The Y-axis gives the energy (upper panel) and the flux density (lower panel). Open dots indicate photometric data, the solid dark line spectroscopic data, and the solid light line a model template of M82 (kindly made available by D. Elbaz). The dashed line shows a power law due to synchrotron emission (see Section 5.1.2), whose emission is dominant in the radio galaxy M87 (Virgo A) down to the infrared domain. The dotted line shows two modified black bodies ($\beta = 2$) at 40 and 20 K (see Section 4.1).

be determined through the analysis of the colors. Broad and narrow band photometry can be combined with spectroscopy to construct spectral energy distributions, provided that the data are taken within similar apertures [193] (otherwise aperture corrections are required) and that the analyzed source is not confused. Photometric data (magnitudes or fluxes) must be transformed to rest frame monochromatic fluxes (flux densities) to account for the transmissivity of the adopted filter.

Figure 9.1 shows the observed UV to centimeter radio spectral energy distribution of three representative galaxies in the nearby universe, the elliptical M87, the spiral M100 and the starburst M82 (other examples of SEDs of starburst and normal galaxies can be found in Figure 12.2). The difference in the observed spectral energy distribution of these three galaxies is striking. The elliptical galaxy M87 emits most of its energy through stellar emission in the optical-near-infrared bands, despite the fact that this object is a powerful radio galaxy (Virgo A). As all ellipticals, M87 is characterized by a very red optical spectrum with no emission lines, and by a very weak UV emission. Devoid of dust, as most elliptical galaxies are, and hosting a powerful radio source, its near-infrared to centimeter radio emission is dominated by synchrotron emission [194]. The spiral galaxy M100 has a relatively blue stellar spectrum with prominent emission lines and an important UV emission. This galaxy emits $\sim$ 30% of its energy in the infrared domain ($\sim$ 5 µm–1 mm). Its infrared spectrum is characterized by the presence of PAHs in the 5–15 µm spectral range. For $\lambda > 1$ cm the spectral energy distribution has the power spectrum shape typical of synchrotron emission. The starburst galaxy M82 has a UV, optical, and near-infrared spectrum similar to that of M100, with a relatively blue color and prominent emission lines. Its energy output, however, is dominated by the dust emission in the infrared domain (> 50%). This strong infrared emission suggests that most of the emitted stellar light is absorbed by dust. Because dust extinction is more important in the ultraviolet than in the optical or near-infrared (see Chapter 12), the underlying stellar population of M82 is certainly much bluer and younger than that of M100, despite the fact that their observed stellar spectral energy distributions look similar.

9.1
The Emission in the UV to Near-Infrared Spectral Domain

The continuum emission of galaxies in the spectral range 912 Å $< \lambda <$ 5 µm is dominated by the stellar emission. Modified by the presence of dust which absorbs the stellar light (see Chapter 12), the shape of the spectral energy distribution in this spectral domain is strongly tied to the relative weight of the underlying stellar populations composing galaxies which, in turn, depend on their past star formation history [195].

The mean rest frame 2000 Å–200 µm spectral energy distributions of different types of extragalactic sources are shown in Figure 9.2. This figure shows that the spectral energy distribution of normal galaxies becomes bluer going from quiescent systems (ellipticals, lenticulars) to star-forming spirals (Sa-Sd) and starbursts

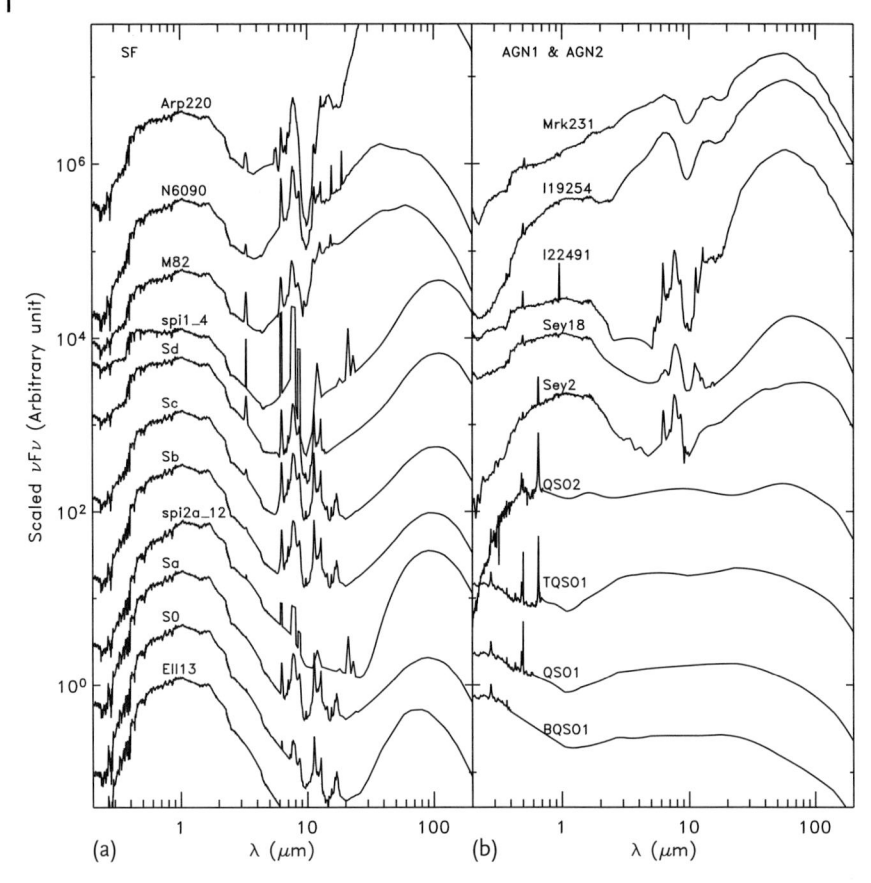

Figure 9.2 The optical to far-infrared spectral energy distributions of different kinds of extragalactic sources in units of $\nu S(\nu)$ vs. λ, from [186]. (a) From top to bottom, starbursts, spirals of decreasing morphological type, lenticulars, and ellipticals; (b) from top to bottom, six type 2 and three type 1 AGNs. Reproduced by permission of the AAS.

such as M82 or Arp 220. The shape of the SED does not significantly change from galaxy to galaxy in the near infrared, but it does at optical and UV wavelengths. In early-type objects, the emission peaks at about 6000–10 000 Å, and drops significantly at $\lambda \leq 4500$ Å (in energy units, the energy output $\nu S(\nu)$ is maximal at $\sim 1.6\,\mu$m). On the contrary, late-type galaxies have blue spectra. Another obvious difference between early- and late-type galaxies is the presence of emission lines in star-forming systems.

Compared to normal galaxies, quasars are bluer in optical and ultraviolet bands and redder in the near-infrared. This is because their spectra obey a power law

$$S(\lambda) \propto \lambda^{\alpha_\lambda} \propto \lambda^{-(\alpha_\nu+2)} \tag{9.1}$$

with a frequency spectral index of $\alpha_\nu = -0.44$ in the spectral range 1300 Å $< \lambda <$ 5000 Å, and $\alpha_\nu = -2.45$ for 5000 Å $< \lambda <$ 9000 Å, as shown in Figure 9.3 [85]. The

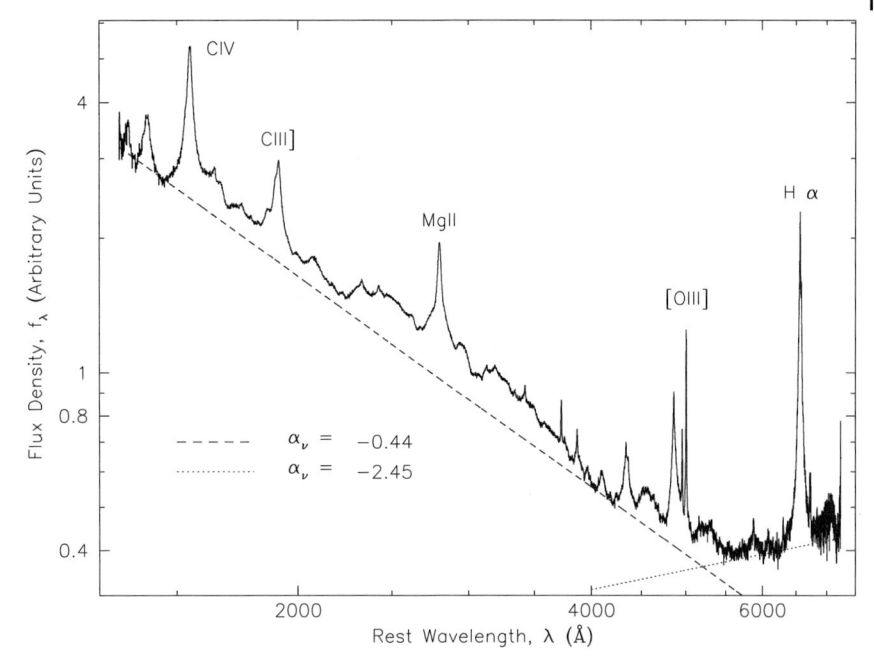

Figure 9.3 The rest frame composite quasar spectrum with a resolution of ∼ 1 Å obtained by combining the spectra of more than 2200 quasars in the SDSS, from [85]. Reproduced by permission of the AAS.

abrupt change in slope observed at ∼ 5000 Å is due to the increased contamination by the stellar emission of the host galaxy at long wavelengths with respect to the emission of the accretion disk [85]. Seyfert galaxies, where the emission of the central active nucleus is combined with that of the stellar population inhabiting the host galaxy, have spectral energy distributions in between that of normal galaxies and quasars. The observed dispersion and difference in the spectral indices with respect to those determined in other works might indicate an intrinsic variation of these parameters in different categories of galaxies [85].

9.1.1
UV, Optical, and Near-IR Colors

UV, optical, and near-infrared colors, defined as the ratio of the fluxes observed within different bands, are generally expressed in magnitudes as

$$m_1 - m_2 = -2.5 \log \frac{F_1}{F_2} \quad (9.2)$$

where F_1 and F_2 are the fluxes within filter 1 and 2, respectively. Colors can be used to quantitatively trace the shape of the galaxy spectral energy distribution. Galaxies predominantly composed of young stellar populations (spirals, irregulars, BCD) have blue colors while those dominated by evolved stars (elliptical, lenticulars) are

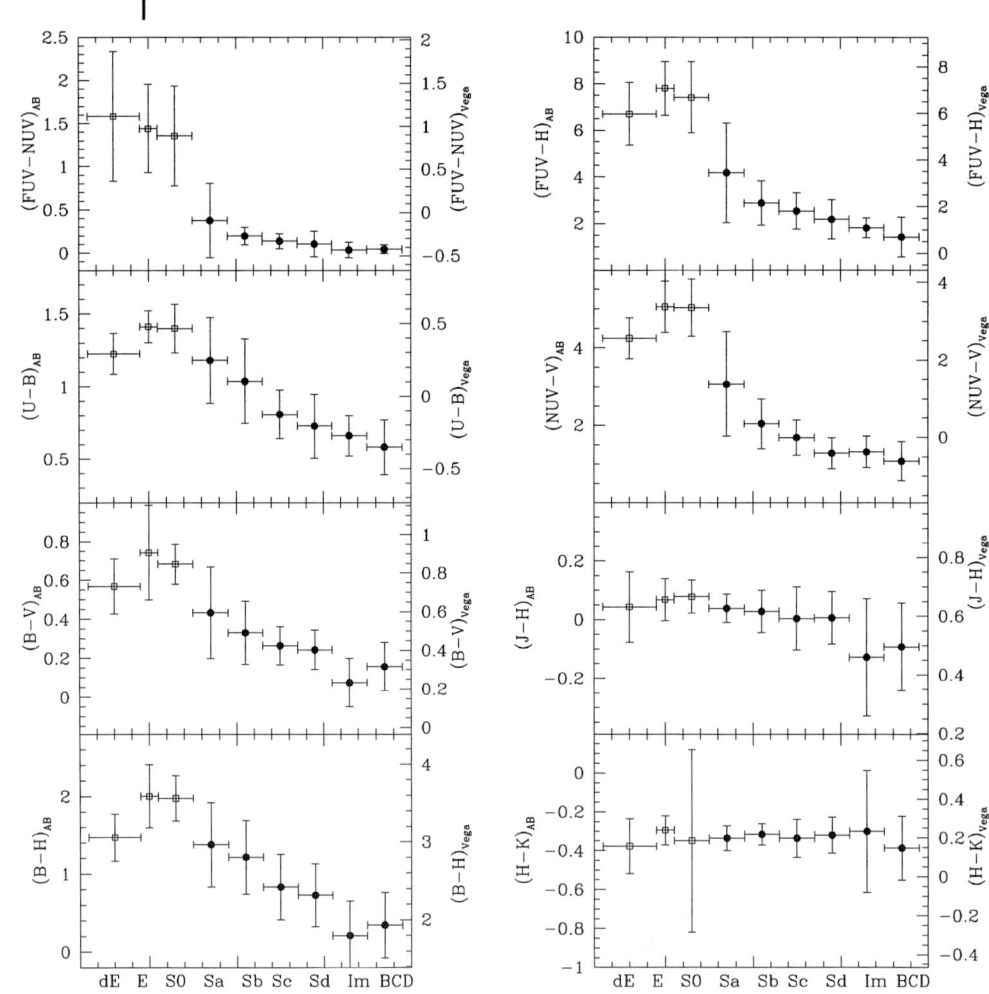

Figure 9.4 The mean extinction-corrected UV, optical, and near-infrared colors (in AB and Vega systems) of nearby galaxies of different morphological classes. UBVJHK are in the Johnson photometric system, FUV and NUV are in the GALEX ultraviolet bands.

generally red [37, 195–197]. Figure 9.4 shows the average UV, optical, and near-infrared colors of normal galaxies of different morphological types. Differences between the various morphological classes are more evident whenever the color index samples a large spectral range (UV to near-infrared). Early-type spirals (Sa-Sab) are generally redder than late-type spirals (Sc-Sd) or irregulars (Im, BCD). Similarly, dwarf ellipticals and spheroidals are bluer than giant ellipticals and lenticulars, except when observed in the far-UV band (FUV), where the emission of ellipticals is boosted by the presence of evolved stars (UV upturn [78]).

Mean colors for different morphological classes in other color systems can be found in [197]. Once separated into quiescent (E-S0-S0a, dE-dS0) and star-forming systems (Sa-Sm-Im-BCD), however, UV, optical, and near-infrared colors (and

SEDs) of galaxies are more tightly related to the stellar mass than to morphological type [189, 198]. The variation in color with stellar mass is more important for star-forming systems (Sa-Sm-Im-BCD) than for quiescent objects (E-S0-S0a, dE-dS0). In late-type galaxies the variation in color is principally due to different star formation histories, while in early-types the change in color results from the combined effects of age and metallicity. Color diagrams can also be used to distinguish stars from quasars, whose compact nature make them morphologically similar [199]. Some confusion with white dwarfs is possible, and selection biases may arise because bright QSO emission lines may perturb the color–color distribution in some filter combinations for given redshift ranges.

9.1.2
Fitting SEDs with Population Synthesis Models

In recent years, the study of stellar evolution has made great progress, making it possible to construct stellar population synthesis models which are suitable for fitting and reproducing the stellar spectral energy distribution of galaxies (for an extended review on this topic we refer the reader to the recent work of [200]). Models of stellar evolution have been used to compute stellar evolutionary tracks that cover the largest possible range in the parameter space of masses and metallicities. They sample all of the main evolutionary phases, from the main sequence (MS) to the asymptotic giant branch (AGB), RGB, HB, and white dwarf populations. Among these, the most widely used are those obtained by the Padova group [79, 201–203] and the Geneva group [204–206].

Evolutionary tracks trace the variation in luminosity, effective temperature, and gravity as a function of time for stars of different mass and metallicity. Combined with observed or synthetic libraries of stellar spectra, they can be used to predict the UV to near-infrared spectra of a star of any mass and metallicity at a given epoch of its evolution. Observed spectra [207–210], which are preferred to synthetic spectra for their high spectral resolution ($R \sim 2000-40\,000$, corresponding to $\sim 2.5-0.1$ Å), are generally limited to the optical window, and are still undersampled for the most massive O stars or low-metallicity objects. Synthetic stellar spectra of model atmospheres [80, 211], which generally have a lower spectral resolution than observed spectra ($R \sim 200$, corresponding to $\sim 10-100$ Å depending on the wavelength; see however the POLLUX database, [212]), have the advantage of covering the whole UV to near-infrared spectral range of stellar emission.

Assuming an initial mass function and a given star formation history, population synthesis models sum up the contribution of stars of different mass, metallicity, and age to create the composite spectrum of a given galaxy. By tracing stellar evolution, they reproduce the time evolution of the UV to near-infrared spectral energy distribution of a model galaxy. Several of these population synthesis models are available on the net: among these, the most widely used are GALAXEV [213], GRASIL [160], Maraston [77, 214, 215], PEGASE [216], for which there exists a high resolution version PEGASE-HR [217], and Starburst99 [218, 219].

Population synthesis models do not always account for the effects of dust (i.e., reddening of the spectra). For this reason, the observed SED must be first corrected for dust attenuation as prescribed in Chapter 12. Extinction-corrected spectral energy distributions of the observed galaxies can be fitted assuming different star formation histories. In starburst galaxies, the star formation history is generally assumed to have two components: constant on relatively long time scales related to the secular evolution of the host galaxy, with superimposed one (or more) instantaneous or short duration episodes of strong activity (burst). In normal galaxies, the star formation activity is generally represented with an exponentially declining law,

$$\mathrm{SFR}(t, \tau_{\exp}) = \frac{1}{\tau_{\exp}} e^{-\frac{t}{\tau_{\exp}}} \tag{9.3}$$

or with a delayed exponentially declining law, called a Sandage law [189, 220]

$$\mathrm{SFR}(t, \tau_{\mathrm{San}}) = \frac{t}{\tau_{\mathrm{San}}^2} e^{-\frac{t^2}{2\tau_{\mathrm{San}}^2}} \tag{9.4}$$

Here, t is the age of the galaxy and $\tau_{\exp}$ and τ_{San} are the characteristic time scales for star formation for an exponentially declining and a Sandage star formation law, respectively. The time dependence of the exponentially declining and Sandage relations are shown in Figure 9.5 for different characteristic time scales. Once the age of the galaxy is fixed (generally between 12 and 13.5 Gyr), the characteristic time scale for star formation ($\tau_{\exp}$, τ_{San}, or the age and the duration of the burst) can be determined by fitting the extinction-corrected UV to near infrared spectral energy distribution of the observed galaxies. A wide sampling of the SED is required for constraining the content of both the young (UV) and old (near infrared) stellar populations. Delayed exponentially declining star formation laws are preferred because they are more realistic (they reproduce the expected growing of activity in the protogalaxy and have shapes similar to those obtained by more sophisticated models of galaxy evolution [221]) and because they are less sensitive to the degeneracy in the t over $\tau_{\exp}$ ratio typical of exponentially declining star formation laws [189].

9.2
The Dust Emission in the Infrared Domain

The importance of the infrared domain explored by IRAS, ISO, Spitzer, AKARI, and Herschel resides in the fact that dust absorbs an important fraction of the stellar light of galaxies, modifying their spectral energy distribution. The absorbed energy is re-radiated by dust in the infrared domain. In normal galaxies, $\sim$ 30% of their total energy is emitted in the infrared, while in strong starbursts such as Arp 220 this fraction can be as high as 90%. Fitting infrared SEDs is necessary for measuring the dust properties such as total infrared luminosity, mass, temperature, fraction of PAHs, hardness of the interstellar radiation field (ISRF), and so

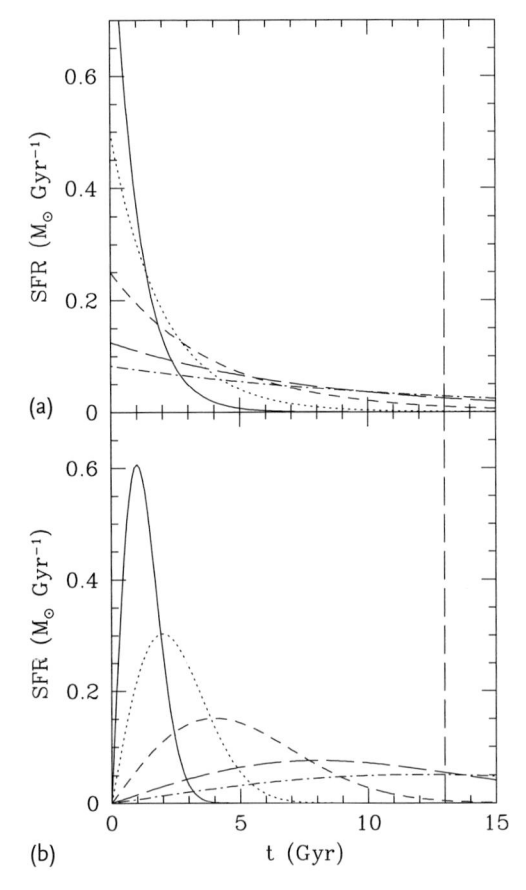

Figure 9.5 The variation of (a) exponentially declining and (b) delayed exponentially declining (or Sandage) star formation laws as a function of time. The different curves represent star formation laws with a characteristic time scale of 1 (solid), 2 (dotted), 4 (short-dashed), 8 (long-dashed), and 12 (dotted-dashed line) Gyr, respectively. The vertical long dashed line indicates the typical age of galaxies in the nearby universe (13 Gyr).

on, all crucial ingredients in the study of the physical processes within the ISM (e.g., [162]; see Chapter 18).

The infrared SED of different extragalactic sources has been determined by several authors [134, 135, 140, 153, 164, 190] and recently, using Herschel data, by [222] (AGNs), [223] (lensed galaxies), [224] (high-redshift galaxies), [194] (local, normal galaxies). As an example, [135] determined the typical spectral energy distribution of galaxies using observational data for a large variety of nearby objects (see Figure 9.6). These SEDs differ for three different reasons: (a) the relative weight of PAH features in the mid-infrared domain, (b) the peak wavelength of the far-infrared emission, and (c) the infrared luminosity. Using the $S(60\,\mu m)/S(100\,\mu m)$ flux density ratio to characterize the different SEDs, they have shown that galaxies with the highest infrared luminosities (ULIRGs) are dominated by a warm dust

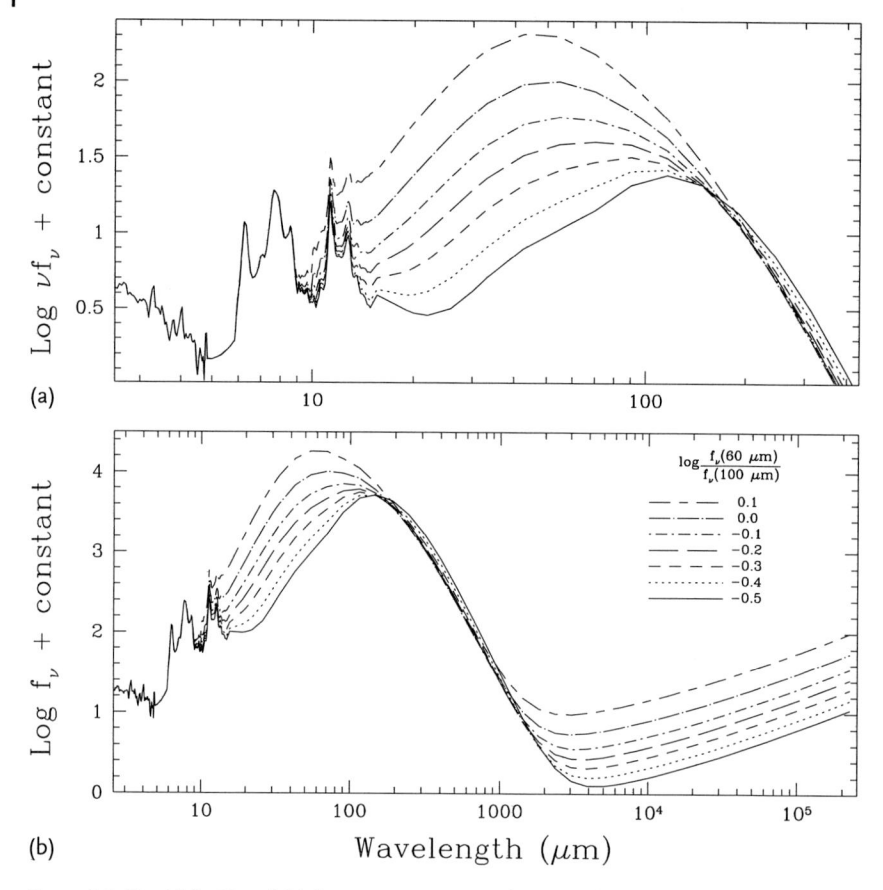

Figure 9.6 The (a) far-IR and (b) far-IR to centimeter radio spectral energy distributions (normalized at the 6.2 μm feature) of galaxies of different far-IR 60/100 μm color, from [135]. The Y-axis is given in (a) energy and (b) flux densities. Reproduced by permission of the AAS.

component whose emission peaks at ∼ 40–50 μm, while in normal galaxies with low infrared luminosities the dust emission is dominated by a cold component whose emission peaks at $\lambda \sim 100$–150 μm. Furthermore, in infrared quiescent objects the contribution of PAHs is important, while it is negligible in ULIRGs.

9.2.1
Mid- and Far-Infrared Colors

The 5–20 μm spectral range of galaxies is characterized by the presence of broad emitting features due to the presence of PAHs, as described in Section 4.2.1. These features dominate the emission in normal galaxies, while they are relatively weak in AGNs probably because the intense UV radiation field destroys PAHs. The low resolution 5–38 μm Spitzer IRS spectra of the inner few kiloparsec of the SINGS galaxies shown in Figure 9.7 clearly show that PAHs are partly or totally destroyed

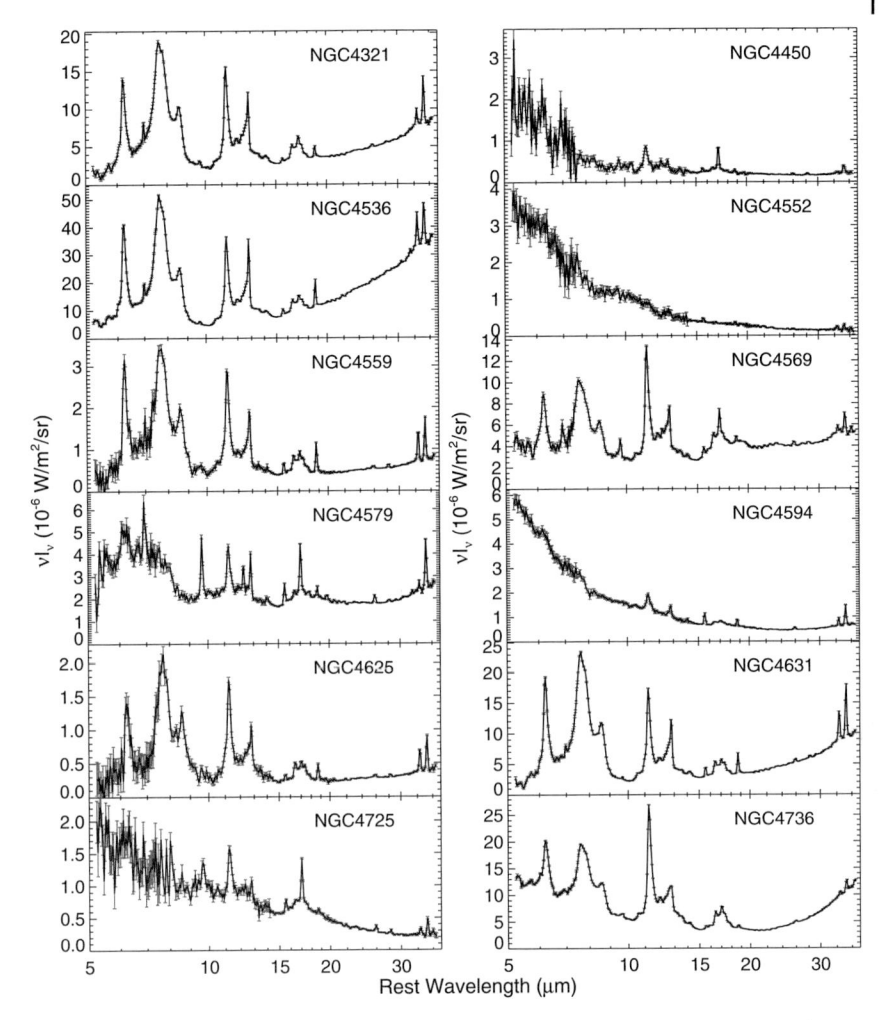

Figure 9.7 The low resolution 5–38 μm Spitzer IRS spectra of the inner few kiloparsec of 12 SINGS galaxies. PAHs are fully or partly destroyed in Seyfert galaxies (NGC 4552, 4569, 4579, 4594, 4725) and LINERs (4321, 4450, 4736), but are dominant in normal, star-forming galaxies (NGC 4536, 4559, 4625, 4631), from [44]. Reproduced by permission of the AAS.

in active galaxies (Seyfert and AGNs), and are dominant in normal, star-forming galaxies [44].

For this reason, color plots made using broad band imaging in the near-infrared can be used to identify the presence of an AGN even in highly obscured environments [109, 225–227]. Because AGNs are characterized by a power law spectrum in the near-infrared [85], they have [3.6 μm]–[4.5 μm] color indices (Spitzer bands) that are significantly redder than those of star-forming galaxies, where the emission is dominated by the Rayleigh–Jeans tail of the cold stellar population (whose peak is at 1.6 µm rest frame) for redshift $z \leq 1$. In the $0 \leq z \leq 1$, however, galaxies span a much larger range in the [5.8 µm]–[8.0 µm] color index than AGNs due to the presence of strong 6.2 and 7.7 µm PAH features in their spectra (see Figure 9.8). For this reason, AGNs can be identified as those objects with near to mid infrared colors in the range

$$([5.8]-[8.0]) > 0.6 \wedge ([3.6]-[4.5]) > 0.2 \times ([5.8]-[8.0]) + 0.18$$
$$\wedge ([3.6]-[4.5]) > 2.5 \times ([5.8]-[8.0]) - 3.5 \qquad (9.5)$$

as clearly shown in Figure 9.8. An alternative method based on the 24 over 8 µm flux ratio proposed for galaxies at redshift $z > 0.6$ is given by [227]. The far-infrared spectral domain, up to 500 µm, has been recently covered by the Herschel space mission [194, 222, 228]. Combined with Spitzer data, S_{250}/S_{70} vs. S_{70}/S_{24} color diagrams separate star-forming galaxies from AGNs, where the 24 µm flux density is dominated by the hot torus emission [222]. The 25–500 µm colors of normal galaxies of different morphological types are shown in Figure 9.9. In star-forming galaxies, the flux density ratios S_{60}/S_{500}, S_{25}/S_{250}, or S_{100}/S_{250} are strongly correlated with the generally used IRAS color index S_{60}/S_{100} (panels a, b, and c). The dynamic range covered by S_{60}/S_{500}, however, is a factor of about 30 larger than that covered by the S_{60}/S_{100} flux density ratio, and is thus a much clearer tracer of the

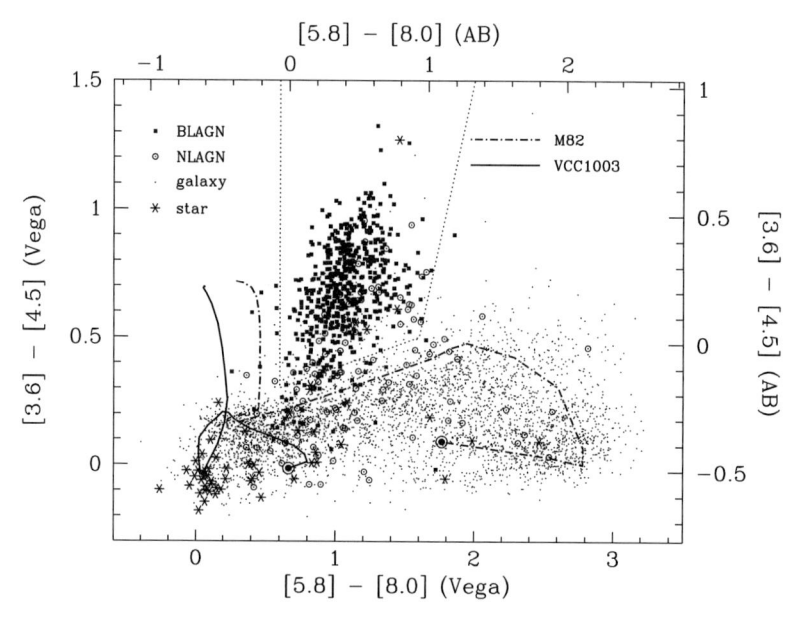

Figure 9.8 The Spitzer [3.6]–[4.5] vs. [5.8]–[8.0] color plot for stars (star), normal galaxies (dot), narrow-line AGN (NLAGN, o-dot), and broad-line AGN (filled squares). The solid and the dotted-dashed lines indicate the $0 \leq z \leq 2$ color tracks for two nonevolving galaxies with spectra similar to that of the S0/Sa NGC 4429 and the starburst M82, respectively ($z = 0$ at the dark bull's eyes), from [226]. Reproduced by permission of the AAS.

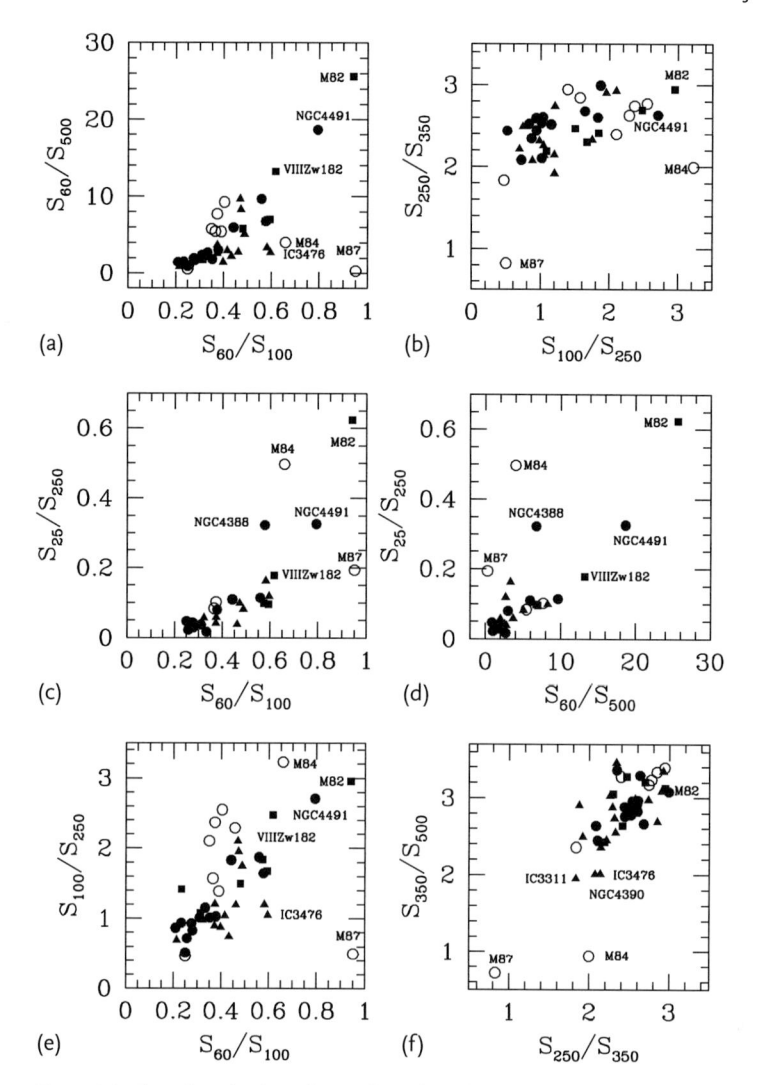

Figure 9.9 The infrared colors of typical nearby galaxies, from [194]. Galaxies are coded according to their morphological type: empty circles for E-S0a, filled circles for Sa-Sb, triangles for Sbc-Scd, squares for Sd, Im, BCD, and Irr galaxies.

average temperature of the bulk of the dust component. Starburst galaxies, defined here as those objects with $S_{60}/S_{100} > 0.5$ [229], have a S_{60}/S_{500} that spans from ∼ 3 to ∼ 30. The prototype starburst galaxy in the local universe, M82, has a S_{60}/S_{500} of ∼ 26. Early-types with synchrotron-dominated IR emission (M87, M84) are well separated in all IRAS-SPIRE or SPIRE color diagrams, with respect to the other dust-dominated E-S0a.

9.3
The Thermal and Nonthermal Radio Emission

The observations of normal, nearby galaxies [120, 167, 230] have shown that the synchrotron spectral index α_{syn} is ~ 0.8, which is consistent with a power spectrum slope for the electrons of cosmic rays $\gamma \sim 2.6$. Flatter slopes have been found in dwarf irregulars ($\alpha_{syn} \sim 0.5$; [231]). Concerning the core emission of normal galaxies, spirals have slopes of $\alpha_{syn} \sim 0.8$, while early-types (E/S0) have much flatter ones ($\alpha_{syn} \sim 0$; [165]). At the same time, it has been shown that the contribution of the thermal component to the total radio continuum emission of nearby late-type galaxies is generally $\lesssim 10\%$ at 1 GHz (20 cm), and ~ 20–30% at 10 GHz (2 cm) [120].

Radio galaxies generally have steep spectral indices ($\alpha_{syn} \sim 0.7$), while radio quasars typically have flat spectra ($\alpha_{syn} < 0.5$), except for those seen at large angles with respect to the jet (steep-spectrum radio quasars), whose $\alpha_{syn} > 0.5$. Differences in the spectral indices are seen between FRI and FRII radio galaxies, and between the core and the diffuse emission, the former being flatter than the latter probably because of self-absorption. Starburst galaxies have a large variety of spectral indices, but all observations consistently indicate that $\alpha_{syn} > 0$ [232].

10
Spectral Features

The spectra of galaxies are characterized by several features generally associated with emission or absorption lines. These features are of paramount importance for the study of the physical and chemical properties of galaxies. They can be used for detecting the presence of a central AGN, for quantifying the amount of internal attenuation (see Chapter 12), and for measuring the metallicity and the age of the gaseous and stellar components of any kind of galaxy. They are crucial for determining the very nature of the underlying stellar population, and for quantifying several critical parameters in the study of the physical and chemical properties of the interstellar medium of galaxies. At the same time, these features are generally used to measure the redshift of galaxies in spectroscopic surveys.

10.1
Galaxy Characterization through Emission and Absorption Lines

Galaxies can be classified according to their spectral properties. The different nature of various physical processes responsible for exciting the interstellar medium are revealed by characteristic line emission properties, while characteristic absorption features can be used to reveal the nature of the underlying stellar populations. Star-forming galaxies have emission lines similar to those observed in HII regions, where the gas is ionized by hot stars. In AGNs, the gas is ionized by fast shocks and nonthermal or power law continua. AGNs can be divided into Seyfert galaxies and LINERS. In Seyfert galaxies, the emission lines come from highly ionized gas, while in LINERs are from low ionized gas [233, 234]. LINERs are very frequent in massive galaxies: they have nuclear optical spectra dominated by low ionization species such as [OII]λ3726, λ3729, [OI]λ6300, and [SII]λ6717, λ6731 [235]. Furthermore, the broadening (and the variability) of some emission lines observed in the nucleus of several galaxies indicates the presence of an active galactic nucleus powered by a massive black hole.

Thus, emission lines can be used to discriminate different types of galaxies. Different line diagnostics have been proposed to separate Seyfert galaxies from LINERs and normal star-forming galaxies dominated by HII regions. Those most commonly used are based on optical line ratios.

At the same time, absorption features such as strong Balmer lines are used to identify post-starburst objects such as those populating nearby clusters [236], while the $D_n(4000)$ break, the strongest discontinuity in the optical spectrum, is generally used to age-date the last episode of star formation [96].

10.1.1
Classification of the Nuclear Activity

Active galaxies are divided into different subclasses according to the presence of different features in their emission line spectra. Seyfert galaxies have a nucleus characterized by strong emission lines [237]. According to the standard classification criteria reported in [237] and proposed by [238–240], Seyfert galaxies are divided into Seyfert 1 and Seyfert 2. Broad ($\geq 2000\,\mathrm{km\,s^{-1}}$) permitted lines such as hydrogen and helium, but narrow forbidden lines, are characteristics of Seyfert 1. Seyfert 2 galaxies have permitted and forbidden narrow lines. In the local universe, Seyfert 1 and 2 galaxies are equally frequent, while Seyfert 1 outnumber Seyfert 2 at high redshifts [241]. Intermediate classes between Seyfert 1 and 2 have been divided into five subclasses according to their $R = \mathrm{H}_\beta/[\mathrm{OIII}]5007\,\text{Å}$ flux ratio:

Sy1.0   if $5.0 < R$
Sy1.2   if $2.0 < R < 5.0$
Sy1.5   if $0.33 < R < 2.0$
Sy1.8   if $R < 0.33$ with a broad component visible in H_α and H_β
Sy1.9   broad component visible in H_α but not in H_β
Sy 2   no broad component visible

Among galaxies with active nuclei (AGNs), we can also mention QSO, BL Lac, and LINERs. Originally identified as radio sources, quasi stellar objects or quasars (now generally called QSO) are extragalactic sources with a very compact morphology, and with spectroscopic characteristics similar to that of the nuclei of Seyfert galaxies (in Seyfert galaxies, the emission of the AGN is contaminated by the stellar populations inhabiting the galaxy). They can be divided into radio-quiet and radio-loud according to their radio luminosity, the former being a factor of 10–30 more frequent than the latter. BL Lac are compact radio sources with stellar morphologies, but have weak absorption line features in their spectra which indicates that they are extragalactic sources. They do not have emission lines, and are variable sources with luminosities that can change by a factor of 2 in very short time scales (hours). LINERs (see below) are galaxies with ionization properties similar to Seyfert 1, but with narrow Balmer emission lines, with a possible mix of AGN-type and stellar photoionization characteristics (see below).

10.1.2
Classification of Post-Starburst and Post-Star-Forming Galaxies

Passive galaxies dominating rich clusters, whose star formation history has probably been perturbed by their interaction with the hostile environment, have been

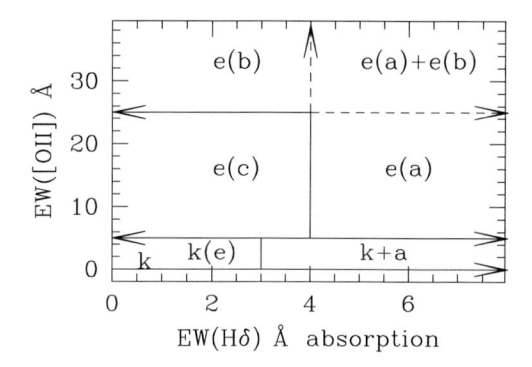

Figure 10.1 The schematic representation of the spectral classification diagram described in the text. The scheme shows, for each spectral type, the range in rest frame [OII] emission and H_δ absorption equivalent width (from [244]). Reproduced by permission of the AAS.

classified according to the presence of several features in their optical spectra [242, 243]. Post-starburst galaxies, objects where the star formation activity has been abruptly truncated after a strong burst episode, are generally indicated as E+A or k+a (and a+k) galaxies, and are characterized by strong Balmer absorption lines with no emission [244]. Galaxies with normal star formation activity brusquely truncated without a starburst event, defined as post star forming galaxies, are characterized by moderate Balmer absorption lines.

The spectral classification of passive galaxies is generally based on the values of the [OII] 3727 Å emission line and H_δ 4101 Å Balmer absorption line, since they are observable in the optical spectra of galaxies in the redshift range $0.0 \leq z \leq 1$. These two lines are used because they are tightly related to the present and past star formation activity of galaxies. The spectral classification scheme, firstly proposed by [236, 245] and later revisited by [244], has been defined to identify galaxies with an ongoing (starburst) and a recent (post-starburst) star formation activity and to distinguish them from passively evolving systems and quiescent star-forming objects. The classification scheme, graphically shown in Figure 10.1, is defined as follows [244]:

Passively evolving galaxies: k class: securely undetected [OII] line in emission and $EW(H_\delta)_A < 3$ Å. Their spectra indicate that these galaxies did not undergo a star formation episode in the last 1–1.5 Gyr. They are classified as "k" galaxies since their spectra is similar to that of a K giant-type star. A less stringent criterion is used for "k(e)" galaxies ($EW[OII]_E < 5$ Å).

Post-star-forming/post-starburst galaxies: k+a class: $EW[OII]_E < 5$ Å and $EW(H_\delta)_A > 3$ Å. Their spectra indicate that the star formation activity in these objects terminated between 5×10^7 and 1.5×10^9 yr ago. This class, originally called "E+A" or H_δ-strong galaxies, is generally indicated with "k+a" because their spectra resembles a mixture of K and A type stars. Among these galaxies, those with $EW(H_\delta)_A > 8$ Å are purely post-starburst galaxies, sometimes indicated with "a+k". An indicative value of $EW(H_\delta)_A \sim 5$ Å can be taken to separate post-star-forming from post-starburst galaxies: it should be noticed, however, that while post-star-forming galax-

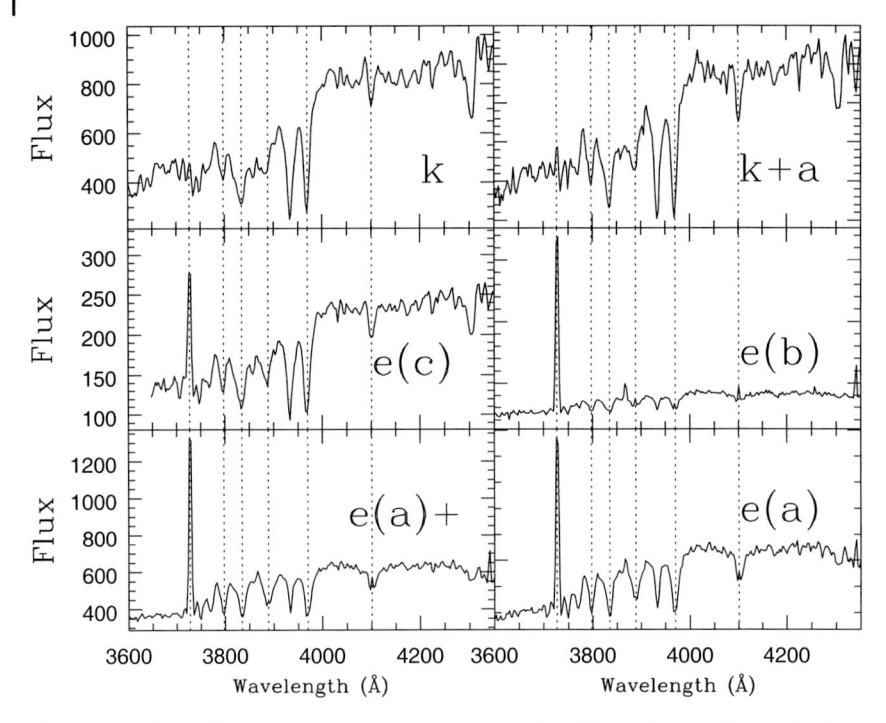

Figure 10.2 The rest frame composite spectra of the different classes described in the text (with a spectral resolution of 3 Å). The vertical dotted lines indicate the main lines of interest, in order of increasing wavelength: [OII] 3727 Å, H_θ 3798 Å, H_η 3835 Å, H_ζ 3889 Å, H_ϵ 3970 Å, and H_δ 4101 Å (from [244]). Reproduced by permission of the AAS.

ies never reach $EW(H_\delta)_A = 5$ Å, in evolved poststarbursts H_δ drops to values lower than 5 Å [246–249].

Quiescent star-forming galaxies: e(c) class: $5 < EW[OII]_E < 25$ Å and $EW(H_\delta)_A < 4$ Å. These spectra are similar to those of local Sa-Sd galaxies. The star formation activity is relatively constant, with no sudden variation due to a starburst event or to a truncation.

Dusty starburst candidates: e(a) and e(a)+ classes: $EW[OII]_E > 5$ Å and $EW(H_\delta)_A > 4$ Å (with some emission in the "+" class). These spectral features are rare in nearby normal galaxies but frequent in dusty starbursts and luminous infrared objects, even though their optical spectral properties do not necessarily indicate the presence of a large amount of dust. They can also indicate post-starburst galaxies with a residual star formation activity.

Strong emission line starbursts: e(b) class: $EW[OII]_E > 25$ Å. The strong [OII] emission line indicates galaxies with strong ongoing star formation activity, such as dust-poor starburst galaxies, very late spirals, or BCDs.

Figure 10.2 shows the composite spectra of galaxies belonging to the different classes.

10.1.3
Line Diagnostics

10.1.3.1 Optical Lines

Combining spectroscopic observations of galactic nuclei with photodissociation models, several authors have proposed various line diagnostic diagrams to identify active galaxies [234, 250–252]. Thanks to the spectroscopic SDSS survey, these classification schemes have been recently tested and updated on statistically significant samples [253, 254]. Here we follow the classification proposed by [254], which has been determined using a combination of stellar population synthesis models and detailed self-consistent photoionization models [233], combined with observed data from the SDSS DR4. This classification has been defined to limit the regions sampled by the different galaxy populations within standard optical line ratio diagnostic diagrams (Figure 10.3).

This classification is based on the relationship between the line ratios of [OIII]λ5007 to H$_\beta$, [NII]λ6584 to H$_\alpha$, [SII]λ6717, λ6731 to H$_\alpha$, and [OI]λ6300 to H$_\alpha$ (Figure 10.3). Whenever emission lines at significantly different wavelengths are used, they must first be corrected for dust extinction using the Balmer decrement, as described in Section 12.2.1. Balmer emission lines should also be corrected to take into account the underlying Balmer absorption [252]. All emission lines must have sufficiently high signal to noise ratios (the authors use a S/N > 6), and aperture effects might be important [254].

These diagnostic diagrams enable the separation of galaxies into four different categories: normal star-forming systems (indicated as HII galaxies), AGNs, which are separated into Seyfert galaxies and LINERs, and composite galaxies (Comp), that have nuclear spectra in between AGNs and HII regions. Composite galaxies have line ratios [OIII]/H$_\beta$ vs. [NII]/H$_\alpha$ in between the theoretical maximum star-

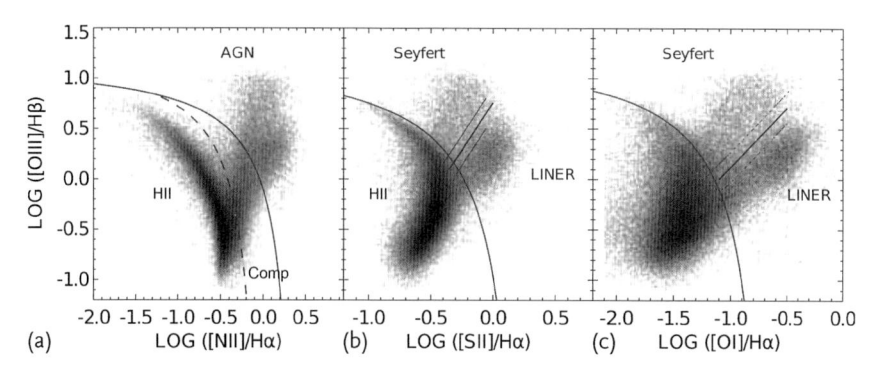

Figure 10.3 The three line diagnostic diagrams used to discriminate star-forming (HII) from AGNs. The later are separated into LINERs and Seyfert galaxies. The long solid line shows the extreme starburst classification of [233], the dashed line shows the pure star-forming line of [253]. Galaxies in between these lines are defined as composite HII-AGN types (Comp). The short solid line separates Seyfert galaxies from LINERs. Gray dots are SDSS galaxies (from [254]). Reproduced with the permission of John Wiley & Sons Ltd.

burst line, defined by [233] with pure stellar photoionization models (solid line), and the empirical curve proposed by [253] to rule out any possible AGN contamination (dashed line)[1]. Star-forming galaxies form a tight sequence from low (low [NII]/H$_\alpha$, high [OIII]/H$_\beta$) to high (high [NII]/H$_\alpha$, low [OIII]/H$_\beta$) metallicities. The [NII]/H$_\alpha$ line ratio is more sensitive to the presence of a low-level AGN than [SII]/H$_\alpha$ and [OI]/H$_\alpha$. The [SII]/H$_\alpha$ and [OI]/H$_\alpha$ diagrams better separate LINERs from Seyfert galaxies. Analytically, the different categories can be defined as follows:

Star-forming galaxies:

$$\log([\text{OIII}]/\text{H}_\beta) < 0.61/[\log([\text{NII}]/\text{H}_\alpha) - 0.05] + 1.30 \tag{10.1}$$

$$\log([\text{OIII}]/\text{H}_\beta) < 0.72/[\log([\text{SII}]/\text{H}_\alpha) - 0.32] + 1.30 \tag{10.2}$$

$$\log([\text{OIII}]/\text{H}_\beta) < 0.73/[\log([\text{OI}]/\text{H}_\alpha) + 0.59] + 1.33 \tag{10.3}$$

Composite galaxies:

$$0.61/[\log([\text{NII}]/\text{H}_\alpha) - 0.05] + 1.30 < \log([\text{OIII}]/\text{H}_\beta) \tag{10.4}$$

$$0.61/[\log([\text{NII}]/\text{H}_\alpha) - 0.47] + 1.19 > \log([\text{OIII}]/\text{H}_\beta) \tag{10.5}$$

Seyfert galaxies:

$$0.61/[\log([\text{NII}]/\text{H}_\alpha) - 0.47] + 1.19 < \log([\text{OIII}]/\text{H}_\beta) \tag{10.6}$$

$$0.72/[\log([\text{SII}]/\text{H}_\alpha) - 0.32] + 1.30 < \log([\text{OIII}]/\text{H}_\beta) \tag{10.7}$$

$$0.73/[\log([\text{OI}]/\text{H}_\alpha) + 0.59] + 1.33 < \log([\text{OIII}]/\text{H}_\beta) \tag{10.8}$$

or

$$\log([\text{OI}]/\text{H}_\alpha) > -0.59 \tag{10.9}$$

and

$$1.89 \log([\text{SII}]/\text{H}_\alpha) + 0.76 < \log([\text{OIII}]/\text{H}_\beta) \tag{10.10}$$

$$1.18 \log([\text{OI}]/\text{H}_\alpha) + 1.30 < \log([\text{OIII}]/\text{H}_\beta) \tag{10.11}$$

---

1) These classification diagnostics are well adapted for Seyfert 2 galaxies but quite uncertain for Seyfert 1, where Balmer and forbidden lines originate from very different regions (see Section 3.2).

LINERs:

$$0.61/[\log([NII]/H_\alpha) - 0.47] + 1.19 < \log([OIII]/H_\beta) \qquad (10.12)$$

$$0.72/[\log([SII]/H_\alpha) - 0.32] + 1.30 < \log([OIII]/H_\beta) \qquad (10.13)$$

$$\log([OIII]/H_\beta) < 1.89 \log([SII]/H_\alpha) + 0.76 \qquad (10.14)$$

$$0.73/[\log([OI]/H_\alpha) + 0.59] + 1.33 < \log([OIII]/H_\beta) \qquad (10.15)$$

or

$$\log([OI]/H_\alpha) > -0.59 \qquad (10.16)$$

$$\log([OIII]/H_\beta) < 1.18 \log([OI]/H_\alpha) + 1.30 \qquad (10.17)$$

Those objects not belonging to any of these categories are defined as ambiguous galaxies:
These are galaxies with a discrepant classification in one of the three diagrams, where the emission comes from the line of sight superposition of different resolved regions with different spectral characteristics, although the presence of intrinsically strange objects is also possible (X-ray excitation, double nuclei, etc.).

According to this classification scheme, [254] classified their $\sim$ 85 000 SDSS galaxies into star-forming galaxies (75%), Seyfert galaxies (3%), LINERs (7%), and composites (7%), with 8% of ambiguous galaxies.

Kewley and collaborators [254] proposed an alternative diagnostic diagram based only on the [OIII]/[OII] vs. [OI]/H_α line ratios; the former being a sensitive diagnostic of the ionization parameter of the gas, the latter being sensitive to the hardness of the ionizing radiation field. This method requires that line emissions are corrected for dust attenuation. As explained in [254], the ionization parameter is a measure of the amount of ionization that a radiation field can drive as it moves through the gas. Seyfert galaxies have a higher ionization parameter than LINERs or star-forming galaxies, and can thus be differentiated from the LINERs on the Y-axis. Seyfert galaxies and LINERs have hard power law ionizing radiation fields (high [OI]/H_α flux ratios), and are thus separated from spirals on the X-axis (Figure 10.4)[2]. This alternative diagnostic diagram based on the [OIII]/[OII] vs. [OI]/H_α line ratios gives the following classification:

Star-forming and composite galaxies:

$$\log([OIII]/[OII]) < -1.701 \log([OI]/H_\alpha) - 2.163 \qquad (10.18)$$

LINERs:

$$-1.701 \log([OI]/H_\alpha) - 2.163 < \log([OIII]/[OII]) \qquad (10.19)$$

$$\log([OIII]/[OII]) < 1.0 \log([OI]/H_\alpha) + 0.7 \qquad (10.20)$$

2) An alternative recipe for separating Seyfert galaxies from LINERs can be found in [255].

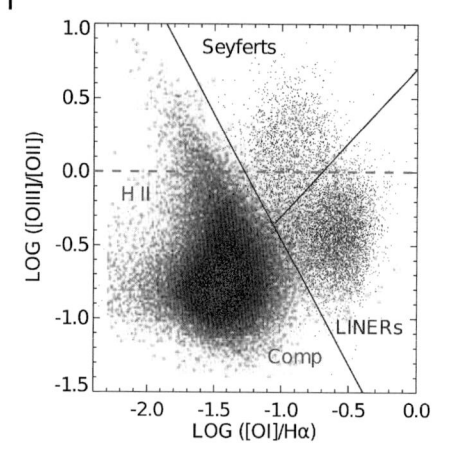

Figure 10.4 The [OIII]/[OII] vs. [OI]/H$_\alpha$ classification diagram of [254] applied to SDSS galaxies. Reproduced with permission of John Wiley & Sons Ltd.

Seyfert galaxies:

$$-1.701 \log([OI]/H_a) - 2.163 < \log([OIII]/[OII]) \tag{10.21}$$

$$1.0 \log([OI]/H_a) + 0.7 < \log([OIII]/[OII]) \tag{10.22}$$

Figure 10.5 shows how this classification method compares to the methods proposed by [234, 250–253]. An alternative classification method especially tuned for the classification of narrow emission line galaxies was proposed by [256–258].

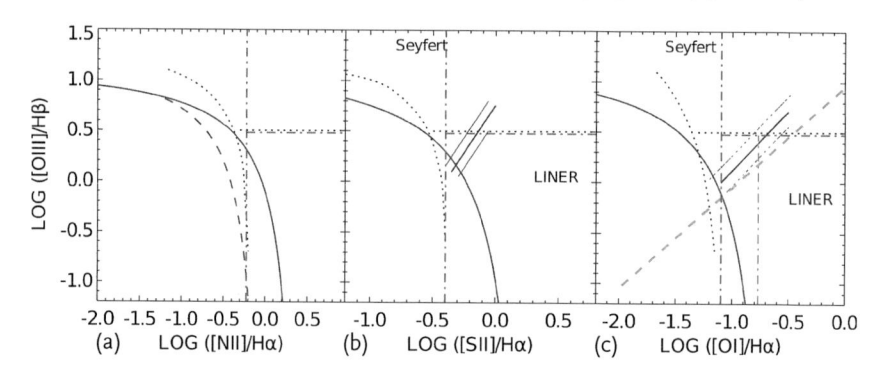

Figure 10.5 Comparison of different classification methods of galaxies based on emission line ratios. The extreme starburst classification of [233] is indicated by the solid line, the pure star formation limit of [253] by the dashed line, the Seyfert-LINER limit of [254] by the short solid line, the [251] limit by the curved dotted line, the [234] LINER limit by the diagonal dashed line, and the [252] classification scheme by the dot-dashed line. (From [254]). Reproduced with permission of John Wiley & Sons Ltd.

Table 10.1 Infrared emission lines in the 10–37 µm range (from [259]). For the molecular hydrogen lines H$_2$ S(2) at 12.28 µm, H$_2$ S(1) at 17.04 µm and H$_2$ S(0) at 28.22 µm see Section 4.2.3.

Line	λ µm	Ionization potential eV
[SIV]	10.51	34.8
[NeII]	12.81	21.6
[NeV]	14.32	97.1
[NeIII]	15.56	41.0
[SIII]	18.71	23.3
[OIV]	25.89	54.9
[FeII]	25.99	7.9
[SIII]	33.48	23.3
[SiII]	34.82	8.2

10.1.3.2 Infrared Lines

The infrared spectra of galaxies are characterized by the presence of several important emission and absorption lines. The most frequently observed emission lines are those in the 10–37 µm range (covered by Spitzer,) and are listed in Table 10.1.

As in the optical, several of these infrared emission lines can be used to disentangle star-forming galaxies from AGNs. The observed emission lines of highly excited ions cannot be produced inside the HII regions by the ionizing radiation emitted by the most massive O stars, but are rather evidence of an AGN. Among these, the most widely used are the neon line [NeV]λ14.32 µm, whose ionization potential is 97.1 eV, and the [OIV]λ 25.89µm with a ionization potential of 54.9 eV [260]. This oxygen line [OIV] can however be excited also by very hot stars such as Wolf–Rayet stars or by shocks associated with a starburst activity [261]. On the contrary, the PAH emission bands which are prominent in star-forming systems are suppressed by the photodissociation of their carriers in the hard radiation field of AGNs [262]. The most widely used PAH bands are the 6.2, the 7.7µm (although its measurement is quite uncertain because of the nearby silicate absorption feature at 9.7 µm) and the weak 3.3 µm band accessible from the ground for bright, nearby sources.

A widely used diagnostic diagram is the [OIV]λ 25.89µm/[NeII]λ 12.81µm vs. the equivalent width of the PAH 6.2 µm [263, 264] band shown in Figure 10.6. Indeed, Figure 10.6 shows that star-forming galaxies (empty circles) have, on average, a stronger PAH emission and a lower [OIV]/[NeII] line ratio than ULIRGs (triangles) and Seyfert galaxies (Sy 1: filled squares, Sy 2: filled pentagons). The [OIV]λ 25.89µm/[SIII]λ 33.48µm vs. [NeIII]λ 15.56µm/[NeII]λ12.81µm diagram (Figure 10.7) has been proposed by [261] as an alternative infrared line diagnostic diagram useful for identifying the principal excitation mechanism in a galaxy. Further discussion on other infrared line diagnostics can be found in [265] (on

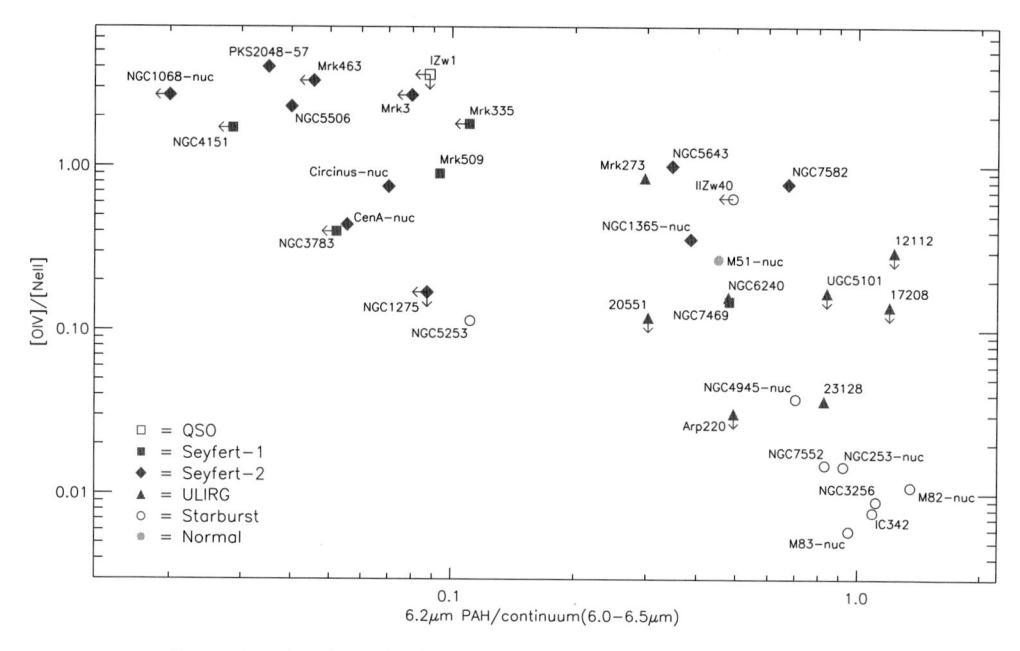

Figure 10.6 The relationship between the [OIV]/[NeII] flux ratio and the equivalent width of the PAH band at 6.2 μm, from [263]. Different symbols are used for different extragalactic sources. Reproduced by permission of the AAS.

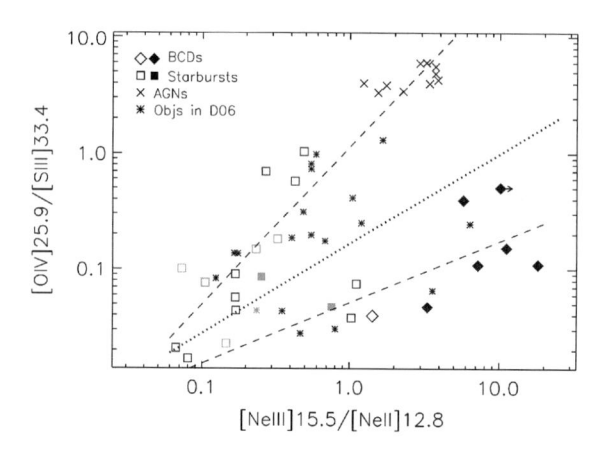

Figure 10.7 The relationship between the [OIV]/[SIII] and the [NeIII]/[NeII] flux ratios for different classes of objects: diamonds are BCD galaxies, squares for starbursts, crosses for AGNs, and stars for star-forming galaxies in [264]. Filled symbols are for galaxies with Wolf–Rayet signatures in their optical spectra. The two branches populated by the different galaxy populations are indicated with the two dashed lines. The dotted line gives the empirical separation between the two populations [261]. Reproduced by permission of the AAS.

10.2
Gas Metallicity from Emission Lines

Extinction-corrected optical emission lines are used to measure the metal abundance (metallicity) of the interstellar gas. As explained in Section 3.2.2, the direct estimate of the heavy element abundances can be derived once the electron density and temperature are known. While electron densities can be estimated from the [OII]$\lambda 3727/\lambda 3729$ or [SII]$\lambda 6716/\lambda 6731$ doublets, the electron temperature is generally measured using the ratio [OIII]$\lambda 4959 + \lambda 5007$/[OIII]$\lambda 4363$, although other lines have been also proposed ([OII]$\lambda 3726 + \lambda 3729$/[OII]$\lambda 7320 + \lambda 7330$; [NII]$\lambda 5755$/[NII]$\lambda 6584$ [273]). Metallicity calibrations based on electron temperature estimates (direct calibrations) have been determined thanks to photoionization models of HII regions [274].

The [OIII]$\lambda 4363$, [NII]$\lambda 5755$, and [OII]$\lambda 7320 + \lambda 7330$ lines, however, are very weak and difficult to observe in galaxies. Furthermore, [OIII]$\lambda 4363$, the principal tracer of the gas temperature, is detected only in metal-poor extragalactic sources [275]. For this reason, several indirect line emission–metallicity relations have been proposed in the literature. These have been calibrated using nearby HII regions by comparing accurate metallicity estimates based on electron temperature measurements with other emission line ratios. The line ratios generally used are [OII]$\lambda 3727/\lambda 3729$, [OIII]$\lambda 4959 + \lambda 5007$, and [NII]$\lambda 6548 + \lambda 6584$ [276, 277]. The calibrations have been obtained using photodissociation models [278–282] or a combination of the two methods [283]. All these indirect methods have their own weaknesses, and it is quite impossible to quantify the uncertainty in the derived metallicities for each of them. Comparing various calibrations [284] has shown that the differences in $12 + \log(O/H)$ can be up to a factor of 0.7 dex, as shown in Figure 10.8. This value should thus be considered as the uncertainty on the absolute metallicity determined using indirect calibrations.

However, as stated in [284], the relative error on $12 + \log(O/H)$ is only ~ 0.15 dex when metallicities are measured using a single calibration. When comparing metallicity estimates from different sources it is thus crucial to use consistent calibrations. For this reason, [284] gave different analytical relations useful for transforming $12 + \log(O/H)$ values from one calibration system to another.

The SDSS team proposed a new calibration based on the R_{23} index, defined as

$$R_{23} = ([OII]\lambda 3727 + [OIII]\lambda\lambda 4959, 5007)/H_\beta \tag{10.23}$$

through the relation [281]

$$12 + \log(O/H) = 9.185 - 0.313x - 0.264x^2 - 0.321x^3 \tag{10.24}$$

where $x = \log R_{23}$. This widely used relation is valid for relatively bright galaxies with metallicities $12 + \log(O/H) > 8.5$ (for comparison, the solar neighborhood

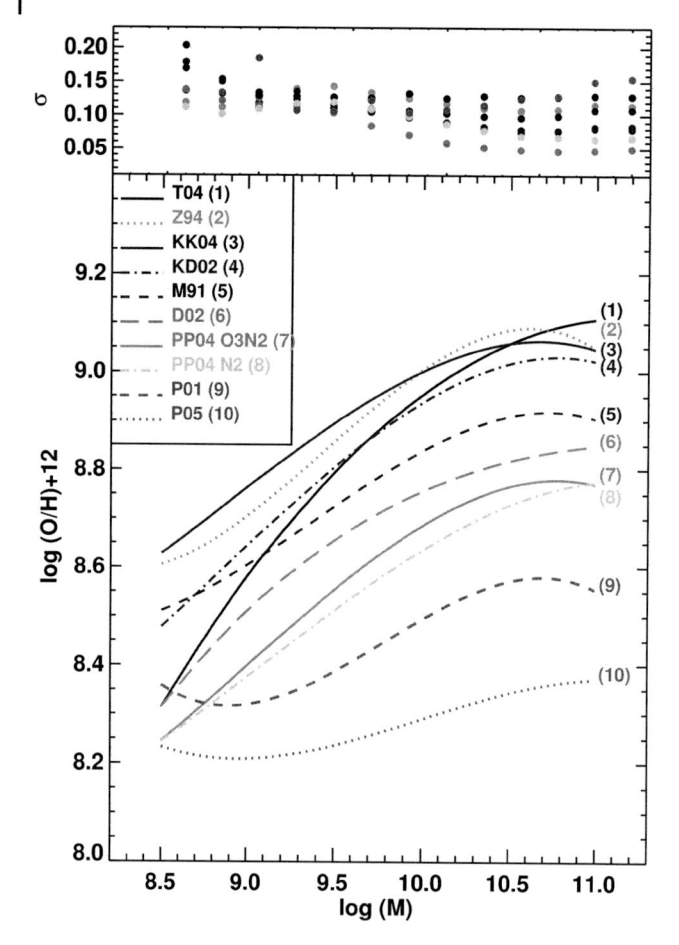

Figure 10.8 The metallicity–mass relation for SDSS galaxies with metallicities $12 + \log(O/H)$ measured using different calibrations, from [284]. The different calibrations are: 1 (T04) [281], 2 (Z94) [279], 3 (KK04) [282], 4 (KD02) [280], 5 (M91) [278], 6 (D02) [283], 7 (PP04 O3N2) and 8 (PP04 N2) [276], 9 (P01) [285], 10 (P05) [277]. Reproduced by permission of the AAS.

has a metallicity of $12 + \log(O/H) = 8.69 \pm 0.05$ [286]). After a critical analysis of the different techniques, [284] suggests the use of either the [280] or the [276] calibrations. These calibrations give low residual discrepancies in relative metallicities, and a low residual discrepancy after other metallicities have been converted into these two methods. The calibration of [280], revised by [284], is

$$\log([NII]/[OII]) = 1106.8660 - 532.15451 Z$$
$$+ 96.373260 Z^2 - 7.8106123 Z^3 + 0.23928247 Z^4 \quad (10.25)$$

where $Z = 12 + \log(O/H)$, for $\log([NII]/[OII]) > -1.2 (12 + \log(O/H) > 8.4)$. For lower metallicities, the calibration has a more complex analytical form since it takes into account variations in the ionization parameter (see [284]). The one proposed by [276] is

$$12 + \log(O/H) = 9.37 + 2.03 N2 + 1.26 N2^2 + 0.32 N2^3 \tag{10.26}$$

where N2 = $\log([NII]\lambda 6584/H_\alpha)$, or

$$12 + \log(O/H) = 8.73 - 0.32 O3N2 \tag{10.27}$$

where O3N2 = $\log\{([OIII]\lambda 5007/H_\beta)/([NII]\lambda 6584/H_\alpha)\}$. If the first of these two relations has the advantage of being sensitive to the presence of AGNs, it can be used only for metallicities smaller than solar. The second relation has a lower dispersion and is preferred for metallicities larger than $\sim 1/4$ solar. Other calibrations based on the sulfur [SII] and [SIII] [280, 287] lines or on the infrared oxygen [OIII]88 μm [288] line have been also proposed in the literature.

10.3
Stellar Age and Metallicity from Absorption Lines

The spectra of galaxies are marked by the presence of several absorption features whose intensities generally depend on the mean age and metallicity of the underlying stellar populations. As mentioned in Section 3.3, these features are due to the absorption of photons in stellar atmospheres[3]. Population synthesis models with sufficient spectral resolution can be used to predict the variation of the intensity, equivalent width, or any other defined feature, as a function of the mean age and metallicity of the underlying stellar population. This is made possible by relating the observed line indices (measured using a standard spectral resolution) of different stellar libraries to stellar atmosphere parameters. Using this technique, absorption lines can also provide individual element abundances.

The spectral indices generally used in the literature are those of the Lick/IDS system [93, 94], defined in [90, 95, 290] and first measured on a large sample of stars with different spectral types using the same instrumental configuration at the Lick Observatory, with a spectral resolution of 8–10 Å in the 4000–6400 Å range (25 indices). The relationships between the strength of the Lick/IDS indices and the age, metallicity, and individual chemical abundances has been calibrated by several authors using both single stellar bursts [89, 215, 291] or more complicated star formation histories [213]. The mean age and metallicity of the observed galaxies is determined either by plotting an age sensitive spectral index, such as H_β, versus a metallicity sensitive index (Mgb, $\langle Fe \rangle$) and comparing the data to model grids (see Figure 10.9), or by simultaneously fitting population synthesis models with as many Lick/IDS indices as possible with a χ^2 technique [292].

The analysis done so far, however, has shown that both methods are limited by an age–metallicity degeneracy mostly due to the fact that Lick/IDS indices are broadly

---

3) In the UV spectral domain, however, absorption lines might be modified by the presence of the interstellar medium, [289].

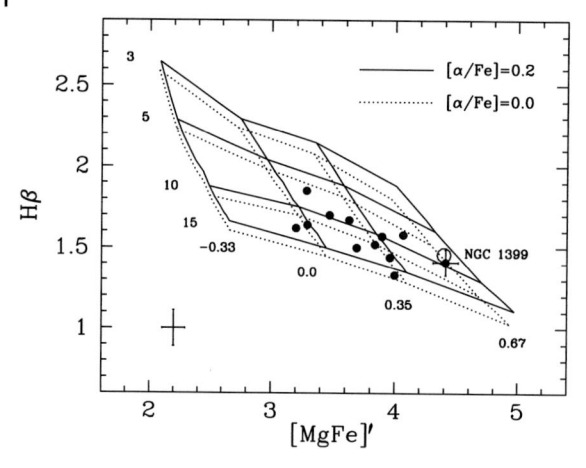

Figure 10.9 The relationship between the age-sensitive indicator H$_\beta$ and the metallicity-sensitive [MgFe]' Lick/IDS index measured in elliptical galaxies compared to the grid models of [293] for different ages (3, 5, 10, 15 Gyr), and metallicities ([Z/H] = −0.33, 0.0, 0.35, 0.67) for two different α/Fe ratios (0.0, dotted line; 0.02 solid line). Reproduced with the permission of John Wiley & Sons Ltd.

defined lines ($\sim$ 40 Å) which include several absorption features from various elements, and are thus difficult to be translated into element abundances or stellar ages [215, 292, 293]. These effects, combined with the fact that the stellar library used in the stellar population models to compute Lick/IDS indices are dominated by stars with sub-solar metallicities, can be important and should be considered in the analysis of galaxy spectra. As a direct consequence of this effect, [293] have shown that the H$_\gamma$ and H$_\delta$ Balmer absorption lines, contrary to H$_\beta$, are very sensitive to the α/Fe at super-solar metallicities typical of massive ellipticals (see Figure 10.10).

The α/Fe element ratio is a key parameter in the study of the star formation history of early-type galaxies. This is due to the fact that while the elements N and Na, and the α-elements – O, Ne, Mg, Si, S, Ar, Ca and Ti – that are built up with α-particle nuclei, are produced principally by Type II supernovae, the elements Fe and Cr come from the delayed explosions of Type Ia supernovae. The massive progenitor of Type II supernovae have shorter lifetimes than the mean age at which binary systems evolve into a Type Ia supernovae. Thus, the ratio α/Fe is tightly related to the time scale over which star formation occurred [294]. Thomas et al. [215] defined a new index based on the Lick/IDS system, [MgFe]', completely independent of the α/Fe ratio and thus perfectly adapted for measuring the mean metallicity of galaxies (see Figures 10.11 and 10.12). This index is defined as [215]:

$$[\text{MgFe}]' = \sqrt{Mgb(0.72 \times \text{Fe}5270 + 0.28 \times \text{Fe}5335)} \tag{10.28}$$

Lick/IDS indices are sensitive to line broadening due to both the stellar velocity dispersion within the galaxies and to instrument resolution. Before any comparison can be made between data and model predictions, the observed spectra should

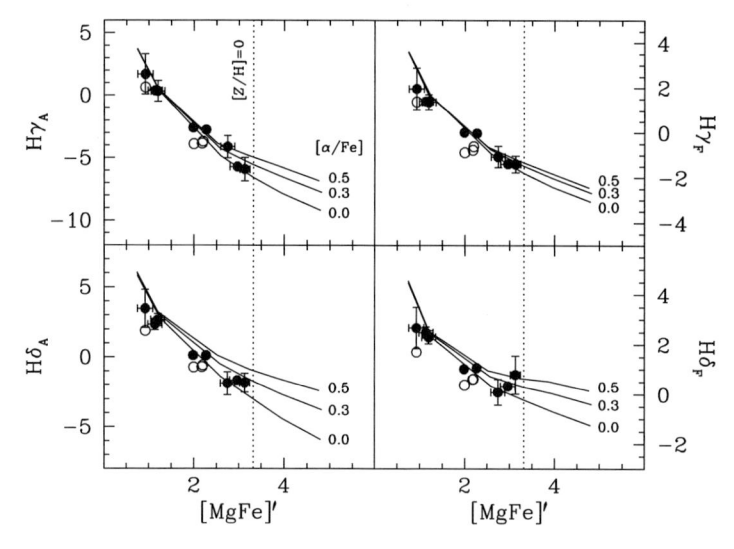

Figure 10.10 The variation of the Lick/IDS Balmer absorption indices H_γ and H_δ (two different definitions A and F) vs. the (α/Fe independent) metallicity index [MgFe]' for three different abundance ratios $[\alpha/Fe] = 0.0, 0.3,$ 0.5 and a fixed age of 12 Gyr, as determined from the models of [293] (solid lines), compared to the observed data of galactic globular clusters (dots). Reproduced with the permission of John Wiley & Sons Ltd.

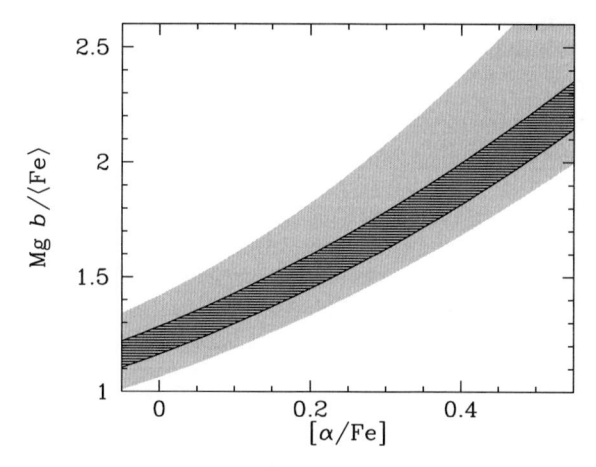

Figure 10.11 The variation of the Lick/IDS ratio Mg$b/\langle$Fe$\rangle$ as a function of the element abundance ratio α/Fe, from [215]. The dark area shows models in the metallicity range $-1.35 \leq [Z/H] \leq 0.35$ and with ages between 8 and 13 Gyr, the light-gray area shows models with the same metallicity but ages between $3\,\text{Gyr} \leq t \leq 15\,\text{Gyr}$. Reproduced with the permission of John Wiley & Sons Ltd.

be corrected to the same resolution. Furthermore, the data should be transformed into the spectrophotometric system of the Lick/IDS stellar libraries to limit the uncertainties related to the poorly known absolute flux calibration of the Lick/IDS stars. This can be done by observing a sample of calibration stars selected from the

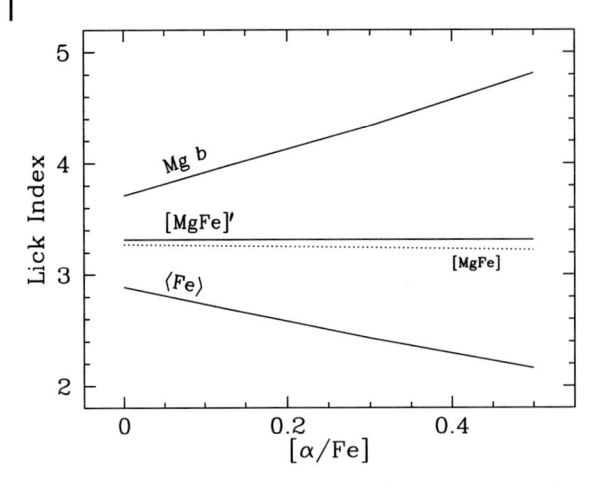

Figure 10.12 The variation of different Lick/IDS indices as a function of α/Fe at a fixed solar metallicity and 12 Gyr age, from [215]. The dotted line is for a slightly different definition of the [MgFe] index. Reproduced with the permission of John Wiley & Sons Ltd.

Lick/IDS library using the same telescope configuration as the science targets, and then deriving possible small offsets between the indices measured in those stars and the ones given in the Lick/IDS system. Some Lick indices are affected by emission, as is the case for the Balmer H_β absorption line and, to a minor extent, H_γ and H_δ. The contribution of the line emission should be removed before comparing the measured indices to the model predictions. This could be done by either fitting the relation between the equivalent width of the Balmer emission line and other emission lines, providing a statistical correction that can be applied whenever other emission lines are detected, or by fitting an optimal spectral template and subtracting the emission directly from the residuals [210].

Spectral indices other than those defined in the Lick/IDS system have been proposed in the literature. Those of [295] in the spectral range 3800–4400 Å (13 indices) where particularly useful to show the existence of a large population of dwarfs in M32 [295] as well as to study the effects of the environment on the stellar populations of early-type galaxies inhabiting rich clusters [296, 297]. Another important spectral index is the CaII triplet in the near-infrared ($\lambda\lambda 8498$, 8542, 8662 Å), proposed by [298] as a useful tool for studying the K and M giant populations in galaxies. The availability of high resolution stellar libraries is now allowing the construction of new population synthesis models and the definition of new line index systems optimized for limiting the contamination of various lines of different nature, thus minimizing the age-metallicity degeneracy effect. The MILES library [292], for instance, is based on INT spectra with a resolution of 2.3 Å of 985 stars in the spectral range 3540–7410 Å.

11
Gas Properties

The cold gaseous phase is the principal feeder of star formation and is thus of paramount importance in the study of galaxy formation and evolution. The primordial gas component, which is continuously accreted into the disk of galaxies during their life (infall), is located in a gas reservoir extending beyond the optical disk of the galaxy and collapses inside molecular clouds to form new stars. At the same time, evolved stars produce and inject recycled gas into the interstellar medium through stellar winds. The column density, mass, and temperature of the gaseous component of the interstellar medium can be inferred through the observation of several emission and absorption lines in the spectra of galaxies. While the mass of the atomic gas component can be measured using the 21 cm HI emission line, the mass of the molecular phase is mainly determined indirectly by means of several CO emission lines in the millimeter domain. Absorption features are generally used to quantify the column density of the gas in absorbing systems.

11.1
Gas Density, Mass, and Temperature

The gas phase is the principal component of the interstellar medium. The atomic hydrogen issued from the primordial nucleosynthesis, which is $\sim 75\%$ of the total baryonic mass of the universe, mixes with dust in the ISM of spiral galaxies and forms molecular hydrogen that later collapses into molecular clouds to form stars. Quantifying the gas content of galaxies is thus critical for understanding the matter cycle in galaxies (see Chapter 18).

The gas component of the interstellar medium of galaxies is composed of both atomic and molecular gas, and can be neutral or ionized. The atomic phase is generally the dominant component of the ISM. There are two phases of the atomic ISM. One is cold (60–100 K) and relatively dense ($n(HI) \sim$ some tens of atoms cm^{-3}) and dominates the absorption. The other is warm (several thousands degrees), and of low density ($n(HI) \sim 0.1$–0.3 cm^{-3}). In the Milky Way, the warm and the cold gas components have $\sim$ a similar mass [75].

A Panchromatic View of Galaxies, First Edition. Alessandro Boselli.
© 2012 WILEY-VCH Verlag GmbH & Co. KGaA. Published 2012 by WILEY-VCH Verlag GmbH & Co. KGaA.

The molecular hydrogen is the dominant component of the gas phase of the ISM in molecular clouds: here, the density of the gas is on the order of 10^2–10^3 mol cm^{-3}, and can reach densities up to 10^4–10^6 mol cm^{-3} in the densest cores, where the gas has very low temperatures (~ 10 K). In galaxies, the total amount of gas M_{gas}, which includes both the atomic and molecular gas phases, can be determined by the simple relation

$$M_{gas} = h_e(M(HI) + M(H_2)) \tag{11.1}$$

where $M(HI)$ and $M(H_2)$ are the atomic and molecular gas components, respectively, and h_e is the fraction of helium, generally taken to be $\sim 30\%$ ($h_e = 1.3$).

11.1.1
The Atomic HI Mass

11.1.1.1 21 cm Emission Line
Observed HI line emission fluxes at 21 cm, obtained by integrating the HI emission profile in the velocity space,

$$SHI_0 = \int S(V) dV \tag{11.2}$$

(see Figure 11.1), can be used to derive the total atomic hydrogen mass, $M(HI)$, of a galaxy. The HI emission of galaxies can be determined using either single beam or interferometric observations.

Single beam observations might need to be corrected for beam attenuation (c_1) and pointing offsets due either to pointing inaccuracies (c_2) or uncertainties in the galaxy coordinates (c_3) if the size of the observed galaxy and the uncertainty of its coordinates are comparable or larger than the half power beam width (HPBW) of the telescope. This is generally not the case, although the Arecibo radio telescope is

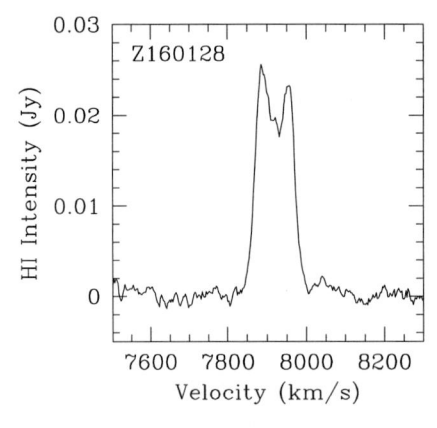

Figure 11.1 The 21 cm HI line emission of the spiral galaxy CGCG 160128 in the Coma supercluster, from [299]. Reproduced with permission (c) ESO.

a possible exception, where the beam size is HPBW $\sim$ 3.5′. If the HI distribution in a galaxy is partially resolved by the telescope beam, the observed HI flux SHI_0 underestimates the true flux, SHI, by an amount that depends on how much the HI distribution of the emitting source fills (or overfills) the beam of the telescope. When the HI distribution is unknown, the beam attenuation correction c_1 can be estimated using a model HI distribution for the target galaxy combined with the telescope beam power pattern. These corrections are important whenever the optical angular size of the observed galaxies is comparable to the HPBW of the telescope. The data reduction pipelines of interferometric data, like those taken at the VLA or Westerbork telescope, generally account for beam attenuation and pointing offsets. This is also the case for HI blind surveys such as ALFALFA [31], where galaxies are entirely mapped.

Another correction to the observed data can be made for HI self-absorption (c_4). In the densest regions of the galactic disk, the interstellar medium can be opaque at 21 cm [128]. Considering that the filling factor of the densest regions in the disk is small, the HI self-absorption can be measured by multiplying the observed HI flux by the factor $c_4 = (a/b)^{0.12}$, where a and b are the major and minor optical diameters of the galaxy [300]. Alternative correction recipes for self-absorption are used for the Hyperleda database (see [301]). The corrected HI flux is thus given by the relation

$$SHI = c_1 c_2 c_3 c_4 SHI_0 \quad (11.3)$$

These corrections are generally small – the average correction measured by [300] for a compilation of 9000 galaxies with single beam observations is $\sim$ 25% ($c_1 c_2 c_3 c_4 = 1.25$). Self-absorption corrections are not systematically applied to the data since model-dependent.

The uncertainty on the flux $\sigma(SHI)$ is given by the relation

$$\sigma(SHI) = \sqrt{\sigma_{\text{abs cal}}^2 + \sigma_{\text{cor}}^2 + \sigma_{\text{stat}}^2} \quad (11.4)$$

where $\sigma_{\text{abs cal}}$ is the uncertainty on the absolute flux calibration, generally determined by observing a noise diode or calibrating objects in the sky at regular intervals. The uncertainty introduced by the noise diode calibration was important ($\sim$ 10%) in the past, but has been significantly reduced in modern telescopes ($\sim$ 2%), as is the case for Arecibo after the installation of the Gregorian optical system. σ_{cor} is the uncertainty introduced by the application of different corrections, and can be quantified as $\sigma_{\text{cor}} = (c_1 c_2 c_3 c_4 - 1) SHI_0 / 3$. The statistical uncertainty, σ_{stat}, can be measured via the relation

$$\sigma_{\text{stat}} \, [\text{Jy km s}^{-1}] = 2 \, rms \, [\text{Jy}] \sqrt{W_{\text{HI}} \, [\text{km s}^{-1}] \delta V_{\text{HI}} \, [\text{km s}^{-1}]} \quad (11.5)$$

where rms is the root mean squared measured on the spectral baseline outside the line (in Jy), W_{HI} is the HI line width (that can be measured at either 50 or 20% of the peak), and δV_{HI} is the velocity resolution of the spectrum, both in km s^{-1}. For undetected galaxies, upper limits to their SHI_0 fluxes can be determined using the relation

$$SHI_0 \,[\text{Jy km s}^{-1}] = 2\,rms\,[\text{Jy}]\sqrt{300\sin(i)\delta\,V_{HI}\,[\text{km s}^{-1}]} \quad (11.6)$$

where i is the inclination of the galaxy. Three hundred km s^{-1} is generally used for massive galaxies, while lower values can be taken for low luminosity systems.

The atomic hydrogen mass of the galaxy is then given by the relation

$$M(\text{HI})\,[M_\odot] = 2.356 \times 10^5\,SHI\,[\text{Jy km s}^{-1}]D^2\,[\text{Mpc}] \quad (11.7)$$

where D is the distance of the galaxy, in Mpc, and SHI is its total corrected HI flux, in Jy km s^{-1}. The resulting uncertainty on the HI gas mass is on the order of $\sim 10\%$.

HI masses in normal, late-type galaxies generally span from $\sim 10^{10}\,M_\odot$ in massive Sc to $\sim 10^7\,M_\odot$ in dwarf systems [302, 303]. The ALFALFA survey has recently shown that a significant fraction of early-type galaxies is not totally devoid of HI gas, but have HI masses of $\sim 10^7$–$10^8\,M_\odot$ [304].

The study of the statistical properties of isolated galaxies defined an important parameter, the HI-deficiency parameter [305], often used to trace the effects of the environment on the late-type galaxy population. The HI-deficiency parameter, HI-def, is defined as the logarithmic difference between the average HI mass of a reference sample of isolated galaxies of similar type and linear dimension, and the HI mass actually observed in individual objects:

$$HI\text{-}def = \log M(\text{HI})_{\text{ref}} - \log M(\text{HI})_{\text{obs}} \quad (11.8)$$

According to [306],

$$\log M(\text{HI})_{\text{ref}}\,[M_\odot] = c + d\log(h\,diam\,[\text{kpc}])^2 - 2\log(h) \quad (11.9)$$

where c and d are weak functions of the Hubble type, as shown in Table 11.1, $diam$ is the optical isophotal diameter (in kpc) of the galaxy, and $h = H_0/100\,\text{km s}^{-1}\,\text{Mpc}^{-1}$. Given the nonlinearity of this relation, the calibration coefficients depend on the adopted cosmological model [307]. Galaxies with an HI-deficiency parameter equal to zero have a normal HI mass content while objects with $HI\text{-}def = 1$ have ten times less HI than galaxies of similar morphological type and linear dimension.

11.1.1.2 Gas Density and Temperature from Absorption Lines

Gas clouds located along the line of sight of bright QSO and radio galaxies can absorb the emitted radiation, producing absorption features in the spectra of the emitting sources. Absorption lines have Voigt profiles, which result from the convolution of a Lorentz and Gauss profile [92]. The Lorentz profile is due to the natural quantum mechanical broadening (damping) of an absorption line due to the intrinsic uncertainty on the ΔE energy of the upper atomic level, expressed by the

Table 11.1 The calibration of the HI deficiency parameter.

Type	c	d	Ref.
E-S0a	6.88	0.89	HG84 [306]
Sa-Sab	7.75	0.59	S96 [308]
Sb	7.82	0.62	S96 [308]
Sbc	7.84	0.61	S96 [308]
Sc	7.16	0.87	S96 [308]
Scd-Im-BCD	7.45	0.70	BG09 [307]

Notes: $h = H_0/100\,\text{km}\,\text{s}^{-1}\,\text{Mpc}^{-1}$; calibrations determined using HI fluxes not corrected for self-absorption; an updated version of these coefficient will be soon provided by the ALFALFA survey.

Uncertainty Principle, $\Delta E \Delta t \approx \hbar$. The Gauss profile is due to the internal motion of the gas in the cloud, which is related to both the thermal and turbulent motions. The Voigt profiles are characterized by the width of the line, which is quantified by the Doppler parameter b, the HI column density, and the velocity with respect to the mean redshift of the absorbing cloud. The Doppler parameter is defined as

$$b = \sqrt{2}\sigma = \sqrt{b^2_{\text{thermal}} + b^2_{\text{turbulent}}} = \sqrt{\frac{2kT}{m} + b^2_{\text{turbulent}}} \quad (11.10)$$

where σ is the characteristic width of the Gauss profile, k is the Boltzmann constant, and m the mass of the hydrogen atom [92]. Whenever the spectral resolution is not sufficient to fit the Voigt profile to the data, the curve of growth can be used to determine $N(HI)$. The curve of growth relates the observed equivalent width, W (Å), to the column density of the gas (see Figure 11.2). The observed equivalent width is defined as the integral of the line normalized by the nearby continuum [97]:

$$W(\lambda)\,[\text{Å}] = \int \left[1 - \frac{I(\nu)}{I(\nu,0)}\right] d\lambda$$
$$= \frac{\lambda^2}{c} \int [1 - e^{-\tau(\nu)}]\,d\nu \quad (11.11)$$

where $I(\nu)$ is the intensity of the line and $I(\nu,0)$ is the continuum emission under the line.

Figures 11.2 and 11.3 show that, whenever the column density of the gas is $N(HI) < 10^{13}\,\text{cm}^{-2}$ (the linear part of the curve of growth), the gas is optically thin and the equivalent width does not depend on the Doppler parameter b. The relationship between the density of the absorbing gas, N, and the rest frame equivalent width of the absorption line, $W_0(\lambda)$, is then given by [97]

$$N\,[\text{cm}^{-2}] = 1.13 \times 10^{20} \frac{W_0(\lambda)\,[\text{Å}]}{\lambda_0^2\,[\text{Å}]\,f_{\text{osc}}} \quad (11.12)$$

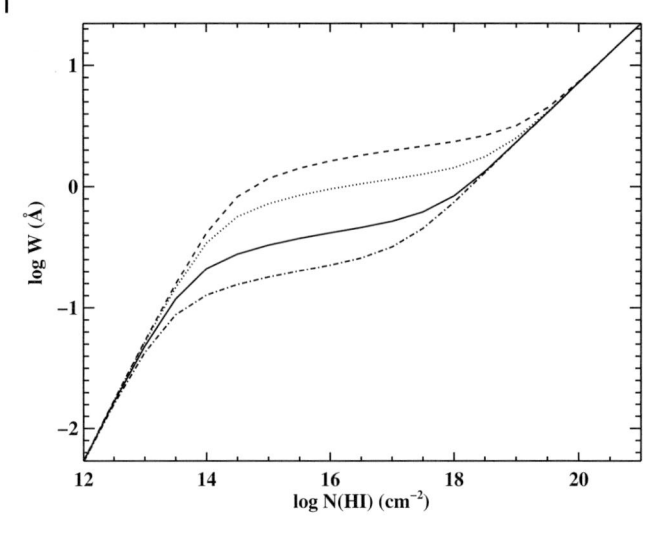

Figure 11.2 The relationship between the logarithm of the equivalent width W (in Å) and the neutral hydrogen column density $N(HI)$ given by the curve of growth for a Lyα absorption line. The four different curves represent different Doppler parameters $b = 10$ (dashed-dotted), 20 (solid), 50 (dotted), and 90 (dashed) km s^{-1}. For $N(HI) < 10^{13}$ cm^{-2}, known as the linear part of the curve of growth, the equivalent width does not depend on b (Figure 11.3). In the column density range 10^{13} cm^{-2} < $N(HI)$ < 10^{19} cm^{-2}, the flat part of the curve of growth, profiles are saturated and the equivalent width increases with b for constant $N(HI)$ (Figure 11.3). For $N(HI) > 10^{20}$ cm^{-2}, the profile has damping wings and the HI column density scales as the square of the equivalent width (courtesy of M. Fumagalli).

where f_{osc} is the oscillator strength. This relation reduces to

$$N(HI) \, [cm^{-2}] = 1.84 \times 10^{14} \, W_{rest} \, [\text{Å}] \tag{11.13}$$

for the determination of $N(HI)$ using the Lyman-α line. In the flat part of the curve of growth (Figure 11.2), for gas column densities in the range 10^{13} cm^{-2} < $N(HI)$ < 10^{19} cm^{-2}, the Lyα absorption line is saturated (Figure 11.3) and the broadening of the line is dominated by the Doppler contribution. Thus, the gas column density depends on W and b:

$$W_0(\lambda) \, [\text{Å}] \sim \frac{2 b \lambda_0 \, [\text{Å}]}{c} \sqrt{\ln\left(\frac{\pi^{0.5} e^2 \, [\text{eV}] \, N(HI) \, [cm^{-2}] \, f_{osc}}{m_e \, [g] c b}\right)} \tag{11.14}$$

with m_e = mass of the electron, and for Lyα this reduces to

$$N(HI) \, [cm^{-2}] = \frac{b \, [km \, s^{-1}]}{7.578 \times 10^{-16}} e^{\left(\frac{123.3 \, W(\lambda) \, [\text{Å}]}{b \, [km \, s^{-1}]}\right)^2} \tag{11.15}$$

In this range of densities, higher order Lyman series lines, which are characterized by low oscillator strengths f_{osc} and thus lie in the linear regime, are required to remove the degeneracy between $N(HI)$ and W that is due to the unknown Doppler

11.1 Gas Density, Mass, and Temperature

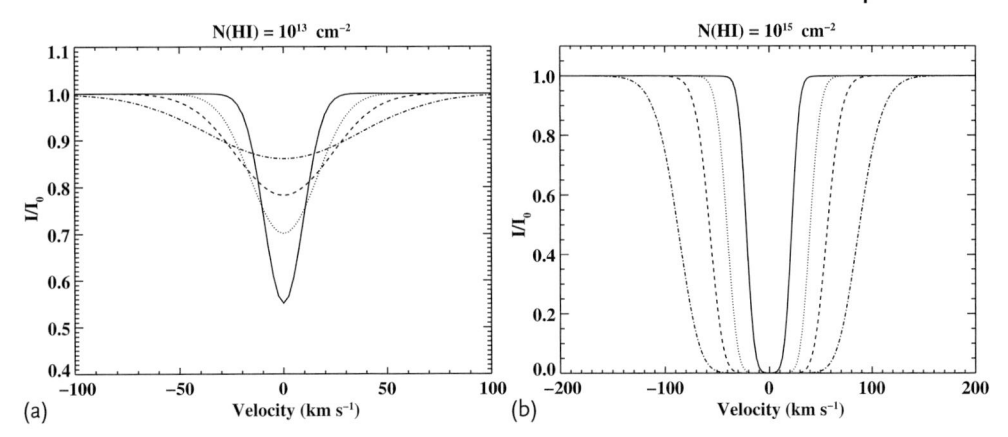

Figure 11.3 The absorption profile of the Ly$_\alpha$ line for a gas column density of (a) $N(\text{HI}) = 10^{13}$ cm^{-2} and (b) $N(\text{HI}) = 10^{15}$ cm^{-2}, with Doppler parameters $b = 10$ (solid line), 20 (dotted), 30 (dashed), and 50 (dotted-dashed) km s^{-1}. For low column densities (a), the equivalent width of the Ly$_\alpha$ absorption line does not depend on the Doppler parameter b (linear part of the curve of growth in Figure 11.2). For intermediate column densities (b), (10^{13} cm^{-2} < $N(\text{HI})$ < 10^{19} cm^{-2}) the equivalent width of the Ly$_\alpha$ absorption line is saturated and the equivalent width increases with b for a constant $N(\text{HI})$. This figure corresponds to the flat part of the curve of growth shown in Figure 11.2 (courtesy of M. Fumagalli).

parameter. This is clearly illustrated in Figure 11.4, where the Ly$_\alpha$ and the Ly$_\beta$ absorption profiles determined for the same Doppler parameters and gas densities are compared. While Ly$_\alpha$ is saturated at densities $N(\text{HI}) > 10^{14}$ cm^{-2}, Ly$_\beta$ saturates at gas column densities that are an order of magnitude higher ($N(\text{HI}) > 10^{15}$ cm^{-2}). Thus, Ly$_\beta$ is ideal for quantifying the gas density for HI column densities in the range 10^{14} cm^{-2} < $N(\text{HI})$ < 10^{15} cm^{-2}.

For systems with higher column densities (Lyman-limit systems; $N(\text{HI}) > 1.6 \times 10^{17}$ cm^{-2}), however, the column density of the gas can be determined (with a large uncertainty) by fitting the UV profile shortward of $\lambda_{\text{rest}} < 912$ Å with the relation

$$S(\lambda) = S_{\text{cont}}(\lambda > 912 \text{ Å}) e^{-\tau(\lambda)} = S_{\text{cont}}(\lambda > 912 \text{ Å}) e^{-\tau_{912\text{Å}} \left(\frac{\lambda [\text{Å}]}{912}\right)^3} \quad (11.16)$$

where $S_{\text{cont}}(\lambda > 912 \text{ Å})$ is the UV continuum at wavelengths longer than the Lyman break. Considering that, for an optically thick gas,

$$\tau = N(\text{HI}) [\text{cm}^{-2}] \sigma [\text{cm}^2] \quad (11.17)$$

where σ is the cross-section for ionization of hydrogen and $\sigma = 6.3 \times 10^{-18} (E_\gamma/13.6 \text{ eV})^{-3}$ cm^2 [309]

$$N(\text{HI}) [\text{cm}^{-2}] = \frac{\tau_{912}}{6.3 \times 10^{18}} \left(\frac{\lambda [\text{Å}]}{912}\right)^3 \quad (11.18)$$

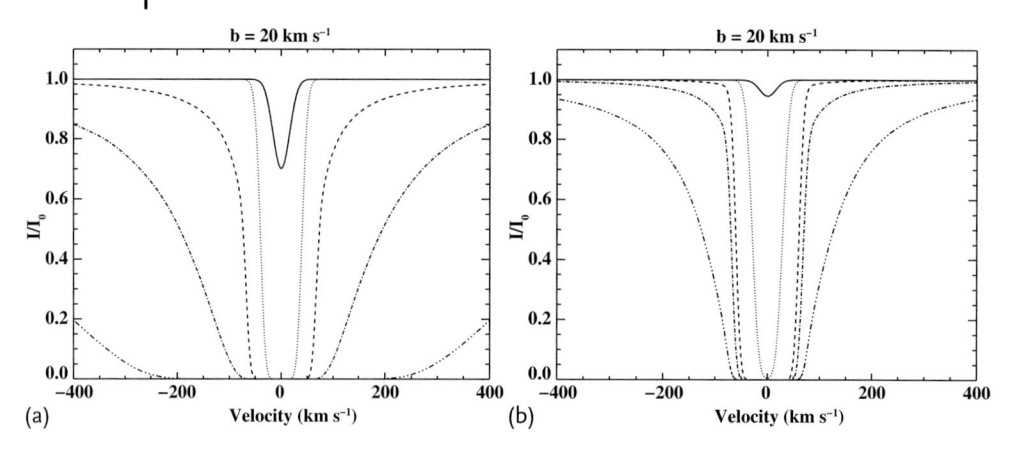

Figure 11.4 The absorption profiles of the (a) Ly$_\alpha$ and (b) Ly$_\beta$ lines for a Doppler parameter $b = 20\,\text{km s}^{-1}$ for neutral hydrogen column densities N(HI) of 10^{12} (solid line), 10^{15} (dotted), 10^{18} (dashed), 10^{19} (dotted-dashed), and 10^{20} (dashed-three dotted) cm^{-2}. The Ly$_\alpha$ absorption line is saturated for densities N(HI) > 10^{14} cm^{-2}, the Ly$_\beta$ line for N(HI) > 10^{15} cm^{-2}, an order of magnitude larger than the Ly$_\alpha$ line. The damped Ly$_\alpha$ profile is evident for N(HI) > 10^{19} cm^{-2} (courtesy of M. Fumagalli).

In the damping part of the curve (N(HI) > 10^{19} cm^{-2}), the lines are saturated but dominated by the Lorentzian damping wings. Here the column density of the gas is given by [310]

$$N(\text{HI})\,[\text{cm}^{-2}] = 1.88 \times 10^{18}\, W_{\text{rest}}^2\,[\text{Å}] \tag{11.19}$$

HI clouds detected through the absorption features in the UV spectrum of bright QSOs are divided into different classes according to their HI column density: the Lyman-α forest is composed of low column density systems with $10^{12} < N(\text{HI}) < 1.6 \times 10^{17}$ cm^{-2} [92]. At HI column densities larger than $N(\text{HI}) = 1.6 \times 10^{17}$ cm^{-2}, the atomic gas becomes optically thick to the Lyman continuum radiation, producing a sharp break in the UV spectrum of the QSO shortward of 912 Å. These systems (1.6×10^{17} cm$^{-2} < N(\text{HI}) < 2 \times 10^{20}$ cm^{-2}) are generally called Lyman-limit systems. Damped Lyman-α systems are those with HI column densities of $N(\text{HI}) > 2 \times 10^{20}$ cm^{-2} [311].

Whenever the gas column density is high, on the order of $\sim 10^{21}$ cm^{-2}, and the background QSO is a powerful radio source, then the HI 21 cm line can be observed as an absorption line. The line is not saturated, and thus its column density is given by the relation [129]

$$N(\text{HI})\,[\text{cm}^{-2}] = 1.84 \times 10^{18} \int \tau T_s\, dV \tag{11.20}$$

where τ is the optical depth

$$\tau = -\ln\left(1 - \frac{\Delta S}{S\, c_f}\right) \tag{11.21}$$

c_f is the covering factor of the radio continuum source by the HI gas cloud, ΔS is the 21 cm HI absorption line depth, S is the radio continuum flux density, dV is the velocity width, and T_S is the spin temperature. For point-like radio sources, the covering factor is $c_f = 1$. For small optical depths in the 21 cm HI absorption line, $\int \tau dV \sim W(\text{HI})$, where $W(\text{HI})$ is the equivalent width of the 21 cm HI line (expressed in km s^{-1}), and Eq. (11.20) becomes

$$N(\text{HI}) \, [\text{cm}^{-2}] = 1.84 \times 10^{18} \, W(\text{HI}) \, [\text{km s}^{-1}] T_S \, [\text{K}] \tag{11.22}$$

Comparing HI column densities measured in damped Lyman-α systems (Eq. (11.19)) with those measured using 21 cm HI absorption lines gives an estimate of the spin temperature of the HI cloud. Whenever the HI cloud is associated with a stellar disk, as is the case in the Milky Way or in the Magellanic clouds, the determined spin temperature is an average value. Mebold et al. [312] give a more accurate prescription for determining the spin temperature of the cold and warm gas phases simultaneously. Absorption line measurements in the UV rest frame can also be used to determine the column density of species other than hydrogen. Thus, the abundance of other elements in the gaseous phase of the ISM of intergalactic clouds can be inferred.

11.1.2
The Molecular H_2 Mass

11.1.2.1 H_2 from CO Emission Lines Observations

Because of its symmetric structure, molecular hydrogen (H_2) has no permanent electric dipole moment. This makes dipole rotational transitions strongly forbidden, while the very low intensity quadrupole rotational transitions are hardly observable since they are very weak [75]. The molecular hydrogen interstellar medium of normal, late-type galaxies is generally cold ($T_R \sim 10$ K), and is thus unobservable through its vibrational or electronic emission lines. For this reason, the molecular hydrogen mass is generally determined by observing several rotational lines of carbon monoxide (CO), the second most abundant molecule in the cold ISM. This technique is based on the assumption that CO is a good tracer of the mass of giant molecular clouds [313]. In spite of its high optical depth, it has been shown that the dynamical mass of galactic giant molecular clouds is proportional to the intensity of the CO line multiplied by $\sqrt{n(H_2)} T_R^{-1}$, leading to the relations

$$\frac{N(H_2)}{I(\text{CO})} \left[\frac{\text{cm}^{-2}}{\text{K km s}^{-1}} \right] = X = c_1 \sqrt{n(H_2) \, [\text{cm}^{-3}]} T_R^{-1} \, [\text{K}] \tag{11.23}$$

or equivalently,

$$\frac{M(H_2)}{L_{\text{CO}}} \left[\frac{M_\odot}{\text{K km s}^{-1} \text{pc}^2} \right] = \alpha = c_2 \sqrt{n(H_2) \, [\text{cm}^{-3}]} T_R^{-1} \, [\text{K}] \tag{11.24}$$

where $N(H_2)$ is the H_2 column density (defined as the mean molecular gas volume density, $n(H_2)$, whose value is ~ 200 cm^{-3} in normal galaxies, multiplied by the

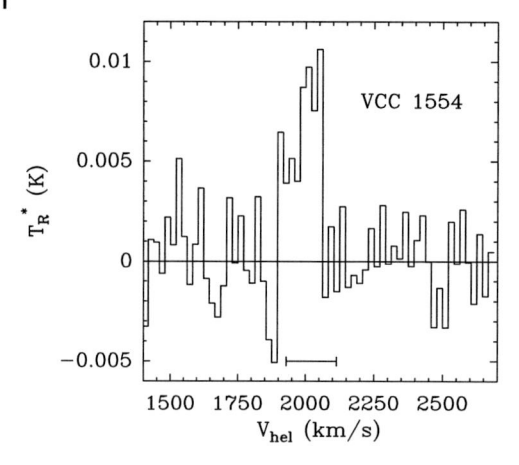

Figure 11.5 The ^{12}CO($J = 1-0$) line emission of the spiral galaxy VCC 1554 (NGC 4532) in the Virgo cluster (from [315]). Reproduced with permission (c) ESO.

path-length along the line of sight through the gas), $I(CO)$ is the intensity of the CO line, $M(H_2)$ is the molecular gas mass (including helium), L_{CO} the luminosity of the CO line, and T_R is the equivalent Rayleigh–Jeans brightness temperature of the (optically thick) CO line. The relatively constant value of $\sqrt{n(H_2)}\,T_R^{-1}$ observed in different giant molecular clouds in the Milky Way, in nearby late-type galaxies (e.g., [314]), or in starbursts such as M82, justify this assumption [313]. Provided that separate giant molecular clouds along the line of sight do not overlap in velocity, then the CO lines are a tracer of the molecular gas mass in external galaxies.

The most commonly used method to determine the molecular hydrogen mass is based on the observation of the ^{12}CO($J = 1-0$) rotational line at 2.6 mm (115 GHz; see Figure 11.5), assuming a standard conversion factor between the CO line intensity $I(CO)$ and the H_2 gas column density previously determined in the Milky Way or in nearby galaxies. This conversion factor has been determined using a variety of independent techniques to measure the molecular gas mass: the virial equilibrium of giant molecular clouds [313], the line ratios of different CO isotopes [316], assuming a metallicity-dependent gas to dust ratio [39], or relying on γ-ray data [317]. The most updated estimates of X in the Milky Way (X_G) range between 1 and 3×10^{20} cm^{-2}/(K km s^{-1}) (or equivalently $\alpha_G = 2.1$ and 6.4 $M_\odot$/(K km s^{-1} pc^2)). The most commonly used value, $X_G = 2.2-2.3 \times 10^{20}$ cm^{-2}/(K km s^{-1}) ($\alpha_G = 4.6-4.8$ $M_\odot$/(K km s^{-1} pc^2)) [318], is generally defined as the "standard" CO-H_2 conversion factor.

However, both theoretical considerations and observations have shown that X is not "standard". Rather, it changes nonlinearly with the physical properties of the ISM such as the hardness of the UV radiation, the cosmic ray density, the metallicity, and the gas density. This is well illustrated in Figure 11.6, where the CO conversion factor is plotted versus the total gas cloud column density or the A_V extinction for spherical clouds of different volume density, metallicity, and far-ultraviolet radiation fields ([114]; other model predictions can be found in [319]).

11.1 Gas Density, Mass, and Temperature

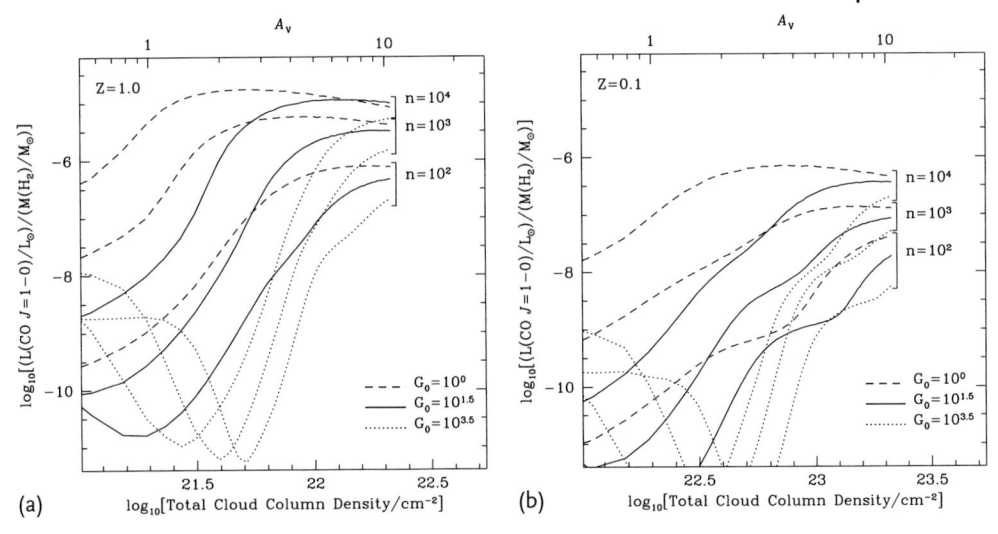

Figure 11.6 The variation of the ^{12}CO($J = 1-0$) line emission with respect to the H$_2$ mass (conversion factor) as a function of the cloud gas column density or extinction A_V predicted by the spherical photodissociation region models of [114]. Results are presented for clouds with volume densities $n = 10^2$, 10^3, and 10^4 cm^{-3} and far-UV intensities $G_0 = 10^0$, $10^{1.5}$, and $10^{3.5}$ (where G_0 is the far-ultraviolet flux in units of the local interstellar value), for (a) solar, and (b) 10% solar metallicities. Reproduced by permission of the AAS.

The X conversion factor can vary by up to a factor of ten between the low density diffuse medium, where the UV radiation can easily dissociate the carbon monoxide but not the self shielded molecular hydrogen, and the high-density regions in the core of molecular clouds. These variations have been observed in the Milky Way, and in nearby galaxies such as M31 and the Magellanic clouds [320–323]. Variations of X with respect to metallicity have been reported by [315, 324–326], although these results have been recently questioned by [327]. In [315], we give an empirical calibration of the X conversion factor as a function of metallicity, galaxy luminosity, or intensity of the UV radiation field (see Table 11.2 and Figure 11.7).

The intensity of the CO line is given by the integral (see Figure 11.5)

$$I(\text{CO}) = \int T dv \, [\text{K km s}^{-1}] \tag{11.25}$$

The brightness temperature can be expressed in different scales: T_R^* (source antenna temperature corrected for atmospheric, ohmic, and spillover losses), T_A^* (the antenna temperature corrected for atmospheric attenuation, resistive losses, and rearward spillover and scattering) or T_{mb} (main beam temperature) (see [328, 329]). The main beam temperature scale is appropriate for galaxies with angular sizes that are comparable with, or smaller than, that of the telescope main beam. The different temperature scales are linked by the following relations: $T_{\text{mb}} = T_A^*/\eta_{\text{mb}}$ and $T_R^* = T_A^*/\eta_{\text{fss}}$, where η_{mb} is the main beam efficiency and η_{fss} is the foreword

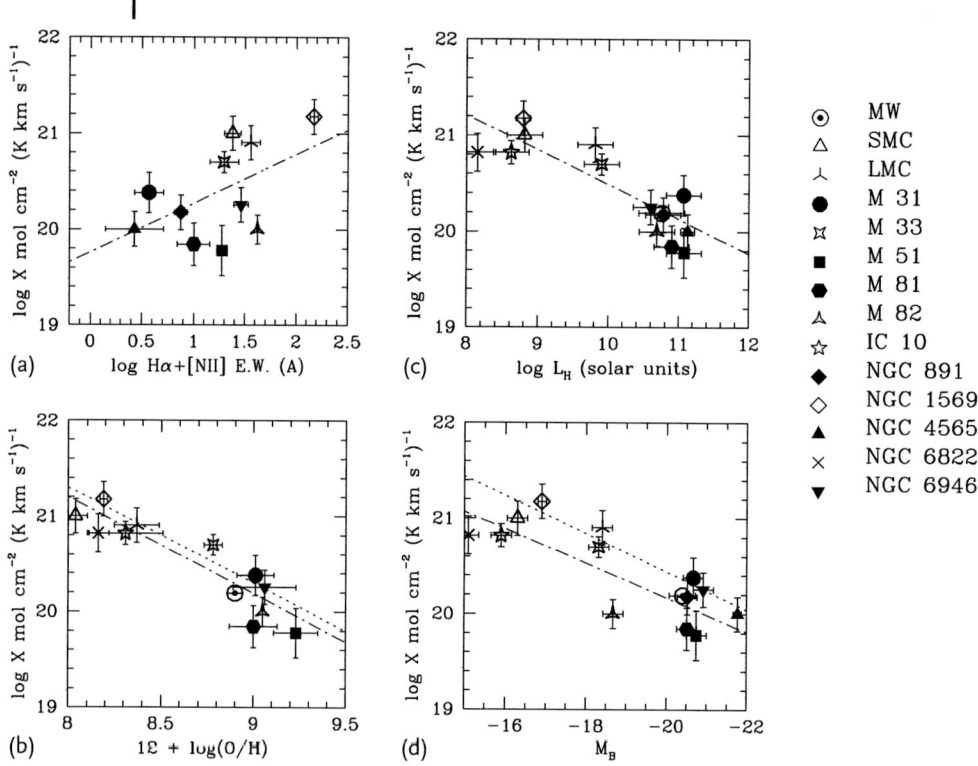

Figure 11.7 The relationship between the X conversion factor and (a) the $H_\alpha +$[NII] equivalent width, which is a tracer of the hardness of the interstellar radiation field (ionizing radiation); (b) the metallicity index $12+\log(O/H)$; (c) the H band luminosity, which is a proxy of the total stellar mass, and (d) the absolute B band magnitude for 14 nearby galaxies (from [315]). The dotted dashed-line is the best fit to the data given by [315] and reported in Table 11.2, the dotted line the best fit given by [325]. Reproduced with permission (c) ESO.

scattering and spillover efficiency. These parameters depend on the telescope used during the observations.

For undetected galaxies, upper limits to their I(CO) fluxes can be determined using the following relation

$$I(CO)\,[\text{K km s}^{-1}] = 2\,rms\,[\text{K}]\sqrt{W_{HI}\,[\text{km s}^{-1}]\delta\,V_{CO}\,[\text{km s}^{-1}]} \quad (11.26)$$

where rms is the root mean squared noise of the spectrum, W_{HI} is the HI line width (here we make the reasonable assumption that the molecular gas has the same kinematical properties of the atomic gas), and $\delta\,V_{CO}$ is the spectral resolution. For galaxies where W_{HI} is not available (or any other high-resolution emission line giving the observed rotational velocity), the HI line width can be estimated assuming the standard relation $W_{HI} = 300\sin(i)\,\text{km s}^{-1}$, where i is the galaxy inclination, or $W_{HI} = 50\,\text{km s}^{-1}$ if $i=0$ (lower values of W_{HI} can be assumed for low luminosity systems). The error on the intensity of the CO line, $\sigma(I(CO))$, can

Table 11.2 The relationships between the X conversion factor and different galaxy properties. X is defined as: log X = slope × Variable + constant (in units of mol cm^{-2} (K km s^{-1})$^{-1}$), adapted from [315].

Boselli et al. 2002

Variable	Slope	Constant	No. of objects	R^{2a}
log H_α + [NII]E.W.[b]	0.51 ± 0.26	19.76 ± 0.43	11	0.29
$12 + \log(O/H)$	-1.01 ± 0.14	29.28 ± 0.20	12	0.83
log L_H[c]	-0.38 ± 0.06	24.23 ± 0.24	14	0.75
M_B	0.18 ± 0.04	23.77 ± 0.28	14	0.67

Literature

Variable	Slope	Constant	Reference
$12 + \log(O/H)$	-1.00	29.30	[325]
$12 + \log(O/H)$	-0.67 ± 0.10	26.43 ± 0.86	[324]
$12 + \log(O/H)$	-1.00	29.20	[318]
M_B	0.20	24.44	[325]

[a] Regression coefficient.
[b] In logarithmic scale (in Å).
[c] In logarithmic scale (in solar units).

be computed as

$$\sigma(I(CO)) \, [\text{K km s}^{-1}] = 2 \, rms \, [K] \sqrt{W_{CO} \, [\text{km s}^{-1}] \delta \, V_{CO} \, [\text{km s}^{-1}]} \quad (11.27)$$

where rms is the root mean squared noise of the spectrum (in K), W_{CO} is the CO line-width, and δV_{CO} is the spectral resolution (both in km s^{-1}).

For single beam observations, when the galaxy's angular size is comparable or smaller than the size of the main beam of the telescope, the molecular gas mass, in solar units, can be determined using the relation

$$M(H_2) \, [M_\odot] = 2.96 \times 10^{-19} X D^2 \, [\text{Mpc}] \, I(CO) \, [\text{K km s}^{-1}] \theta^2 \, [\text{arcsec}] \quad (11.28)$$

with $I(CO)$ expressed in the main beam scale, where θ is the half power beam width (HPBW, in arcsec) of the telescope, and D is the distance to the source, in Mpc. For galaxies with angular sizes larger than the beam, $(\theta/[\text{arcsec}])^2$ should be replaced by $\Omega_{s \times b}$ (in arcsec^2), the solid angle of the source convolved with the telescope beam. It is quite difficult to quantify the uncertainty of the mass of the molecular hydrogen, since the error is dominated by the uncertainty of the X conversion factor.

The ^{12}CO($J = 1-0$) emission line at 115 GHz is easily accessible to ground-based radio telescopes for galaxies in the local universe, since it lies in a spectral range where the atmosphere is quite transparent. Other CO rotational transitions, however, such as ^{12}CO(2–1) (230 GHz) and ^{12}CO($J = 3$–2) (346 GHz), have been observed in nearby objects. It has been claimed that these higher level transitions should be preferred to the ^{12}CO($J = 1$–0) line to trace the dense molecular gas in the core of giant molecular clouds, the gas phase directly associated with the process of star formation. This is due to the fact that the $J = 1$ level is only 5.5 K above the ground state and the $J = 1$–0 transition has a critical density (the density necessary to produce substantial excitation of a rotational transition through collisions with H$_2$) of 1.1×10^3 cm^{-3} in optically thin gas. The $J = 3$ level is 33 K above the ground state and the $J = 3$–2 transition has a critical density of 2.1×10^4 cm^{-3}, thus the ^{12}CO($J = 3$–2) transition traces gas that is, on average, warmer and/or denser than ^{12}CO($J = 1$–0). Furthermore, because of redshift, high CO transition lines, which are in the rest frame at higher frequencies than the ^{12}CO($J = 1$–0) line, fall into the millimeter domain for distant galaxies. For this reason they are becoming the most commonly used tracers of the molecular gas phase in the far universe.

It is generally assumed that the observed CO line is optically thick and thermally excited. Thus, the ratio of the intrinsic surface brightness temperature (or equivalently the luminosity ratio) of the different transitions, defined as [330]

$$r_{J\,J-1} = \frac{I\{CO[J-(J-1)]\}}{I\{CO[(J-1)-(J-2)]\}} \times \frac{(J-1)^2}{J^2} \tag{11.29}$$

is equal to 1 by definition. This means that the CO luminosity, L_{CO}, in K km s^{-1} pc^2, is the same for different CO transitions, and that CO fluxes, in Jy km s^{-1}, scale as J^2. With this assumption, the molecular gas mass, $M(H_2)$ (including the mass of helium), is then given by the relation

$$M(H_2) = \alpha L_{CO} \tag{11.30}$$

for any CO rotational transition. Recent observations combined with models, however, have shown that the molecular gas properties of ULIRGs might be significantly different than those of normal, late-type galaxies. Unlike disk galaxies, where the CO emission comes principally from virialized molecular clouds, in ULIRGs the emission is dominated by the diffuse molecular medium. This implies that the virialized mass of these systems does not only include the molecular phase but rather all the baryonic components, including stars [331]. Furthermore, the molecular gas can be optically thick and thermally excited in ULIRGs, but this is not the case in normal late-type galaxies at any redshift [123, 330], as shown in Figure 11.8. CO luminosities determined using transitions higher than $J \geq 2$ can be strongly underestimated [330]. A further source of uncertainty is related to the fact that, in starburst galaxies, the dust can be optically thick up to several hundreds micrometers, thus affecting the estimate of the observed ratios of high J level CO transitions [332, 333]. In order to compensate for the different behaviors observed in these various classes of extragalactic sources, different α conversion factors have been proposed in the literature [318]. It has been shown that,

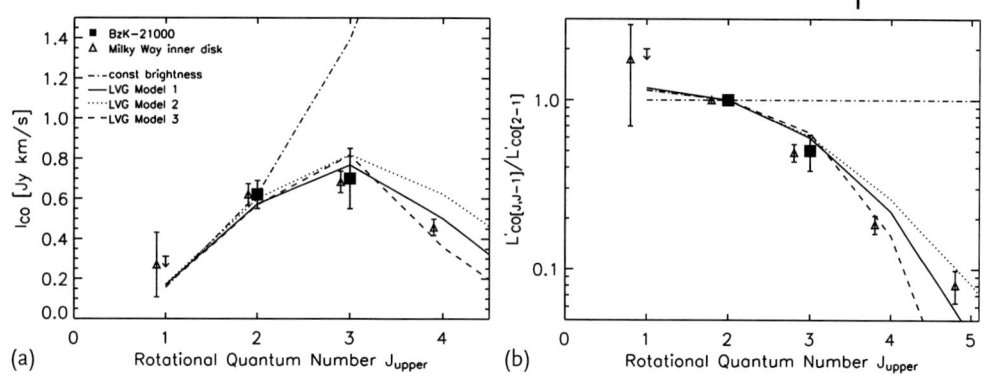

Figure 11.8 The CO spectral line energy distribution (SLED) of the galaxy BzK-21000 at $z = 1.522$, obtained by combining millimeter (Plateau de Bure) and centimeter (Very Large Array) observations, taken from [330] (a). The intensity of different ^{12}CO rotational line transitions of this high redshift galaxy (black filled squares) is plotted as a function of the rotational quantum number J, and compared to those obtained for the inner disk of the Milky Way (empty triangles) and to different large velocity gradient models (LGV). The dotted-dashed line shows the intensity of an optically thick, thermally excited gas, radiated at a constant temperature, and normalized to the ^{12}CO(2–1) emission line. (b) shows the ratios of the CO luminosities normalized to the CO(2–1) transition. $J_{upper} > 2$ transitions can severely underestimate the CO luminosity and thus the molecular gas mass of the observed galaxy. Reproduced by permission of the AAS.

due to the different conditions of the ISM (higher density and warmer temperature), the conversion factor in ULIRGs is $\alpha = 0.8 M_\odot$ (K km s^{-1} pc^2), significantly smaller than the canonical value found in normal, nearby, late-type galaxies ($\alpha = 4.6$–$4.8 M_\odot$ (K km s^{-1} pc^2)). An higher value of α should be used in high-redshift, normal, Milky-Way-like galaxies [330]. Given all these uncertainties, it is clear that relation (11.30) should be used with extreme caution. Late-type galaxies have, on average, molecular hydrogen masses derived from ^{12}CO(1–0) observations on the order of $M(H_2) = 10^7$–$10^9 M_\odot$, corresponding to $\sim 20\%$ of their total gas content [315]. However, these values might be underestimated since they are based on under-sampled CO maps. The fraction of molecular gas does not change with morphological type or galaxy mass. ULIRGs have molecular gas masses on the order of $M(H_2) = 0.4 \times 10^{10}$–$1.5 \times 10^{10} M_\odot$. The molecular gas, the dominant phase of the ISM, is much denser here than in normal, late-type galaxies [334].

11.1.2.2 H$_2$ from Gas to Dust Measurements

An alternative way to measure H$_2$ masses and column densities of galaxies is to consider a constant or metallicity-dependent gas to dust ratio, combined with HI and sub-mm dust emission measurements [39, 335]. The gas to dust ratio in the solar neighborhood is fairly constant, and has a value of ~ 160 [336], while it is $\sim$ a factor of 4 [337] and 10 [338] higher in the Large and Small Magellanic Clouds, respectively. In nearby galaxies, variations of the gas to dust ratio with respect to metallicity have been reported by [339]. Considering the standard relations calibrated on the Milky Way and on the Magellanic Clouds [315],

$$\log(gas/dust) = 10.207(\pm 0.015) - 1.146(\pm 0.024)$$
$$\times (12 + \log(O/H)) + \log(gas/dust)_\odot \quad (11.31)$$

where the gas to dust ratio is given relative to the solar neighborhood. The molecular gas column density can be derived using combined HI 21 cm and sub-mm dust observations, assuming that $gas/dust = 1.3(N(HI) + N(H_2))/N(dust)$, where $N(HI)$, $N(H_2)$, and $N(dust)$ are the atomic, molecular, and dust column densities, respectively (the factor 1.3 is included here to take into account the contribution of helium). Gas and dust column densities of unresolved galaxies can be determined from their mass measurements making some simple assumptions on the gas and dust distribution over their disks, as is done by [315].

11.1.2.3 H$_2$ from Near- and Mid-IR Emission Lines

The direct emission of relatively warm ($T \sim 100-1000$ K) H$_2$ molecules in the ISM of normal [340–342], starburst, and active galaxies [343] has been recently observed by ISO and Spitzer in the 5–30 µm spectral range. These lines are ideal tracers of the molecular gas phase which has been exposed to moderate heating, which includes a large fraction of the molecular clouds. As extensively described in [341], the emission of H$_2$ rotational lines depends on both the temperature and the density of the gas, and this might change in galaxies over a large dynamic range. This degeneracy can be removed by considering several rotational lines and comparing their fluxes to those expected from a gas in thermal equilibrium, for which the density ratio in the excited states is known [115]. Johnstone et al. [115] give an accurate prescription on how near-IR roto-vibrational and pure mid-infrared rotational lines can be combined for measuring H$_2$ column densities and gas masses in normal galaxies. In these objects, [341] show that the fraction of warm molecular gas is between 1 and 30% of the total molecular gas, with temperatures of $\approx 100-150$ K. The warm (~ 150 K) molecular hydrogen accounts for $\sim 10\%$ of the molecular gas phase in starbursts, and for 2–35% in Seyfert galaxies, which have a slightly higher gas temperature (150–180 K; [341, 343]).

The roto-vibrational H$_2$ line $S(1)(v = 1-0, J_u - J_l = 3-1)$ at 2.122 µm traces the molecular gas currently participating in star formation or recently shocked by newly formed stars. This line, which traces gas at temperatures higher than the purely rotational lines in the mid-IR ($T \sim 2000$ K), is due to H$_2$ molecules collisionally excited by shocks, or radiatively excited by the interstellar ultraviolet radiation field or re-processed X-ray photons ([344]). Assuming the Orion vibrational temperature of 2000 K, the mass of the hot molecular hydrogen can be estimated using the relation [344]

$$M(H_2)_{warm} [M_\odot] \simeq 5.08 \frac{I_{v=1 \to 0 S(1)}}{10^{-16} [W\,m^{-2}]} \left(\frac{D}{[Mpc]}\right)^2 \quad (11.32)$$

The fraction of warm to cold molecular gas mass ranges between 10^{-5} and 10^{-7} in normal, late-type galaxies, with larger ratios in more active objects [344]. Colder molecular hydrogen gas with temperatures on the order of $\sim$ 100–400 K, has been detected in early-type and [115] spiral galaxies [342], thanks to the emission of pure rotational lines in the 10–37 μm spectral domain covered by Spitzer.

11.1.2.4 H$_2$ from UV Absorption Lines

The density of the diffuse molecular hydrogen can be directly measured using absorption line measurements of UV photons emitted by background sources. Gas column densities can be derived using the same technique used for HI absorption lines. The molecular hydrogen is so abundant even in the diffuse interstellar medium that the strongest H$_2$ lines in the far-UV are dominated by damped wings, thus column densities can be determined using the same relation (11.19) adopted for measuring HI column densities. The kinetic temperature of the gas can be determined if the spectral resolution is sufficiently high to measure the population of the different rotational levels. The observation of UV absorption lines enables the detection of H$_2$ gas with low column densities (from 10^{14} to 10^{21} cm^{-2}), values much lower than those traced by the CO emission lines, which generally detect molecular gas with column densities in the range 10^{20}–10^{23} cm^{-2}. Direct measurements of the interstellar, diffuse H$_2$ gas using UV absorption lines have been done using FUSE in the Milky Way [345], in the Magellanic Clouds [346], and in the nuclear region of a bunch of starburst galaxies [347]. These observations have shown that, in the nuclei of starburst galaxies, the diffuse molecular hydrogen ($N(H_2) \leq 10^{20}$ cm^{-2}) is several orders of magnitude less abundant than that associated with dense molecular clouds. In the Magellanic Clouds, the fraction of diffuse H$_2$ is $\sim$ a factor of 10 smaller than that estimated from CO line emission measurements [346].

12
Dust Extinction

The electromagnetic radiation emitted by any extragalactic source has to go through the interstellar medium located within the emitting source and our Galaxy before reaching any observer. If observed with ground-based facilities, the emitted light also has to pass through the atmosphere of the Earth. The interstellar medium of galaxies, the Milky Way, and the atmosphere of our planet are not completely transparent and act as a selective filter to the incident radiation. This radiation is absorbed and scattered according to the physical properties of the medium and of the emitted radiation. In particular, the gaseous component of the interstellar medium is opaque to the ionizing radiation ($\lambda < 912$ Å) which is almost completely absorbed within the emitting galaxy. A small fraction of the ionizing photons can, however, escape because of the porosity of the interstellar medium (Lyman escape fraction). Dust grains absorb and scatter the stellar light, particularly light emitted at short wavelengths (UV, optical, and, to a minor extent, near-infrared). The interstellar medium becomes gradually transparent at longer wavelengths. While absorption reddens the emitted light (*reddening*), scattering bluens it (*bluening*). Absorption is more prevalent than scattering, and the observed spectrum of any extragalactic source can thus be significantly reddened by dust extinction (or attenuation). Observed spectra must be corrected for internal attenuation, Galactic extinction, and atmospheric extinction to reconstruct the intrinsic spectra of the emitting sources. The atmospheric extinction is generally measured by observing various calibration stars of different colors at high and low elevation. This is generally done during the initial data reduction and will not be further discussed in the present volume. For an accurate discussion on dust extinction within galaxies we refer the reader to the beautiful book of [348].

If $S_o(\lambda)$ and $S_c(\lambda)$ are the observed and corrected flux densities at a given wavelength λ, then

$$S_o(\lambda) = S_i(\lambda) S_{MW}(\lambda) S_c(\lambda) \tag{12.1}$$

where $S_i(\lambda)$ is the fraction of the flux density attenuated within the emitting source (internal attenuation) and $S_{MW}(\lambda)$ is the fraction attenuated within the Milky Way. When fluxes are expressed in magnitudes (in a given photometric band), the relation (12.1) becomes

$$m_c = m_o - A_i - A_{MW} \tag{12.2}$$

where m_c and m_o are the corrected and observed magnitudes, respectively, A_i is the internal attenuation, and A_{MW} is the Galactic extinction (both expressed in magnitudes). If F_c and F_o are the corrected and observed fluxes within the photometric bands, then

$$A_i + A_{MW} = 2.5 \log \frac{F_c}{F_o} \tag{12.3}$$

12.1
Galactic Extinction

The interstellar medium of the Milky Way acts as a screen to the radiation emitted by any extragalactic source. If $\tau_{MW}(\lambda)$ is the optical thickness at a given wavelength of the line of sight in the interstellar medium of the Milky Way, then the corrected flux density $S_c(\lambda)$ can be obtained by using the relation [349]

$$S_o(\lambda) = S_c(\lambda) e^{-\tau_{MW}(\lambda)} = S_c(\lambda) 10^{-[A_{MW}(\lambda)/2.5]} \tag{12.4}$$

$S_{MW}(\lambda) = e^{-\tau_{MW}(\lambda)}$, where $\tau_{MW}(\lambda) = 0.921 A_{MW}(\lambda)$ (with $A_{MW}(\lambda)$ expressed in magnitudes). $\tau_{MW}(\lambda)$ is proportional to the column density of the dust and logically decreases with increasing Galactic latitude b, reaching a minimum at the north and south Galactic poles.

The first systematic attempt to measure $\tau(\lambda)$ over the whole sky has been made using the HI map of the Milky Way [350] and assuming a constant dust to gas column density ratio [350]. More recently, Schlegel and collaborators determined a full sky map of the Milky Way dust column density using the all sky survey IRAS 100 µm ISSA map combined with the COBE/DIRBE 100 and 240 µm maps [351]. The Galactic column density map has been determined after subtracting the zodiacal light, point sources, and the contribution of the cosmic infrared background. The use of the 100 and 240 µm bands enabled an accurate determination of the dust temperature (ranging between 17 and 21 K) and thus of the dust column density. This technique has provided a full sky map of the dust column density of the Milky Way with the photometric accuracy of DIRBE and the angular resolution of IRAS which, at 100 µm, is on the order of 1.5 arcmin. This new method improved the Galactic extinction line of sight estimates by a factor of ~ 2 with respect to HI based extinction maps, in low and intermediate gas column density regions. The improvement was even greater in high gas density regions. Alternative methods based on varying background galaxy counts with respect to the Galactic latitude were proposed in the past [352] and gave consistent results [353], but are now not common because they are not sensitive to small angular scale variations.

The Galactic extinction $A_{MW}(B)$ is generally given in magnitude units (commonly indicated with A_B), that is, in the Johnson B band. $A_{MW}(B)$ can be converted into a Galactic extinction for other photometric bands or the continuum for any wavelength λ, $A_{MW}(\lambda)$, using a Galactic extinction curve $k_{MW}(\lambda)$ (see Section 12.1.1), and remembering that

$$A_{MW}(\lambda) = k_{MW}(\lambda) E_{MW}(B-V) \tag{12.5}$$

where

$$E_{MW}(B-V) = A_{MW}(B) - A_{MW}(V) \tag{12.6}$$

is the color excess. In the Milky Way,

$$k_{MW}(V) = \frac{A_{MW}(V)}{E_{MW}(B-V)} = 3.1 = R_{MW}(V) \tag{12.7}$$

and

$$k_{MW}(B) = \frac{A_{MW}(B)}{E_{MW}(B-V)} = 4.1 \tag{12.8}$$

Thus,

$$A_{MW}(\lambda) = \frac{k_{MW}(\lambda) A_{MW}(B)}{4.1} \tag{12.9}$$

The Galactic extinction curve $k_{MW}(\lambda)$ can also be expressed as a color excess relative to $E(B-V)$, $E(\lambda-V)/E(B-V)$, where

$$E(\lambda-V) = A(\lambda) - A(V) \tag{12.10}$$

and

$$k_{MW}(\lambda) = \frac{E_{MW}(\lambda-V)}{E_{MW}(B-V)} + R_{MW}(V) \tag{12.11}$$

12.1.1
Extinction Curve

The UV to near-infrared extinction curve (or extinction law), $k(\lambda)$, has been carefully measured only in the Milky Way [354–358], the LMC [359, 360], and the SMC [338, 360]. The extinction curves of these three galaxies are given in Figure 12.1 and Table 12.1. The three extinction curves are very similar for wavelengths larger than $\lambda > 5000$ Å, and yet are very different in the UV spectral range. In particular, while the Galactic extinction law is characterized by a prominent bump at 2175 Å, this is almost absent in the SMC. The observed variations in the extinction curves can be ascribed to the different dust content and composition of these three objects of differing metallicity. It is generally assumed that the extinction laws of

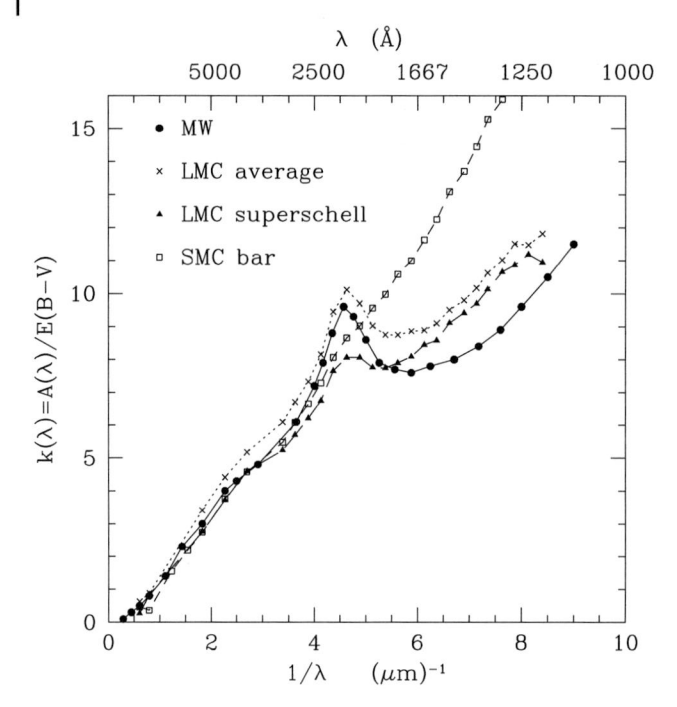

Figure 12.1 The UV to near-infrared extinction curves of the Milky Way, the Large Magellanic Cloud (average and supershell), and the Small Magellanic Cloud (bar), constructed using the data given in Table 12.1.

other galaxies will be similar to one of these three objects, according to their metallicity or star formation activity.

The extinction laws of the Galaxy and Magellanic Clouds are commonly fitted in the optical-UV spectral domain with the function [355]

$$k(\lambda) = R_V + c_1 + c_2 x + c_3 D(x, \gamma, x_0) + c_4 F(x) \tag{12.12}$$

where

$$D(x, \gamma, x_0) = \frac{x^2}{(x^2 - x_0^2)^2 + x^2 \gamma^2} \tag{12.13}$$

and

$$F(x) = \begin{cases} 0.5392(x - 5.9)^2 + 0.05644(x - 5.9)^3 & \text{for } x \geq 5.9 \ \mu m^{-1} \\ 0 & \text{for } x < 5.9 \ \mu m^{-1} \end{cases} \tag{12.14}$$

where $x = \lambda^{-1}$ (in μm). The coefficients of this equation for the Magellanic Clouds [360] are given in Table 12.2. A modified analytical expression of this law

12.1 Galactic Extinction

Table 12.1 Empirical extinction curves.

λ Å	MW			LMC average			LMC supershell			SMC bar		
	$1/\lambda$ μm^{-1}	$\frac{E(\lambda-V)}{E(B-V)}$	$k(\lambda)$	$1/\lambda$ μm^{-1}	$\frac{E(\lambda-V)}{E(B-V)}$	$k(\lambda)$	$1/\lambda$ μm^{-1}	$\frac{E(\lambda-V)}{E(B-V)}$	$k(\lambda)$	$1/\lambda$ μm^{-1}	$\frac{E(\lambda-V)}{E(B-V)}$	$k(\lambda)$
47 620	0.21	−3.02	0.06									
34 480	0.29	−2.91	0.17									
22 220	0.45	−2.76	0.32									
16 390	0.61	−2.58	0.50									
12 500	0.80	−2.23	0.85									
9 009	1.11	−1.60	1.48									
6 993	1.43	−0.78	2.30									
5 495	1.82	0.00	3.08									
4 405	2.27	1.00	4.08									
4 000	2.50	1.30	4.38									
3 436	2.91	1.80	4.88									
2 740	3.65	3.10	6.18									
2 500	4.00	4.19	7.27									
2 398	4.17	4.90	7.98									
21 980				0.45	−3.31	0.10	0.45	−2.48	0.28	0.45	−2.696	0.04
16 500				0.61	−2.78	0.63	0.61	−2.49	0.27	0.61	−2.277	0.46
12 500				0.80	−2.53	0.88	0.80	−1.93	0.83	0.80	−2.381	0.36
8 100				1.23			1.23			1.23	−1.186	1.55
6 500				1.54			1.54			1.54	−0.545	2.19
5 500				1.82	0.00	3.41	1.82	0.00	2.76	1.82	0.000	2.74
4 400				2.27	1.00	4.41	2.27	0.96	3.72	2.27	1.025	3.76
3 700				2.70	1.77	5.18	2.70	1.84	4.60	2.70	1.841	4.58
2 960				3.38	2.68	6.09	3.38	2.48	5.24	3.38	2.740	5.48
2 760				3.62	3.30	6.71	3.62	2.94	5.70	3.62	3.343	6.08
2 580				3.88	3.92	7.33	3.88	3.45	6.21	3.88	3.913	6.65
2 420				4.13	4.74	8.15	4.13	3.99	6.75	4.13	4.551	7.29
2 290				4.37	6.04	9.45	4.37	4.90	7.66	4.37	5.335	8.07
2 160				4.63	6.71	10.12	4.63	5.30	8.06	4.63	5.921	8.66

Notes: $k(\lambda)$ are determined assuming $R_{MW}(V) = 3.08$ [361], $R_{LMC\,average}(V) = 3.41$ [360], $R_{LMC\,supershell}(V) = 2.76$ [360], and $R_{SMC}(V) = 2.74$ [360].

Table 12.1 (continued)

λ Å	1/λ μm⁻¹	MW $\frac{E(\lambda-V)}{E(B-V)}$	$k(\lambda)$	λ Å	1/λ μm⁻¹	LMC average $\frac{E(\lambda-V)}{E(B-V)}$	$k(\lambda)$	λ Å	1/λ μm⁻¹	LMC supershell $\frac{E(\lambda-V)}{E(B-V)}$	$k(\lambda)$	λ Å	1/λ μm⁻¹	SMC bar $\frac{E(\lambda-V)}{E(B-V)}$	$k(\lambda)$
2299	4.35	5.77	8.85	2050	4.88	6.29	9.70	2050	4.88	5.30	8.06	2050	4.88	6.283	9.02
2188	4.57	6.57	9.65	1950	5.13	5.61	9.02	1950	5.13	5.00	7.76	1950	5.13	6.820	9.56
2101	4.76	6.23	9.31	1860	5.38	5.34	8.75	1860	5.38	4.98	7.74	1860	5.38	7.225	9.97
2000	5.00	5.52	8.60	1780	5.62	5.34	8.75	1780	5.62	5.14	7.90	1780	5.62	7.853	10.59
1901	5.26	4.90	7.98	1700	5.88	5.45	8.86	1700	5.88	5.33	8.09	1700	5.88	8.256	11.00
1799	5.56	4.65	7.73	1630	6.13	5.48	8.89	1630	6.13	5.69	8.45	1630	6.13	8.886	11.63
1701	5.88	4.60	7.68	1570	6.37	5.69	9.10	1570	6.37	5.82	8.58	1570	6.37	9.513	12.25
1600	6.25	4.73	7.81	1510	6.62	6.09	9.50	1510	6.62	6.35	9.11	1510	6.62	10.346	13.09
1490	6.71	4.99	8.07	1450	6.90	6.39	9.80	1450	6.90	6.65	9.41	1450	6.90	10.960	13.70
1393	7.18	5.36	8.44	1400	7.14	6.76	10.17	1400	7.14	6.94	9.70	1400	7.14	11.705	14.45
1316	7.60	5.91	8.99	1360	7.35	7.22	10.63	1360	7.35	7.37	10.13	1360	7.35	12.536	15.28
1250	8.00	6.55	9.63	1310	7.63	7.61	11.02	1310	7.63	7.90	10.66	1310	7.63	13.138	15.88
1176	8.50	7.45	10.53	1270	7.87	8.10	11.51	1270	7.87	8.11	10.87	1270	7.87	13.903	16.64
1111	9.00	8.45	11.53	1230	8.13	8.07	11.48	1230	8.13	8.43	11.19	1230	8.13	14.514	17.25
				1190	8.40	8.41	11.82	1190	8.40	8.19	10.95	1190	8.40	14.895	17.63
				1160	8.62			1160	8.62			1160	8.62	16.418	19.16

Table 12.2 Coefficients of the extinction law fitting functions for LMC and SMC.

Galaxy	Equation	R_V	c_1	c_2	c_3	c_4	c_5	x_0 μm^{-1}	γ μm^{-1}
MW	(12.15)	3.00	−0.175	0.807	2.991	0.319	6.097	4.592	0.922
LMC (Average)	(12.12)	3.41 ± 0.06	−0.890 ± 0.142	0.998 ± 0.027	2.719 ± 0.137	0.400 ± 0.036	–	4.579 ± 0.007	0.934 ± 0.016
LMC (Supershell)	(12.12)	2.76 ± 0.09	−1.475 ± 0.152	1.132 ± 0.029	1.463 ± 0.121	0.294 ± 0.057	–	4.558 ± 0.021	0.945 ± 0.026
SMC (Bar)	(12.12)	2.74 ± 0.13	−4.959 ± 0.197	2.264 ± 0.040	0.389 ± 0.110	0.461 ± 0.079	–	4.600 ± 0.000	1.000 ± 0.000
SMC (Wing)	(12.12)	2.05 ± 0.17	−0.856 ± 0.246	1.038 ± 0.074	3.215 ± 0.439	0.107 ± 0.038	–	4.703 ± 0.018	1.212 ± 0.019

Notes: MW from [358], LMC and SMC from [360].

has been recently proposed by [358] for the Galactic extinction law:

$$k_{MW}(\lambda) = \begin{cases} R_V + c_1 + c_2 x + c_3 D(x, \gamma, x_0) & \text{if } x \leq c_5 \\ R_V + c_1 + c_2 x + c_3 D(x, \gamma, x_0) + c_4(x - c_5)^2 & \text{if } x > c_5 \end{cases} \quad (12.15)$$

The coefficients for the Galactic extinction law are listed in Table 12.2, for which an extension to the near-infrared exists. These analytical functions can be used to determine $k(\lambda)$ for any λ in the UV to near-infrared range. It is worth noting that the average LMC extinction law, characteristic of relatively quiescent regions, is similar to that observed in the Milky Way, while the supershell LMC extinction curve, measured in active star-forming regions, has a different shape. As suggested by [360], the sequence MW, LMC average, SMC wing, LMC supershell and SMC bar can be representative of regions with increasing star formation activity.

12.2
Internal Attenuation

The electromagnetic radiation emitted by any source within a galaxy is attenuated (absorbed and scattered) by its ISM located along the line of sight to the observer. Attenuation within a galaxy depends on several factors, namely:

1. The wavelength λ of the emitted radiation: UV photons are more efficiently absorbed and scattered than optical or near-infrared ones, thus the light emitted by recently formed massive stars is more affected than that of the old stellar population.
2. The properties of the ISM, such as the metallicity, gas, and dust column densities, which characterize the galaxy extinction law. Extinction is more important in dust-rich starbursts than in normal galaxies, is weak in metal-poor dwarfs, and often negligible in early-type galaxies devoid of their ISM.
3. The relative geometry of the emitting sources and of the absorbing material on large and small scales. The optical images of edge-on galaxies such as NGC 891 or NGC 4565 clearly show that a large fraction of the stellar light in the disk is absorbed by the dust located in the plane of the galaxy. Recently formed young stellar populations are embedded inside a clumpy interstellar medium along a thin disk of thickness comparable to that of dust. For this reason their emission is more affected by attenuation than that of old stellar populations, which had time to migrate outside the high-density, dust-rich regions in the thin dust layer. On small scales, the ISM can be well mixed with the emitting sources (diffuse medium) or make a screen in front of them (clumpy medium). The latter is the most efficient geometrical configuration for attenuating the emitted light. On even smaller scales, the porosity of the ISM can allow a fraction of the ionizing radiation to escape where it is otherwise totally absorbed by the gaseous component (Lyman continuum escape fraction).

The attenuation of the stellar continuum and of the line emission are generally treated with different techniques, although in some cases the continuum attenuation is scaled to the line attenuation (Calzetti's law).

12.2.1
Attenuation of the Emission Lines

The extinction of hydrogen recombination lines is generally determined using the Balmer decrement: as shown in Section 3.2.1, for densities such as those found in the ISM of galaxies, the relative intensity of the hydrogen recombination lines depends only weakly on density and temperature. For this reason, their expected ratio should be fairly constant within galaxies, with values close to those given in Table 3.1 for a medium of density $n_e = 10^4$ cm^{-3} and effective temperature $T_e = 10^4$ K, when all the emitted photons are immediately absorbed by the diffuse gas located around the emitting star (case B; see Section 3.2.1). Whenever these conditions are satisfied, which is generally the case within galaxies, the expected intensity ratio of the H_α and H_β lines should be $I(H_\alpha)/I(H_\beta) = 2.85$. Since extinction is more efficient for H_β ($\lambda = 4861$ Å) than H_α ($\lambda = 6563$ Å), observed ratios $I_o(H_\alpha)/I_o(H_\beta) > 2.85$ indicate that both lines have been attenuated by the presence of dust along the line of sight (geometrical effects should be negligible since the gas emitting in H_α and H_β is ionized by the same source). We can thus derive the extinction of the H_β line using the relation [168]

$$\log\left[\frac{I_o(H_\alpha)}{I_o(H_\beta)}\right] = \log\left[\frac{I(H_\alpha)}{I(H_\beta)}\right] - C(H_\alpha) + C(H_\beta)$$

$$= \log\left[\frac{I(H_\alpha)}{I(H_\beta)}\right] - C(H_\beta) f(H_\alpha) \quad (12.16)$$

where

$$f(H_\alpha) = \frac{C(H_\alpha) - C(H_\beta)}{C(H_\beta)} = \frac{k(H_\alpha) - k(H_\beta)}{k(H_\beta)} \quad (12.17)$$

is the Galactic reddening function normalized to H_β. Once the extinction is determined for the Balmer lines, it can be determined for other lines, knowing that

$$C(\lambda) = \log\left[\frac{I(\lambda)}{I_o(\lambda)}\right] \quad (12.18)$$

is the logarithmic extinction coefficient, $I(\lambda)$ and $I_o(\lambda)$ are the extinction-corrected and the observed line fluxes, respectively. The extinction can also be expressed in magnitudes $A(\lambda)$, where

$$A(\lambda) = 2.5\,C(\lambda) \quad (12.19)$$

Thus, $C(\lambda)$ is related to the extinction law as:

$$C(\lambda) = \frac{k(\lambda)\,E(B-V)}{2.5} \quad (12.20)$$

Table 12.3 Extinction coefficients for optical and near-infrared emission lines.

Line	λ Å	$k_{MW}(\lambda)$[a]	$f(\lambda)$[a,b]
[OII]	3 727/29	4.751	+0.324
H$_\delta$	4 101	4.418	+0.231
H$_\gamma$	4 340	4.154	+0.158
[OIII]	4 363	4.128	+0.151
H$_\beta$	4 861	3.588	0.000
[OIII]	4 959	3.497	−0.025
[OIII]	5 007	3.452	−0.038
[OI]	6 300	2.642	−0.263
[NII]	6 548	2.524	−0.297
H$_\alpha$	6 563	2.517	−0.294
[NII]	6 584	2.507	−0.301
[SII]	6 716	2.444	−0.319
[SII]	6 731	2.437	−0.321
P$_\beta$	12 820	0.832	−0.768
P$_\alpha$	18 750	0.451	−0.874

[a] $k(\lambda)$ and $f(\lambda)$ are determined assuming $R_{MW}(V) = 3.08$.
[b] $f(\lambda)$ is relative to H$_\beta$.

Table 12.3 gives the Galactic extinction coefficients for several optical and near-infrared emission lines determined using the extinction law given by [354].

12.2.2
Attenuation of the Stellar Continuum

Dust attenuation reddens the spectra at short wavelengths, deeply modifying the spectral energy distribution of galaxies. Since the UV radiation is preferentially emitted by young and massive stars that are generally more affected by attenuation from surrounding dust clouds than the evolved stellar populations, star formation activities and histories determined using rest frame UV-optical observations can be highly inaccurate if the data are not properly corrected for dust extinction.

12.2.2.1 The Far-Infrared to UV Flux Ratio

Radiative transfer models have shown that the UV attenuation can be accurately determined assuming that the energy of the emitted light absorbed by dust is re-radiated in the infrared, as graphically shown in Figure 12.2. The energetic balance between the absorbed UV and emitted IR radiation can be quantified using the IR to UV flux ratio. The strength of this method, presently estimated as the most accurate method for correcting UV data, resides in the fact that it is almost completely

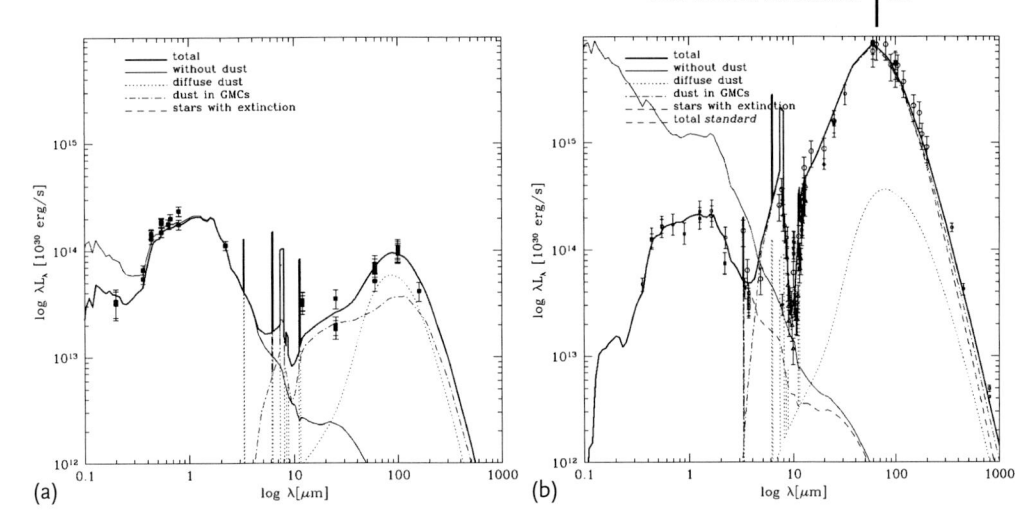

Figure 12.2 The UV to submillimeter spectral energy distribution of (a) the star-forming galaxy M100 and (b) the starburst Arp 220, from [160]. The UV to near-infrared emitted stellar radiation (light solid line) is partly absorbed by dust and re-radiated in the infrared (dotted and dotted-dashed lines) to produce the observed spectral energy distribution (dark solid line). The observed stellar SED (dashed line) is mainly modified in the UV domain. The energy absorbed by dust and re-emitted in the infrared is a small fraction of the total emitted energy in normal, spiral galaxies such as M100, while it is dominant in strong starbursts such as Arp 220. Reproduced by permission of the AAS.

independent on the geometrical configuration of the system and on the extinction law [362–364]. This is due to two main reasons: a) in active star-forming systems, the emitting dust is principally heated by the UV radiation produced by young stars that are physically located inside HII regions, where dust and stars are well mixed, and b) the infrared emission is not self-absorbed by dust, thus the energy is isotropically emitted in the infrared. At the same time, the energetic balance implies that galaxies are considered to be integrated objects: the energy emitted in one region can be absorbed by dust located elsewhere, thus loosing the spacial information on dust attenuation. Obviously, the far-infrared to UV flux ratio is an accurate tracer of dust attenuation only in those objects where the far-infrared emission comes from dust heated by the stellar component, as is the case in normal, star-forming galaxies and starbursts, but not in objects hosting an AGN where the infrared emission is due to the accretion disk.

The calibration of the attenuation law relating the $A(UV)$ extinction to the infrared to UV flux ratio has been determined either using different radiative transfer models especially tuned to reconstruct the geometrical configuration of emitting stars and absorbing dust in galaxies [160, 161, 363, 365–370], or empirical relations between the observed total infrared luminosity and the UV emission using different, observed (IR) and synthetic (stellar) SEDs of galaxies. The synthetic spectra are attenuated using an attenuation curve scaled to a given extinction law (generally Galactic or LMC), then subtracted from the unobscured spectra to estimate the total energy absorbed by dust which, assuming an energetic balance, equals the total energy emitted in the infrared (F_{TIR}) by the galaxy. The observed UV flux within a filter is then estimated by convolving the reddened SED with the UV filter response, thus allowing the $F_{TIR}/F(UV)$ flux ratio for any $A(UV)$ extinction to be calculated. This model-independent technique has the advantage of providing an accurate zero-point correction to the UV data, but it cannot be used to reconstruct the shape of the dust-free UV spectral profile which strongly depends on the presence of the UV bump at 2175 Å in the assumed extinction curve.

Presently, the most widely used empirical calibration valid for actively starforming galaxies, given by [371], is

$$A(NUV) = -0.0495 \left[\log\left(\frac{F_{TIR}}{F(NUV)}\right)\right]^3 + 0.4718 \left[\log\left(\frac{F_{TIR}}{F(NUV)}\right)\right]^2$$
$$+ 0.8998 \left[\log\left(\frac{F_{TIR}}{F(NUV)}\right)\right] + 0.2269 \tag{12.21}$$

for the GALEX NUV (2310 Å) band, and

$$A(FUV) = -0.0333 \left[\log\left(\frac{F_{TIR}}{F(FUV)}\right)\right]^3 + 0.3522 \left[\log\left(\frac{F_{TIR}}{F(FUV)}\right)\right]^2$$
$$+ 1.1960 \left[\log\left(\frac{F_{TIR}}{F(FUV)}\right)\right] + 0.4967 \tag{12.22}$$

for the GALEX FUV (1530 Å) band, where F_{TIR} is the total infrared flux defined in Eq. (7.7), and $F(NUV)$ and $F(FUV)$ are defined as $F(NUV) = \nu(NUV)S(NUV)$ and $F(FUV) = \nu(FUV)S(FUV)$, all expressed in units of $W\,m^{-2}\,Hz^{-1}$, while $A(\lambda)$ is expressed in magnitudes. Other calibrations can be found in [372] (for starburst galaxies), and [367, 373, 374] (for normal, star-forming galaxies). As emphasized in [371], this calibration is not appropriate for strong starburst galaxies, which are better represented by the relation [372]

$$A(1600\,\text{Å}) = 2.5 \log\left[\frac{F_{TIR}}{0.9F(1600)1.75} + 1\right] \tag{12.23}$$

where $F(1600)$ is the 1600 Å IUE flux defined as $\nu(1600)S(1600)$.

These calibrations are not valid for quiescent spirals such as those populating the very nearby universe, where the infrared emission can be dominated by the cirrus heated by the general interstellar radiation field [373]. In these objects, the $A(\lambda)$ vs. $F_{TIR}/F(FUV)$ calibration depends on the shape of the spectral energy distribution of the underlying stellar population. In galaxies with low star formation activity, the dust is efficiently heated by the old stellar population which dominates the interstellar radiation field. For these objects, [375] gives different recipes according to their star formation history:

$$A(FUV) = a_1 + a_2 \left[\log\left(\frac{F_{TIR}}{F(FUV)}\right)\right] + a_3 \left[\log\left(\frac{F_{TIR}}{F(FUV)}\right)\right]^2$$
$$+ a_4 \left[\log\left(\frac{F_{TIR}}{F(FUV)}\right)\right]^3 + a_5 \left[\log\left(\frac{F_{TIR}}{F(FUV)}\right)\right]^4 \tag{12.24}$$

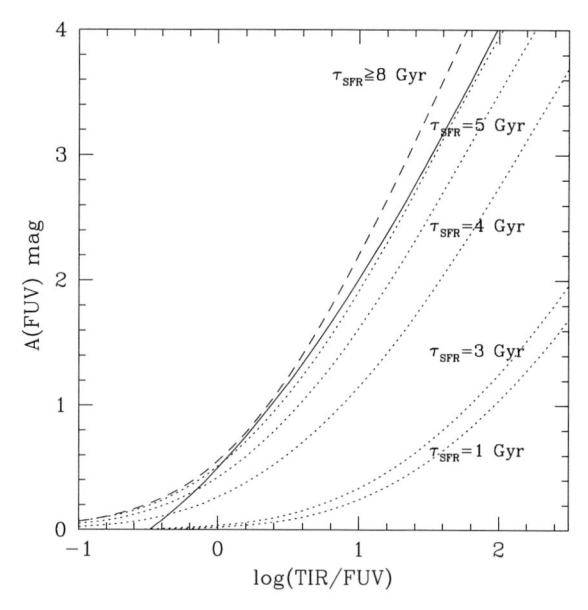

Figure 12.3 The relationship between the attenuation A(FUV) and the $F_{TIR}/F(FUV)$ flux ratio. The dotted lines show the models of [375] for galaxies characterized by a star formation history of the type SFR$(t, \tau_{SFR}) = 1/\tau_{SFR}^2 \exp(-t^2/(2\tau_{SFR}^2))$, where t is the age of the galaxy and τ_{SFR} is a characteristic time scale for star formation, which is small ($\sim$ 2 Gyr) for red galaxies which formed most of their stars in the past and large ($\sim$ 8 Gyr) for blue galaxies still very active in forming new stars (see Section 9.1.2). The solid line represents the calibration of [371] for star-forming galaxies and the dashed line that of [372] for starbursts. Reproduced with the permission of John Wiley & Sons Ltd.

where the a_i coefficients changes with the star formation history. The later is traced by the τ_{SFR} star formation time scale parameter (Table 12.4). Given the strong relationship between the star formation history of galaxies and their UV-optical-near-infrared color, these different calibrations can be scaled to observed colors (Table 12.4). These recipes should be preferred for early-type spiral galaxies or for those star-forming objects with red colors for which the calibrations of [371] or [376] would drastically overestimate the dust attenuation, as shown in Figure 12.3. As defined, however, these calibrations are also appropriate for any normal galaxy active in star formation.

For correcting stellar SEDs or UV, optical, or near-infrared colors, the A(FUV) needs to be transformed into $A(\lambda)$ for any λ. Cortese et al. [375] proposed a recipe in order to take into account the large scale geometrical configuration of disk galaxies. Consider a simple sandwich model to represent a thin layer of dust of thickness ζ embedded in a thick layer of stars where the dust to stars scale height ratio $\zeta(\lambda)$ depends on wavelength (with λ in units of Å) [152] according to

$$\zeta(\lambda) = 1.0867 - 5.501 \times 10^{-5} \times \lambda \tag{12.25}$$

Stars are formed in a thin layer and migrate with time to high galactic latitudes, producing a z-scale layer which increases with wavelength. The internal attenua-

Table 12.4 The a_i coefficients of relation (12.24) for different values of τ_{SFR} and observed colors (all magnitudes are in AB system).

τ_{SFR} (Gyr)	a_1	a_2	a_3	a_4	a_5	FUV-H	FUV-i	FUV-r	FUV-g	FUV-B
≤ 2.6	0.02025	0.06107	0.07212	0.10588	−0.01517	9.1	7.5	7.3	6.7	6.4
2.8	0.02355	0.06934	0.08725	0.10339	−0.01526	8.6	7.0	6.9	6.3	6.0
3	0.03404	0.09645	0.12452	0.09679	−0.01548	8.2	6.6	6.5	5.9	5.7
3.2	0.05822	0.15524	0.17801	0.08664	−0.01593	7.8	6.2	6.1	5.6	5.3
3.4	0.09944	0.24160	0.23161	0.07580	−0.01671	7.4	5.9	5.8	5.3	5.0
3.6	0.15293	0.33799	0.27713	0.06638	−0.01792	7.0	5.5	5.4	5.0	4.7
3.8	0.20982	0.42980	0.31431	0.05909	−0.01957	6.7	5.2	5.1	4.7	4.4
4	0.26302	0.51013	0.34522	0.05377	−0.02164	6.4	4.9	4.8	4.4	4.2
4.2	0.30899	0.57732	0.37157	0.05000	−0.02399	6.1	4.6	4.6	4.2	3.9
4.4	0.34695	0.63224	0.39438	0.04739	−0.02650	5.8	4.3	4.3	3.9	3.7
4.6	0.37760	0.67674	0.41420	0.04555	−0.02900	5.5	4.1	4.0	3.7	3.4
4.8	0.40210	0.71272	0.43139	0.04426	−0.03140	5.2	3.8	3.8	3.5	3.2
5	0.42168	0.74191	0.44624	0.04332	−0.03362	4.9	3.6	3.6	3.3	3.0
5.4	0.45013	0.78536	0.47009	0.04210	−0.03745	4.4	3.1	3.1	2.9	2.6
5.8	0.46909	0.81520	0.48787	0.04138	−0.04050	4.0	2.7	2.7	2.5	2.2
6.2	0.48223	0.83642	0.50127	0.04092	−0.04288	3.6	2.3	2.3	2.1	1.9
6.6	0.49167	0.85201	0.51152	0.04060	−0.04475	3.2	1.9	2.0	1.8	1.6
7	0.49867	0.86377	0.51952	0.04038	−0.04624	2.8	1.6	1.6	1.5	1.2
≥ 8	0.50994	0.88311	0.53315	0.04004	−0.04883	< 2.6	< 1.2	< 1.3	< 1.2	< 1.1

tion $A(\lambda)$ is given by the relation [152]

$$A(\lambda) = -2.5 \log \left\{ \left[\frac{1 - \zeta(\lambda)}{2} \right] \left(1 + e^{-\tau_{\text{dust}}(\lambda) \cdot \sec(i)} \right) \right. \\ \left. + \left[\frac{\zeta(\lambda)}{\tau_{\text{dust}}(\lambda) \cdot \sec(i)} \right] \left(1 - e^{-\tau_{\text{dust}}(\lambda) \cdot \sec(i)} \right) \right\} \quad [mag] \quad (12.26)$$

where i is the galaxy inclination, and $\tau_{\text{dust}}(\lambda)$ is the optical depth. $\tau_{\text{dust}}(\text{FUV})$ can be derived by inverting Eq. (12.26):

$$\tau_{\text{dust}}(\text{FUV}) = \left[\frac{1}{\sec(i)} \right] [0.0259 + 1.2002 \times A(\text{FUV}) + 1.5543 \\ \times A(\text{FUV})^2 - 0.7409 \times A(\text{FUV})^3 + 0.2246 \times A(\text{FUV})^4]$$

(12.27)

$\tau_{\text{dust}}(\lambda)$ can then be determined at any wavelength using the relation

$$\tau_{\text{dust}}(\lambda) = \tau_{\text{dust}}(\text{FUV}) \frac{k(\lambda)}{k(\text{FUV})} \tag{12.28}$$

and assuming an extinction curve.

12.2.2.2 The Calzetti's Attenuation Law

Unfortunately, infrared data are not always available for dwarf galaxies in the nearby universe or for optically or UV selected objects in deep cosmological surveys. Other recipes for dust attenuation, empirically calibrated with available multifrequency data on well defined nearby samples, have been proposed in the literature. Among these, the Calzetti's law is the most widely used. This can be generalized to

$$S_c(\lambda) = S_o(\lambda) 10^{0.4 A(\lambda)} = S_o(\lambda) 10^{0.4 E_g(B-V) k^e(\lambda)} \tag{12.29}$$

where $S_c(\lambda)$ is the correct flux density, $S_o(\lambda)$ is the observed flux density, and $E_g(B-V)$ is the color excess derived from the nebular gas emission which is related to the stellar color excess $E_s(B-V)$ via the relation [372]

$$E_s(B-V) = 0.44 E_g(B-V) \tag{12.30}$$

The attenuation law, $k^e(\lambda)$, is given by

$$k^e(\lambda) = 1.17 \left(-1.857 + \frac{1.040}{\lambda} \right) + 1.78$$

for $0.63\ \mu m \leq \lambda \leq 2.20\ \mu m$ \hfill (12.31)

and

$$k^e(\lambda) = 1.17 \left(-2.156 + \frac{1.509}{\lambda} - \frac{0.198}{\lambda^2} + \frac{0.011}{\lambda^3} \right) + 1.78$$

for $0.12\ \mu m \leq \lambda \leq 0.63\ \mu m$ \hfill (12.32)

This attenuation law should not be confused with the extinction curve described in Section 12.1.1. Given that $E_g(B-V)$ can be directly determined from the hydrogen emission line ratios [377],

$$E_g(B-V) = \frac{\log(R_o/R_i)}{0.4[k(\lambda_a) - k(\lambda_b)]} \tag{12.33}$$

where R_o and R_i are the observed and intrinsic[1] ratios of the hydrogen emission lines a and b, respectively, and $k(\lambda_a)$ and $k(\lambda_b)$ are the extinction curves measured at the wavelength of the two emission lines, the Calzetti's attenuation law on the stellar continuum can be scaled on hydrogen line ratios. This leads to the following relations [378]:

$$A(1600\ \text{Å}) = 1.21 A(H_\beta) = 1.78 A(H_\alpha) \tag{12.34}$$

---
1) The intrinsic values of the hydrogen line ratios can be found in Table 3.1 in Section 3.2.1.

and

$$A(2800\,\text{Å}) = 1.29A(H_\alpha) \tag{12.35}$$

where $A(H_\alpha)$ and $A(H_\beta)$ can be determined using the prescriptions given in Section 12.2.1. The Calzetti's law has been theoretically justified by the presence of a turbulent interstellar medium, where dust acts as a foreground screen for the emitting sources [379, 380]. The limit of this relation is that, although widely used, it is valid only for starburst galaxies. When applied to normal star-forming galaxies [371] or nearby early-type spirals, this relation strongly overestimates the extinction [375].

12.2.2.3 The Slope of the UV Spectrum

Whenever far-infrared data are not available, which is generally the case in deep cosmological surveys, dust extinction is commonly measured using the slope of the UV spectrum [372, 376, 378]. This is due to the fact that, whenever an episode of star formation is occurring, the production of a high quantity of massive, OB stars dominates the UV emission of galaxies. The slope of the UV spectrum,

$$S(\lambda) \propto \lambda^\beta \tag{12.36}$$

can be predicted by population synthesis models. The UV spectrum in the range ~ 1000–$2000\,\text{Å}$ is fairly constant and has a slope $\beta \simeq -2$ which does not change significantly as a function of the star formation history, as depicted in Figure 12.4. Since dust extinction is more efficient at short wavelengths, the presence of dust flattens the UV spectrum, making the slope β a direct tracer of dust attenuation. Radiative transfer models, however, have indicated that this tracer strongly depends on the relative geometry of the emitting stars and absorbing dust, and on the assumed extinction law, particularly the presence of the UV bump at $2175\,\text{Å}$ [364]. However, on given, well defined categories of extragalactic sources, where both the

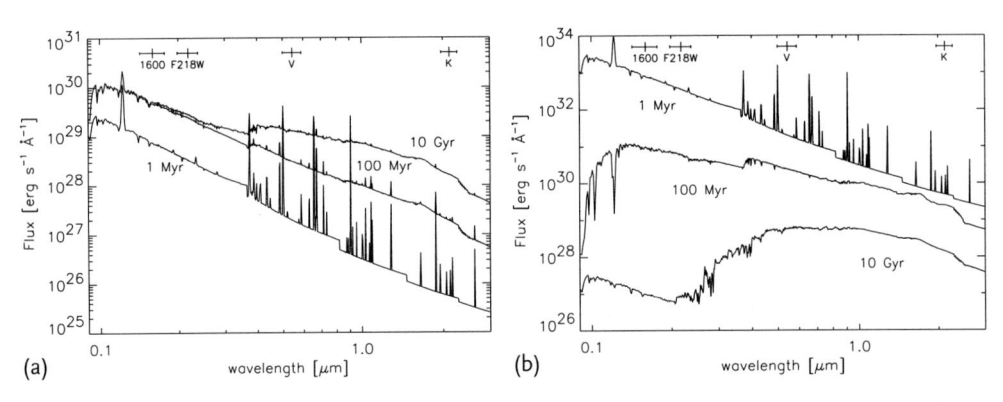

Figure 12.4 Example of extinction-free stellar SEDs for (a) a constant and (b) burst stellar populations of different ages, from [363]. The slope of the UV spectrum in the wavelength range between 1000 and 2000 Å does not significantly change with the history of star formation (constant vs. bursty), or with the age of the burst or the duration of the constant activity. Reproduced by permission of the AAS.

geometry and the extinction law should not drastically change, it can give fair results.

Originally measured in the UV spectra of nearby starbursts [376], the slope of the UV spectrum is now generally measured using the β parameter determined from imaging data, whose definition changes depending on the selected UV bands [378]. Using the GALEX FUV and NUV bands, β_{GALEX} reduces to [381]

$$\beta_{\text{GALEX}} = \frac{0.4[m(\text{FUV}) - m(\text{NUV})]}{\log_{10}(\lambda_{\text{NUV}}/\lambda_{\text{FUV}})} - 2.0 = 2.22[m(\text{FUV}) - m(\text{NUV})] - 2.0$$
(12.37)

where $m(\text{FUV})$ and $m(\text{NUV})$ are the GALEX AB magnitudes (although other slightly different definitions are available in the literature). First empirically determined using the far-infrared to UV flux ratio vs. β relationship in local starbursts using inhomogeneous apertures by [376], its most recent calibration based on integrated UV and total infrared fluxes (TIR) of galaxies hosting a starburst is given by [381]

$$A(\text{FUV}) = 3.85 + 1.96\beta \pm 0.4$$
(12.38)

This relation is widely used in deep cosmological surveys, where massive galaxies had much higher star formation activities than at the present epoch. For nearby late-type galaxies with a normal star formation activity, however, this relation overpredicts the internal attenuation, as shown in Figure 12.5. In these galaxies, and in

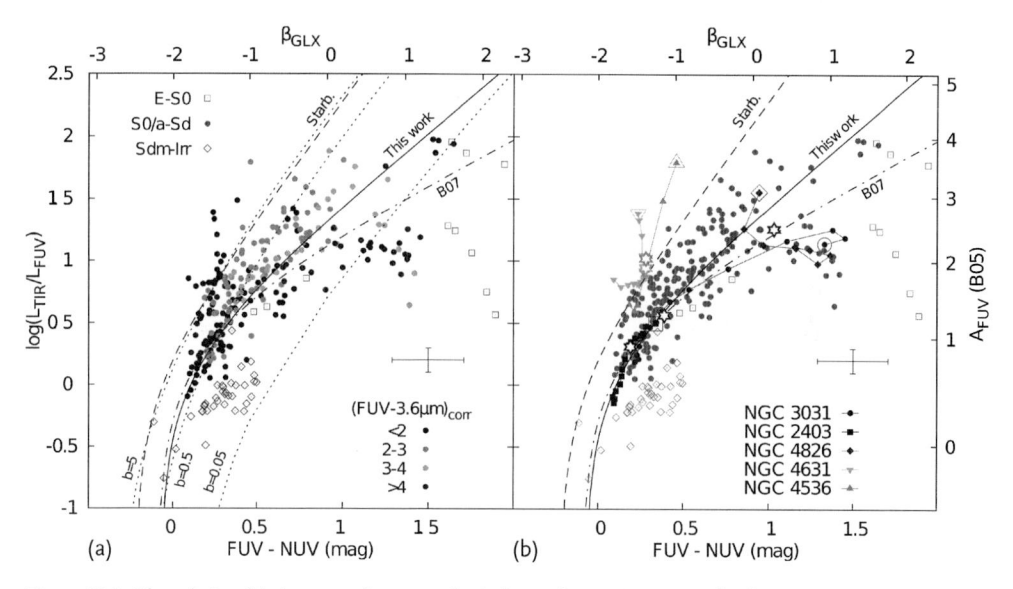

Figure 12.5 The relationship between the total far-infrared to FUV luminosity ratio and the slope of the UV spectrum, β, for SINGS galaxies classified according to their morphological type, from [163]. Normal galaxies in the local universe (solid line) follow a relation significantly different than starburst galaxies (dashed line). Reproduced by permission of the AAS.

all star-forming galaxies not undergoing a starburst phase, other empirical relations should be used. The far-infrared to UV flux ratio vs. β relation in normal galaxies is indeed flatter than in starbursts [163]:

$$\frac{F_{TIR}}{F(FUV)} = 10^{[0.30+1.15(m_{FUV}-m_{NUV})]} - 1.64 \qquad (12.39)$$

13
Star Formation Tracers

The formation and evolution of galaxies can be observationally constrained through the study of their present and past star formation activity. The star formation history of galaxies can be determined by fitting their stellar ultraviolet (UV) to near-infrared spectral energy distribution with population synthesis models, while the direct observation of the youngest high-mass stars is generally used to infer their ongoing activity.

The gas component of the interstellar medium is distributed inhomogeneously within galaxies. Internal or external perturbations such as spiral density waves, gas compression due to supernova explosions or interactions with nearby companions can induce turbulence in the gas, creating high density regions where gravity can break the hydrostatic equilibrium and make the gas collapse. Gas collapses within giant molecular clouds and fragments into smaller systems creating a star cluster. In the densest regions, the temperature can increase sufficiently to start the hydrogen burning process, giving birth to a star. For a complete description of the star formation process we refer the reader to [67] or [75].

13.1
The Initial Mass Function

Stars are formed when a giant molecular cloud fragments into a star cluster with a given mass distribution called initial mass function (IMF). If dN is the number of stars with masses in the range $(M, M + dM)$ formed during the time interval dt, then

$$dN = f(M, t) dt dM = \psi(t) \phi(M) dt dM \qquad (13.1)$$

where $\phi(M)$ is the initial mass function, defined as the amount of mass locked up in a new generation of stars with masses in the interval $(M, M + dM)$, normalized to satisfy the relation

$$\int_{M_{\text{low}}}^{M_{\text{up}}} M \phi(M) dM = 1 \qquad (13.2)$$

A Panchromatic View of Galaxies, First Edition. Alessandro Boselli.
© 2012 WILEY-VCH Verlag GmbH & Co. KGaA. Published 2012 by WILEY-VCH Verlag GmbH & Co. KGaA.

where M_{up} and M_{low} are the upper and lower mass cut-offs of the initial mass function, generally taken to be 0.1 and 100$M_\odot$ respectively. $\psi(t)$ is the star formation rate, the total mass of stars formed per unit time, and often indicated with SFR. The IMF is sometimes expressed in a logarithmic form [382]:

$$\xi(M) = \frac{dN}{d\log(M)} \tag{13.3}$$

The initial mass function $\phi(M)$ is generally represented by a power law:

$$\phi(M) \propto M^{-\alpha} \tag{13.4}$$

Thus,

$$\xi(M) \propto M^{-\alpha+1} \tag{13.5}$$

whose value was first determined by [382] to be $\alpha = 2.35$ (Salpeter IMF). Recent observations of star clusters in the Milky Way or in other nearby galaxies have shown that the shape of the IMF is universal [383–386], with a slope changing in different stellar mass ranges [385]:

$$\begin{aligned}
\alpha &= 0.3 \pm 0.7 \quad \text{if} \quad 0.01 \leq M/M_\odot < 0.08 \\
\alpha &= 1.8 \pm 0.5 \quad \text{if} \quad 0.08 \leq M/M_\odot < 0.50 \\
\alpha &= 2.7 \pm 0.3 \quad \text{if} \quad 0.50 \leq M/M_\odot < 1.00 \\
\alpha &= 2.3 \pm 0.7 \quad \text{if} \quad 1.00 \leq M/M_\odot
\end{aligned} \tag{13.6}$$

Although several indirect observational results on the star formation properties of different types of galaxies at different cosmological epochs can be explained by a variation of the initial mass function, presently there is no direct or convincing evidence against its universality (e.g., [387]). Recently, however, statistical considerations led some authors to conclude that the integrated galactic IMF, generally indicated as IGIMF, which represents the IMF of a whole galaxy obtained by combining the IMF of all the single star clusters composing it, should change with the global star formation activity of the galaxy [388, 389]. These variations are expected to be important for very low star formation activities. The validity of this hypothesis, however, has still to be confirmed observationally.

13.2
The Star Formation Rate

Star formation rates, measured in $M_\odot$ yr^{-1}, are generally estimated by observing the young stellar population (see [390] for a review). Young and massive stars have relatively short lifetimes on the main sequence, thus their presence indicates recent star formation episodes in galaxies. If $L(\lambda)$ is the luminosity relative to the emission of the youngest stars, the star formation rate can be determined through the relation

$$\psi(t) = \text{SFR} = K(\lambda)L(\lambda) \tag{13.7}$$

where $K(\lambda)$ can be inferred from population synthesis models with several assumptions on the shape of the IMF and on the upper and lower mass cut-offs. This widely used technique can be blindly applied only if the initial mass function is universal, thus independent of morphological type, luminosity, and redshift, and if the star formation activity of the target galaxies has been constant for a time greater than or equal to the lifetime of the emitting stars. Stationarity in the star formation process is requested since, at a given wavelength, the number of newly formed emitting stars should equal the number of older emitting stars leaving the main sequence, so that the total emission of a galaxy is proportional to the number of emitting stars formed per year. This stationarity regime is reached only if star formation is constant over a time scale comparable to, or larger than, the lifetime of the emitting stars, as clearly shown in Figure 13.1.

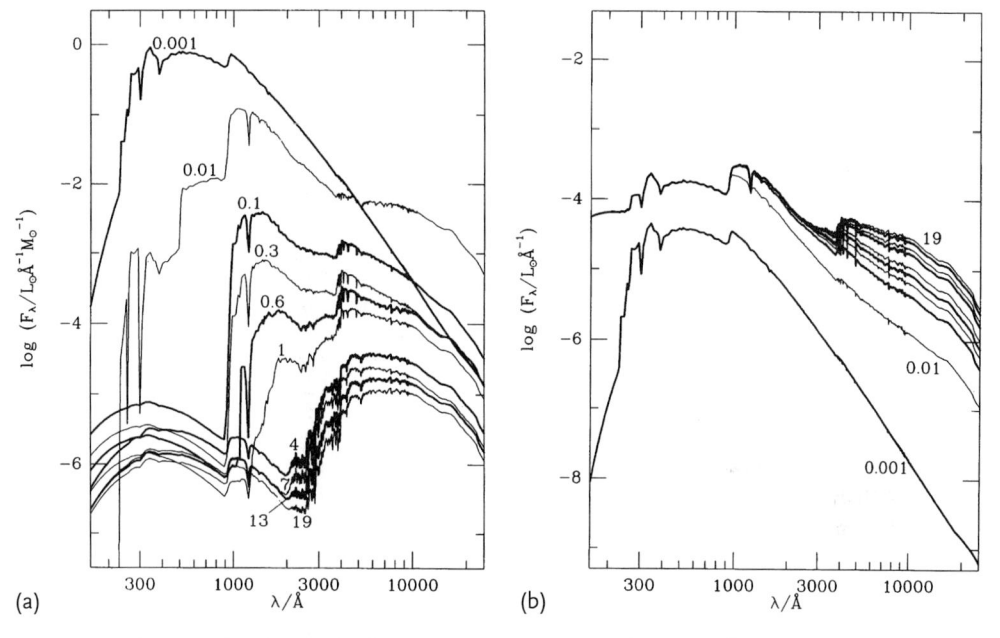

Figure 13.1 The spectral evolution of stellar populations for an instantaneous burst (a) and a constant star formation activity (b), as determined using population synthesis models with a Salpeter IMF (adapted from [391]). The numbers indicate the age (in Gyr) of the single burst and of the beginning of the constant activity. In the instantaneous burst model, the UV flux emitted by the massive stars is at its maximum at the epoch of the burst (0.001 Gyr) and drastically decreases afterwards. In the constant star formation model, the UV flux becomes constant only 10^7 (ionizing radiation, $\lambda < 912$ Å) and 10^8 yr (UV radiation at $\lambda \sim 1500$ Å) after the beginning of the star formation episode. Reproduced by permission of the AAS.

The stationarity condition is quite restrictive since it is well known that the star formation activity of galaxies changes with time because of an inconstant supply of gas or because of possible interactions with the nearby environment. In normal, massive galaxies such as those observed in the nearby universe, star formation can be considered constant on time scales of the order of 10^7–10^8 yr despite the fact that on small scales such as those of single HII regions, the activity changes drastically on time scales smaller than 10^7 yr (the lifetime of a single HII region is on the order of some 10^6 yr). In interacting systems such as the Antennae, or in typical active objects such as those dominating the early universe, stationarity might not be satisfied on time scales on the order of 10^8 yr, thus limiting the number of potential star formation tracers.

13.3
The Birthrate Parameter and the Specific Star Formation Rate

Another useful parameter to characterize the star formation activity of galaxies is the birthrate parameter. The birthrate parameter b, first proposed by [220, 383], is defined as the ratio of the present day star formation activity to the star formation rate averaged over the entire lifetime of the galaxy. Considering galaxies as coeval objects of age t_0, b can be determined using the relation [392]

$$b = \frac{\text{SFR}}{\langle \text{SFR} \rangle} = \frac{\text{SFR}\, t_0 (1 - R)}{M_{\text{star}}} \tag{13.8}$$

where M_{star} is the stellar mass of the galaxy, which can be determined as described in Section 15.1, and R is the fraction of gas that stars re-injected through stellar winds into the interstellar medium during their lifetime (recycled gas fraction). Since stars eject different fractions of gas into the ISM at different epochs of their life, the return parameter R is a function of time and depends on the assumed IMF and birthrate history. Kennicutt et al. [196] have shown that 90% of the returned gas is released in the first gigayear (more than half in the first 200 Myr) of a stellar generation for any assumed IMF. For this reason, an instantaneous recycling approximation for the determination of the birthrate parameter b can be calculated assuming $R = 0.3$ for a Salpeter IMF. The H_α equivalent width, generally quoted as H_α E.W. (or H_α + [NII]E.W. depending whether it has been determined using narrow band imaging including the two nearby [NII] lines), defined as the ratio of the H_α line (current star formation) over the underlying red continuum (evolved stellar population), is an observational entity directly related to the birthrate parameter. The birthrate parameter is also tightly related to the specific star formation rate used in cosmological surveys, defined as [393]:

$$\text{SSFR}\,[\text{yr}^{-1}] = \frac{\text{SFR}}{M_{\text{star}}} = \frac{b}{t_0(I - R)} \tag{13.9}$$

13.4
The Star Formation Efficiency and the Gas Consumption Time Scale

The star formation efficiency is defined as the rate at which the gas content of galaxies is transformed into stars, and is given by the relation [394, 395]

$$\epsilon = \frac{SFR t_{SFR}}{\Sigma_{gas}} \qquad (13.10)$$

where t_{SFR} is the time scale for star formation, which depends on the growth rate of the gravitational instability of the gas and is thus related to the velocity dispersion and to the surface density of the gas, and Σ_{gas} the gas surface density. Given the difficulty of measuring t_{SFR}, the unitless star formation efficiency ϵ is generally replaced by the variable [396]

$$SFE [\text{yr}^{-1}] = \frac{SFR}{M_{gas}} \qquad (13.11)$$

Given that stars are formed within molecular clouds, the total gas mass M_{gas} is sometimes replaced by the molecular gas mass $M(H_2)$. The star formation efficiency is related to the time scale for gas depletion if the fraction of gas ejected by stars and recycled in the interstellar medium is taken into account. This gas consumption time scale, often referred to as the "Roberts time," τ_R [397], is given by the relation [392]

$$\tau_R [\text{yr}] = \frac{M_{gas}/SFR}{1 - R} = \frac{1/SFE}{1 - R} \qquad (13.12)$$

where R is the returned gas fraction.

13.5
Hydrogen Emission Lines

Hydrogen recombination lines are observed only in the presence of the ionizing radiation ($\lambda < 912$ Å) emitted by high-mass ($M > 10 M_\odot$), young ($\leq 10^7$ yr) OB stars [390]. Because of the recombination process (see Section 3.2.1), their intensity is proportional to the global photoionization rate of the ISM, which in turn depends on the number of ionizing stars. Among these lines, the most commonly used are the H_α line at 6563 Å, easily accessible from ground-based facilities for nearby galaxies, and the H_β line at 4861 Å, also observable at relatively high redshift. They can be observed in spectroscopic mode or in imaging mode using narrow band interferential filters. These lines suffer from dust extinction and should be corrected as explained in Section 12.2.1 before they can be used to quantify the

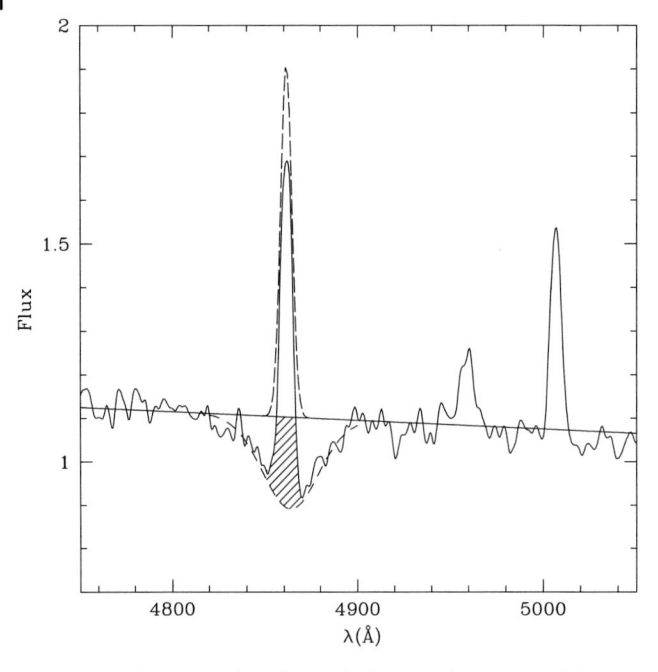

Figure 13.2 The observed H$_\beta$ line (solid line) must be deblended into a corrected emission and an underlying absorption component (dashed lines). The shaded region represents the portion of the absorption line that must be added to the emission line to obtain its correct value (from [84]). Reproduced with permission (c) ESO.

star formation activity of any galaxy[1]. Because of the absorption of the stellar atmosphere of relatively young stars, the H$_\beta$ line is also characterized by a relatively strong underlying absorption feature that must be considered in the determination of the emitted flux (see Figure 13.2).

At the same time, H$_\alpha$ imaging data obtained by using narrow band interferential filters generally include the nearby [NII] lines at 6548 and 6584 Å. Their emission should be removed from the H$_\alpha$ +[NII] flux before using the imaging data to quantify star formation rates. The [NII] contamination is generally measured using integrated spectroscopy, or nuclear spectroscopy combined with aperture corrections. Whenever these are not available, statistical corrections are used, such as those proposed by [399], and depicted in Figure 13.3

$$\frac{[\text{NII}]}{\text{H}_\alpha} = 10^{0.44 \log L(\text{H}) - 4.91} = 10^{0.38 \log M_{\text{star}} - 4.13} \tag{13.13}$$

or [400],

$$\frac{[\text{NII}]}{\text{H}_\alpha} = 10^{0.35 \log L(\text{H}) - 3.85} = 10^{0.30 \log M_{\text{star}} - 3.23} \tag{13.14}$$

---

[1] Alternative recipes for determining the dust-free total H$_\alpha$ luminosity of galaxies have been proposed by [398] by combining H$_\alpha$ imaging data with far-infrared observations.

13.5 Hydrogen Emission Lines

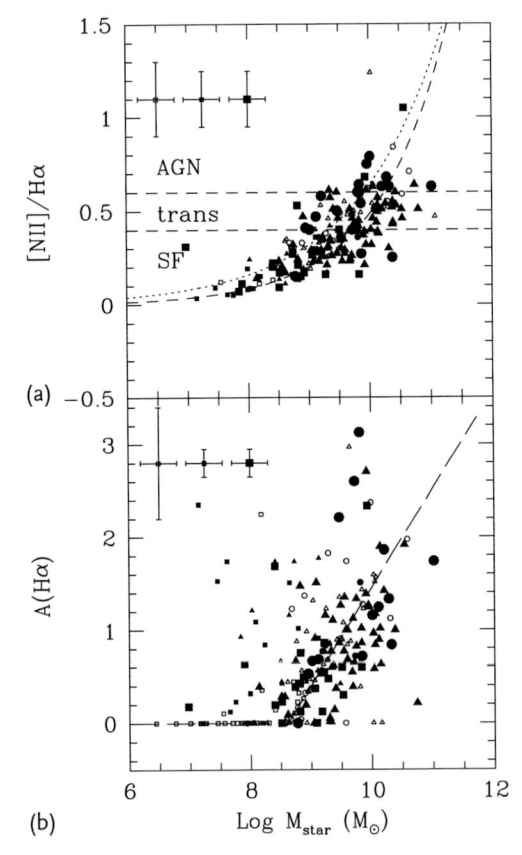

Figure 13.3 The relationship between the [NII]/Hα ratio (a) and A(Hα) (b) and the stellar mass (in logarithmic scale). The horizontal, dashed lines indicate the intervals in [NII]/Hα for AGN, transition, and star-forming galaxies (see Section 10.1.3). The dotted line gives the empirical [NII]/Hα to stellar mass calibration given in [400], the short-dashed line that of [399], while the long-dashed line gives the A(Hα) vs. log M_{star} relation. Circles are for Sa-Sb, triangles for Sbc-Sd and squares for Sm-Im-BCD galaxies. Reproduced by permission of the AAS.

where [NII] is the sum of the 6548 and 6584 Å lines and $L(H)$ is the total H band luminosity (in solar units). Integrated spectroscopy, however, has to be preferred since both aperture corrections [401] or statistical recipes are highly uncertain [387]. Furthermore, an accurate determination of the underlying H$_\beta$ absorption and of the [NII] contamination can only be achieved with a sufficient spectral resolution ($R \sim 1000$).

Two other corrections should be applied to the observed data before they can be used to derive star formation rates. The first one is to account for the fraction of ionizing photons that escape the galaxy without ionizing the gas (escape fraction). This fraction is generally small ($\leq 6\%$) in normal, star-forming galaxies, but can be important in high redshift starburst objects [387]. Also, whenever the gas is mixed with dust, only a fraction of the Lyman continuum photons (f) produced in the star-forming regions contributes to the ionization of the atomic hydrogen. The remaining ($1-f$) is absorbed by dust [402]. Recent observations indicate that f ranges between 0.5 and 1.0 [387].

The constant $K(H_\alpha)$ in Eq. (13.7) needed to transform $L(H_\alpha)$ into star formation rates has different values depending on the adopted IMF, the upper mass cutoff M_{up} and the metallicity. Variations of $K(H_\alpha)$ up to a factor of $\sim$ 1.5–1.6 are obtained by changing the shape of the IMF in a fixed mass range. The commonly used value is [390]

$$K(H_\alpha) = 7.9 \times 10^{-42} \left[\frac{M_\odot \, \text{yr}^{-1}}{\text{erg s}^{-1}} \right] \tag{13.15}$$

determined for a Salpeter IMF in the mass range 0.1–100 $M_\odot$ with solar metallicity. It should be noted that even in the same IMF configuration (Salpeter in the mass range 0.1–100 $M_\odot$) $K(H_\alpha)$ can slightly change when determined using other population synthesis models. Alternative values of $K(H_\alpha)$ for different IMFs can be found in [392], while calibrations for metallicity-dependent relations are found in [403, 404], and IMFs and IGIMFs that are dependent on star formation are found in [388]. Equation (13.7) can be equivalently expressed as [390]

$$\text{SFR}\,[M_\odot \, \text{yr}^{-1}] = 1.08 \times 10^{-53} Q(H^0)\,[\text{s}^{-1}] \tag{13.16}$$

where $Q(H^0)$ is the ionization rate.

Hydrogen recombination lines other than H_α and H_β are now used for determining star formation rates in galaxies, such as the Br_γ line at 2.17 μm, accessible from ground-based facilities, for which [390]

$$K(Br_\gamma) = 8.2 \times 10^{-40} \left[\frac{M_\odot \, \text{yr}^{-1}}{\text{erg s}^{-1}} \right] \tag{13.17}$$

$K(\lambda)$ values for other hydrogen recombination lines can be easily derived using the ratios given in Table 3.1. It is worth mentioning that the Ly_α line (λ1215.67 Å), widely observed in cosmological surveys since it is redshifted into the optical domain for $z > 1.8$, cannot be used for measuring star formation rates for two main reasons [117]:

1. This line is highly extinguished, and
2. Ly_α photons produced in galaxies suffer a large number of resonant scattering. That is, the Ly_α photon emitted inside an HII region is reabsorbed by the first hydrogen atom it encounters and immediately re-emitted as another Ly_α photon in an arbitrary direction. Thus the Lyman α photon propagates in a random walk (resonant scattering) until it escapes the HII region or it is absorbed by dust [75] (see Section 4.2.4).

Although the H_α line is considered to be one of the most reliable star formation tracers in normal galaxies, the uncertainty on the SFR determined with H_α data is relatively important, up to a factor of $\sim$ 50% when high quality spectroscopic data are available for dust extinction and [NII] contamination corrections, and higher otherwise [387, 405]. In heavily obscured objects such as strong starbursts, other dust-free tracers should be preferred.

13.6
UV Stellar Continuum

The UV stellar continuum of galaxies ($\lambda \leq 2000$ Å), due to the emission of the young stellar populations (mostly B and A stars depending on the wavelength range), is often used to determine the present day star formation activity of late-type galaxies. Given the relatively long lifetime of the emitting sources, UV luminosities can be transformed into star formation rates, provided that the star formation activity of the observed galaxies is constant on time scales of a few 10^8 yr (see Figure 13.1). UV data should be corrected for dust extinction following the prescription given in Chapter 12. This dust-extinction-corrected star formation tracer is well tuned for normal, optically or UV selected galaxies[2], for low metallicity systems, and any dust-poor object in the nearby and far universe. However, it is not usable in highly obscured objects such as infrared-selected galaxies or strong starbursts [374]. Kennicutt [390] proposes a calibration constant for transforming UV luminosities into star formation rates that is valid for the spectral range 1500–2800 Å (calibrated for a Salpeter IMF in the mass range 0.1–100$M_\odot$ and solar metallicity) :

$$K(\text{UV}) = 1.4 \times 10^{-28} \left[\frac{M_\odot \text{ yr}^{-1}}{\text{erg s}^{-1} \text{ Hz}^{-1}} \right] \quad (13.18)$$

The validity of $K(\text{UV})$ over such a large spectral range is due to the fact that the UV spectrum of star-forming galaxies is nearly flat, whenever expressed in $S(\nu)$. Alternative values of $K(\text{UV})$ for different IMFs can be found in [403] and [404] (metallicity-dependent values) and in [389] (SFR-dependent values). For UV fluxes measured in the FUV GALEX band at 1530 Å, the star formation rate is required to have been stationary for the past 10^8 yr. The stationarity condition for the NUV GALEX band (several 10^8 years) is quite constraining since it is rarely satisfied in star-forming, nearby galaxies.

---

2) The UV emission of early-type galaxies (E-S0-S0a) is dominated by hot stars in later stages of stellar evolution [78, 181]; thus their emission cannot be used to determine present day star formation rates.

13.7
Infrared

In dust-rich star-forming galaxies, the UV radiation emitted by the young stellar population is absorbed and re-emitted in the far-infrared spectral domain. The total infrared luminosity and the monochromatic flux density are thus tightly related with the activity of star formation and can be transformed, under some conditions, into star formation rates.

13.7.1
Integrated Infrared Luminosity

In highly obscured galaxies such as starburst galaxies, luminous infrared galaxies (LIRGs), and ultra-luminous infrared galaxies (ULIRGs), a large fraction of the ionizing and nonionizing UV radiation emitted by the youngest stars is absorbed and re-emitted by dust in the far-infrared. In these objects, extinction-corrected UV or hydrogen recombination line fluxes are highly uncertain, given the huge attenuation that can easily reach several magnitudes. The far-infrared luminosity has to be preferred for quantifying star formation rates. Kennicutt [390] gives a calibration constant $K(\text{IR})$ for transforming 8–1000 µm infrared luminosities L_{IR} into SFR for a Salpeter IMF in the mass range 0.1–$100 M_\odot$ and solar metallicity:

$$K(\text{IR}) = 4.5 \times 10^{-44} \left[\frac{M_\odot \, \text{yr}^{-1}}{\text{erg s}^{-1}} \right] \quad (13.19)$$

An alternative calibration is given by [406]. As indicated by [390], this calibration is valid only for highly obscured galaxies where most of their infrared emission is in the 10–120 µm spectral range, indicating a strong continuous star formation activity in the last 10^8 yr. As for the other direct (UV, H_α) or indirect (radio continuum) tracers of the young stellar population, the stationarity condition on the last $\sim 10^8$ yr is required. In starburst galaxies, generally identified as those objects with $S(60 \, \mu\text{m})/S(100 \, \mu\text{m}) > 0.6$ [229], this condition should be verified. Since the extinction in the UV bands is on the order of ~ 1 mag even in normal galaxies, the far-infrared radiation contributes to more than 50% of the bolometric emission of the young stellar population (see Figure 9.1). For this reason, this calibration could be extended to selected infrared galaxies and all active, dust-rich, star-forming systems with the exception of low luminosity, low metallicity systems, thus practically to almost all objects detected in deep, high redshift cosmological surveys. We should stress, however, the fact that this calibration cannot be applied to normal quiescent late-type galaxies such as those generally observed in the nearby universe. Their infrared emission not only comes from highly obscured star-forming regions, but from diffuse cirrus regions that are heated by the general interstellar radiation field – thus also from evolved stellar populations [407]. A third potential limit of using this calibration resides in the need to determine the total infrared luminosity L_{IR}, something that is particularly difficult in cosmological surveys when galaxies are detected in a few infrared bands. In these conditions, total infrared luminosities are extrapolated using monochromatic flux densities combined with highly uncertain infrared *K*-corrections.

13.7.2
Monochromatic Infrared Luminosities

To overcome all these technical problems, several authors have proposed to use monochromatic infrared luminosities at relatively short wavelengths (from $\sim 15\text{–}25\ \mu\text{m}$). At these rest-frame wavelengths, dust is principally heated by the UV radiation emitted by the youngest stellar population, with a negligible contribution from the evolved populations, making the monochromatic infrared luminosity a direct tracer of the star formation activity. At the same time, monochromatic luminosities do not need to be extrapolated for the determination of total luminosities, and *K*-corrections might not be necessary for galaxies within a given redshift range. Empirical relations calibrated on nearby samples of star-forming galaxies with both extinction-corrected hydrogen recombination lines and monochromatic infrared luminosities have been proposed in the literature. At 24 μm, the most widely used relations are proposed by [408], calibrated on nearby luminous infrared galaxies:

$$\text{SFR}\,[M_\odot\,\text{yr}^{-1}] = 8.45 \times 10^{-38}\,L(24\,\mu\text{m}\,[\text{erg s}^{-1}])^{0.871} \tag{13.20}$$

Also widely used are those determined using the SINGS nearby galaxy sample [409]:

$$\text{SFR}\,[M_\odot\,\text{yr}^{-1}] = 1.27 \times 10^{-38}\,L(24\,\mu\text{m}\,[\text{erg s}^{-1}])^{0.885} \tag{13.21}$$

$L(\nu)$ is the monochromatic infrared luminosity at 24 μm (see Section 7.1). Although proposed as alternative tracers, monochromatic infrared luminosities at wavelengths shorter than $\sim 15\ \mu\text{m}$ should not be used to quantify star formation rates in late-type galaxies since they are dominated by the emission of PAHs (molecules that can be destroyed in low metallicity environments and hard UV radiation fields [107, 409]). For the same reason, the 24 μm calibrations given in Eqs. (13.20) and (13.21) cannot be used for galaxies at a redshift $z \geq 0.9$; at this redshift, the PAH feature at 12.7 μm falls inside the 24 μm band.

13.8
Radio Continuum

The far-infrared–radio correlation, which relates the total infrared luminosity to the nonthermal 20 cm radio continuum luminosity, is one of the tightest correlations observed in galaxies. It has proved valid for a large variety of objects spanning a large range in luminosity (~ 4 orders of magnitude in L_{IR}), from dwarfs to ULIRGs, including normal galaxies and AGNs [166]. The tightness of this relation is due to the fact that the relativistic electrons spinning in weak magnetic fields responsible for the synchrotron emission, are produced and accelerated in supernova remnants, which are related to the stellar population (young and massive stars) dominating the dust heating in galaxies. For this reason, the radio continuum emission at 1400 MHz (20 cm) has often been proposed as an alternative tracer of star formation in star-forming galaxies. The advantage of this indicator is that it is not sensitive to the presence of dust and can also be used in highly obscured objects.

The first attempt to calibrate the 20 cm radio luminosity of galaxies in SFR in $M_\odot$ yr^{-1} was done by [166]. More recently, [406] proposed a new calibration determined using the far-infrared–radio correlation and assuming a Salpeter IMF in the mass range 0.1–100 $M_\odot$:

$$\text{SFR}\,[M_\odot\,\text{yr}^{-1}] = 5.52 \times 10^{-22} L(1.4\,\text{GHz})\,[\text{W Hz}^{-1}]$$
$$\text{for}\quad L(1.4\,\text{GHz}) > 6.4 \times 10^{21}\,\text{W Hz}^{-1}$$

$$\text{SFR}\,[M_\odot\,\text{yr}^{-1}] = \frac{5.52 \times 10^{-22} L(1.4\,\text{GHz})}{0.1 + 0.9(L(1.4\,\text{GHz})/6.4 \times 10^{21})^{0.3}}$$
$$\text{for}\quad 3 \times 10^{19} \leq L(1.4\,\text{GHz}) \leq 6.4 \times 10^{21}\,\text{W Hz}^{-1} \qquad (13.22)$$

This relation can be easily applied to galaxies at different redshifts since the observed radio luminosities can be transformed into into rest-frame 20 cm luminosities $L(1.4\,\text{GHz})$ by assuming a typical spectral slope $\alpha_{\text{syn}} = 0.8$ (see Chapter 8). This can be done whenever the radio emission is dominated by the nonthermal component, that is, for $\lambda > 10$ cm. For high frequencies, the radio emission is principally thermal and star formation rates can be estimated through the relation (8.2) given in Chapter 8 [168].

This calibration has been used by combining far-infrared and UV data to determine the SFR of galaxies, particularly in low luminosity objects where the far-infrared luminosity alone underestimates SFR. As stated by [406], however, the calibration has a relatively large scatter which can exceed a factor of 2, thus the uncertainty in SFR determined using radio data is quite high. The use of this calibration in radio selected samples should be considered with caution because of the possibly strong contamination of AGNs: although these objects follow the far-infrared–radio correlation, the calibration of their infrared luminosity in terms of star formation rates is highly uncertain because of their peculiar nature. Another source of uncertainty is related to the fact that the nonthermal radio emission not only depends on the density of the relativistic electrons, n_e, but also on the intensity of the magnetic field B (see Eq. (5.5)). It has been shown that galaxy interactions might compress the magnetic field and thus increase the radio continuum emission without affecting the electron density and thus the star formation activity [10]. These effects should be considered when star formation rates are determined using radio continuum data.

13.9
Other Indicators

13.9.1
The X-ray Luminosity

In the absence of a bright active galactic nucleus (AGN), the X-ray luminosity of normal galaxies is due to the emission of its X-ray binary population, with a contribution from the diffuse emission in supernovae remnants and of the hot gas component [54]. X-ray binaries are composed of a compact source, a neutron star or a black hole, which accretes material from a nearby companion. They are generally divided into high-mass (HMXB) and low-mass (LMXB) X-ray binaries, according to the mass of their companion (see Chapter 2). In HMXB, the mass is generally $\geq 10\,M_\odot$ and characterized by a typical lifetime τ on the order of $\sim 10^7$ yr. In LMXB, the mass is $\leq 1\,M_\odot$ with a typical lifetime of $\tau \sim 10^{10}$ yr [410]. The rapid evolution of HMXB makes them a good tracer of star formation [411], while LMXB are connected to the total stellar content of galaxies [412]. The interest in the X-ray emission as a potential tracer of star formation resides in the fact that galaxies are almost transparent to X-rays above ~ 2 keV, except for the densest parts of the most massive molecular clouds [412]. The X-ray luminosity can be used as an alternative estimate of star formation, instead of using tracers that are heavily affected by extinction in dust-rich objects. This, however, is difficult because the X-ray emission of galaxies is dominated by nuclear emission in the presence of a weak AGN, whose contribution should be removed for a correct estimate of the star formation activity. The presence of an AGN can hardly be determined spectroscopically since these objects have X-ray spectra similar to those of X-ray binaries. At the same time, in the absence of an AGN, the contribution of LMXB to the X-ray emission of galaxies might be important. This contribution can be roughly estimated assuming that the emission from LMXB is related to the total stellar or dynamical mass of the galaxy [412].

Empirical relations for the X-ray luminosity as a tracer of star formation have been determined by comparing the X-ray emission of resolved Galactic and extragalactic HMXB combined with the integrated X-ray emission of nearby galaxies to the star formation rate estimated using independent tracers (far-infrared, H_α, UV, radio continuum), the later generally measured using the standard calibrations of [390]. Using ASCA and BeppoSAX observations of nearby galaxies, [413] linear relations between the soft (0.5–2 keV) and hard (2–10 keV) X-ray emission and the star formation rate have been found. Specifically,

$$\text{SFR}\,[M_\odot\,\text{yr}^{-1}] = 2.2 \times 10^{-40}\,L_{0.5-2\,\text{keV}}(X)\,[\text{erg s}^{-1}] \qquad (13.23)$$

and

$$\text{SFR}\,[M_\odot\,\text{yr}^{-1}] = 2.0 \times 10^{-40}\,L_{2-10\,\text{keV}}(X)\,[\text{erg s}^{-1}] \qquad (13.24)$$

with a slightly higher uncertainty in the soft band due to the possible presence of intrinsic absorption. In a similar work based on Chandra and ASCA X-ray data of both resolved HMXB and of unresolved nearby star-forming objects, [412] have shown that both the number of HMXB and the total X-ray emission of late-type galaxies without an AGN is tightly related to their star formation activity. This relation, however, is linear only for SFR $\geq 4.5\,M_\odot\,\mathrm{yr}^{-1}$ ($L_{2-10\,\mathrm{keV}}(X) \geq 3 \times 10^{40}\,\mathrm{erg\,s}^{-1}$):

$$\mathrm{SFR}\,[M_\odot\,\mathrm{yr}^{-1}] = \frac{L_{2-10\,\mathrm{keV}}(X)\,[\mathrm{erg\,s}^{-1}]}{6.7 \times 10^{39}} \qquad (13.25)$$

It is flatter for lower X-ray luminosities:

$$\mathrm{SFR}\,[M_\odot\,\mathrm{yr}^{-1}] = \left(\frac{L_{2-10\,\mathrm{keV}}(X)\,[\mathrm{erg\,s}^{-1}]}{2.6 \times 10^{39}}\right)^{0.6} \qquad (13.26)$$

13.9.2
Forbidden Lines

The hydrogen recombination lines H_α and H_β are redshifted outside the optical range for $z \geq 0.5$ and $z \geq 0.9$, respectively, and are thus not appropriate for quantifying star formation rates using deep optical spectroscopic cosmological surveys. To overcome this problem, several authors tried to empirically calibrate the [OII]λ3727 Å and [OIII] λ5007 Å oxygen forbidden lines to be used as possible indicators of the star formation activity of galaxies [414–417]. The idea behind this is that the oxygen forbidden lines come from HII regions where stars are forming. The main difficulty, however, is that their luminosity is not directly coupled to the number of ionizing photons, or the number of young and massive stars: their excitation is sensitive to metal abundance and to the ionization state of the gas. The use of the [OIII]λ5007 Å line as a star formation tracer was soon dropped since this line is too sensitive to metallicity and ionization [417]. By averaging the empirical calibration of [415] determined using integrated spectroscopy for 90 normal and peculiar galaxies with that of 75 low metallicity irregular galaxies of [414], Kennicutt [390] proposed an empirical calibration valid for a Salpeter IMF in the mass range 0.1–100 $M_\odot$:

$$\mathrm{SFR}\,[M_\odot\,\mathrm{yr}^{-1}] = (1.4 \pm 0.4) \times 10^{-41}\,L([\mathrm{OII}])\,[\mathrm{erg\,s}^{-1}] \qquad (13.27)$$

where the large uncertainty in the constant reflects the difference in the [414, 415] calibrations. The [OII] line is highly extinguished and must be corrected for dust attenuation using the Balmer decrement or, whenever the Balmer lines are not available, some luminosity-dependent empirical relations such as those proposed by [417]. It has been shown that this calibration is not very sensitive to the ionization parameter or to the metal abundance for metallicities $12 + \log(O/H)$ in the range 8.15–8.7, while it changes significantly for $12 + \log(O/H) < 8.15$ or $12 + \log(O/H) > 8.7$ [417] (metallicity-dependent calibrations can be found in [416]). It should be stressed that, if the luminosity-dependent statistical corrections proposed by [417] necessary for determining SFR in high redshift galaxies introduce a statistical uncertainty of a factor of ∼ 2.5 on the SFR determined using the [OII] line which should be added to the already large uncertainty on the H_α–SFR calibration (see previous section), then the uncertainty on the SFR for a single galaxy might be much larger, up to a factor of ∼ 10.

13.9.3
[CII]

The fine-structure line of [CII] at 157.74 μm is emitted primarily by gas exposed to UV radiation in the photodissociation regions and is thus directly related to the star formation activity in galaxies, even in those hosting an AGN. This line is generally the brightest emission line in the far-infrared spectrum of galaxies, with $-3 \leq \log(L([CII])/L_{FIR}) \leq -2$, and is thus observable even in high redshift objects. Given its strong association with the UV radiation emitted by the young stellar population, it has been proposed to use the [CII] line as a direct tracer of the star formation activity of galaxies [418]. An empirical calibration has been determined by [400] assuming a Salpeter IMF in the mass range $0.1-100 M_\odot$:

$$\text{SFR}[M_\odot \text{ yr}^{-1}] = 5.952 \times 10^{-33} 10^{0.788 \log L([CII])} \text{ [erg s}^{-1}] \quad (13.28)$$

This relation is valid for normal galaxies with $10^{8.0} \leq L_{FIR} \leq 10^{10.5} L_\odot$ and $0.2 \leq $ [CII]E.W. ≤ 3.0 μm. The strong decrease of $L([CII])/L_{FIR}$ observed in infrared luminous galaxies ($L_{FIR} > 10^{11} L_\odot$; [419]) probably due to the different physical conditions of their ISM, suggests that this relation is not universal and cannot be blindly used for high redshift objects.

13.9.4
Radio Recombination Lines

Radio recombination lines (RRLs) are due to transitions between highly excited atomic levels ($n \geq 30$) that follow the recombination of electrons with atoms. Among these, the most frequently observed are the hydrogen recombination lines. They are equivalent to the Lyman, Balmer, or Brackett lines in the UV, optical, and near-infrared domain but, because of the highly excited levels of the atoms, they fall in the radio domain.

The interest of these lines as potential tracers of star formation in extragalactic objects has been firstly discussed by [420], showing how the intensity of these lines, in the case of spontaneous emission (normal galaxies with a negligible nonthermal radio continuum emission) or stimulated emission (bright radio sources), depends on the electron density and temperature. Combined with local thermal equilibrium models, they can be used to estimate the number of ionizing photons that, for a given IMF, can be transformed into star formation rates, in solar masses per year [421]. The advantage of the radio recombination lines is that they are not sensitive to dust and are thus perfectly tuned for measuring star formation rates in the core of heavily extinguished starbursts or ultra-luminous infrared galaxies [421]. They are, however, relatively weak therefore not easily accessible for all kind of extragalactic sources.

13.10
Population Synthesis Models

An accurate way of measuring star formation rates in galaxies is that of fitting their extinction-corrected UV to near-infrared spectral energy distribution with population synthesis models, as described in Section 9.1.2. Once the analytical form of the time-dependent star formation rate that best reproduces the observed data is constrained, the present day star formation rate can be determined by simply measuring the SFR at $t = t_0$, where t_0 is the age of the observed galaxy. This technique is subject to the uncertainty introduced by extinction corrections, as are the classical indicators of star formation rates such as the Balmer lines or the UV continuum emission. However, it has the major advantage that it does not need the stationarity condition of star formation to be applied and can be used for any kind of object with a star formation rate which varies on short time scales, such as starbursts, interacting and merging systems. Its major limit, however, resides in the fact that it strongly depends on the assumed star formation history.

13.10.1
Dating a Star Formation Event

Several absorption features visible in the optical spectra of galaxies have been proposed to date the last episode of star formation. The $D_n(4000)$ index and the Balmer absorption lines are those mostly used in the literature.

The stellar spectra of galaxies have their strongest discontinuity at ~ 4000 Å. This feature arises from the accumulation of a large number of spectral lines in a narrow wavelength range (see Section 3.3). As explained by [96], the intensity of the break depends on the opacity due to metals in the atmosphere of hot stars: in the presence of young stellar populations, metals are multiply ionized and the opacity decreases, making the break less pronounced than in the presence of older stellar populations. The $D_n(4000)$ spectral index is defined as the ratio of the average flux density $S(\nu)$ in the band 4000–4100 and 3850–3950 Å (although other definitions are present in the literature [96]). It has been shown that this index changes with time after a single burst of star formation almost independently on metallicity up to ages of $\sim 10^8$ yr [96, 422], as shown in Figure 13.4.

Balmer absorption lines such as H_β, H_γ, and H_δ are related to the presence of warm and young late B to early F type stars. Easily measurable in the spectra of early-type galaxies since they are only marginally contaminated by the nebular gas emission, they indicate the presence of a recent activity. Although they cannot be used to quantify the star formation activity of galaxies, they have been proposed to measure the age of the last star formation event [96], as shown in Figure 13.4, particularly in cluster objects where the activity might have brusquely stopped because

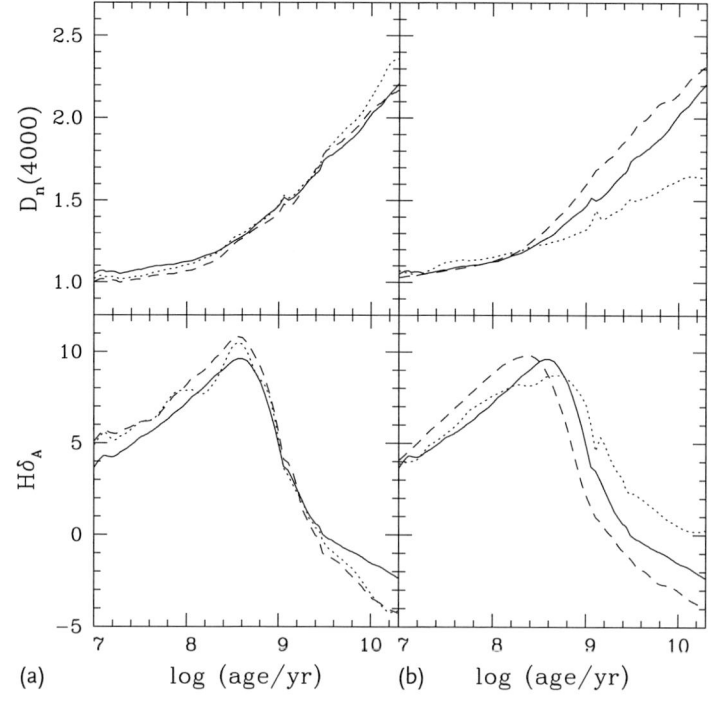

Figure 13.4 The evolution of the $D_n(4000)$ and of the H_δ absorption features as a function of time after an instantaneous burst of star formation. The different lines in (a) show the variation of the two spectral indices as determined using different stellar libraries, while those (b) show those determined using the stellar library STELIB of [209] for solar (solid line), 1/5 solar (dotted line), and 2.5 solar (dashed line) metallicities, from [96]. Reproduced with the permission of John Wiley & Sons Ltd.

of the interaction with the hostile environment [236, 423] (see Section 10.1.2). Detailed studies of these Balmer features using different population synthesis models, however, revealed that these indices suffer from the metallicity–age degeneracy effect and thus can be hardly used without other metallicity indices to date the last episode of star formation [422]. The metallicity dependence is more important for H_γ and H_δ than for H_β [293]. The direct use of these indices for dating the last episode of star formation or for quantifying the ongoing activity of galaxies is also limited by the fact that their intensity depends on the star formation history of the parent galaxy which, except in a few peculiar cases, can be hardly reproduced by a single, instantaneous burst of star formation. These indices are thus very useful when combined with other metallicity-sensitive indices and with population synthesis models tuned to reproduce the different evolutionary histories of galaxies.

14
Light Profiles and Structural Parameters

The reconstruction of the radial brightness distribution of galaxies, often called the radial light profile, generally achieved in optical and near-infrared bands, is of paramount importance for studying their structural properties. It has been claimed that the shape of the inner profile of early-type galaxies is tightly related to the formation process that gave birth to these objects. The presence of a dense inner power law disk in rotating ellipticals is consistent with their formation in gas-rich mergers, while the presence of nuclear cores might be due to the orbital decay of massive black holes that are accreted in mergers [424], or related to a formation from a dry merging event.

Light profiles are also used for measuring total magnitudes and several other structural parameters; i.e., effective radii, surface brightnesses, or concentration indices necessary for determining many major scaling relations, such as the Kormendy relation or the fundamental plane. Two-dimensional images can also be used to estimate the degree of asymmetry of the different underlying stellar populations. These entities are, for instance, of primary importance for quantifying the perturbation induced by the harsh environment on the stellar component of galaxies inhabiting high density regions such as clusters or compact groups.

14.1
The Surface Brightness Profile

14.1.1
Extended Radial Profiles

The surface brightness profile of early- and late-type galaxies can be determined by integrating their images in elliptical annuli of increasing radii aligned along the galaxy's major axis. The ellipticity and position angle can be taken as fixed values or left as free parameters to follow a possible twist of the isophotes. The brightness profiles can also be obtained by using recent two-dimensional fitting tasks such as GIM2D (Galaxy IMage 2D), an IRAF/SPP package [425], GALFIT [426], and BUDDA [427].

A Panchromatic View of Galaxies, First Edition. Alessandro Boselli.
© 2012 WILEY-VCH Verlag GmbH & Co. KGaA. Published 2012 by WILEY-VCH Verlag GmbH & Co. KGaA.

Historically, the surface brightness profiles of early-type galaxies have been fitted with a de Vaucouleurs law [428]:

$$I_{dV}(r) = I(0) \exp\left[-\left(\frac{r}{a}\right)^{1/4}\right] \quad (14.1)$$

Or, alternatively,

$$\mu_{dV}(r) = \mu(0) + 1.086 \frac{r^{1/4}}{a} \quad (14.2)$$

where $I(r)$ and $\mu(r)$ are the radial variation of the light intensity and of the surface brightness, a is a characteristic radius, and $I(0)$ and $\mu(0)$ are the central intensity and surface brightness, respectively. For disk galaxies, the light profile is generally represented by an exponential law,

$$I_{\exp}(r) = I(0) \exp\left[-\frac{r}{h_r}\right] \quad (14.3)$$

where h_r is the exponential disk scale length. These two expressions are a generalization of the Sersic's profile [429]:

$$I_S(r) = I(0) \exp\left[-\left(\frac{r}{a}\right)^{1/n}\right] = I(r_e) \exp\left\{-b_n\left[\left(\frac{r}{r_e}\right)^{1/n} - 1\right]\right\} \quad (14.4)$$

where $n = 4$ for the de Vaucouleurs law and $n = 1$ for the exponential law, as shown in Figure 14.1. The constant b_n is chosen so that a circle of radius r_e (effective radius) includes 50% of the total light of the galaxy. For $n > 1$, $b \simeq 1.9992n - 0.3271$. A detailed discussion on the properties of the Sersic models can be found in [430]. Light profiles of bulge dominated late-type galaxies can be fitted by combining an exponential profile, to trace the light distribution of the disk, and a de Vaucouleurs or a Sersic profile for the bulge, as shown in Figure 14.2. cD galaxies dominating rich clusters also have exponential plus Sersic profiles, where the exponential component probably results from the tidally distorted envelope formed after the accretion of other objects.

14.1.2
The Central Surface Brightness Profile of Early-Type Galaxies

The Sersic model generally provides an accurate description of the global radial light profile of galaxies but it fails to reproduce the inner regions (typically $r \leq 100\,\mathrm{pc}$ for the brightest ellipticals and $r \leq 10\,\mathrm{pc}$ for the dwarfs) of both early-type galaxies brighter than $M_B \leq -20$ (which show central light deficits) and fainter than $M_B \geq -19.5$ (which usually show a central excess) [433]. For this reason, different relations have been proposed in the literature for representing the central brightness profile of early-type galaxies. Among these, the most widely used is the core-Sersic model [434, 435]:

$$I_{cS}(r) = I'\left[1 + \left(\frac{r_b}{r}\right)^{\alpha}\right]^{\gamma/\alpha} \exp\left[-b_n\left(\frac{r^{\alpha} + r_b^{\alpha}}{r_e^{\alpha}}\right)^{1/(\alpha n)}\right] \quad (14.5)$$

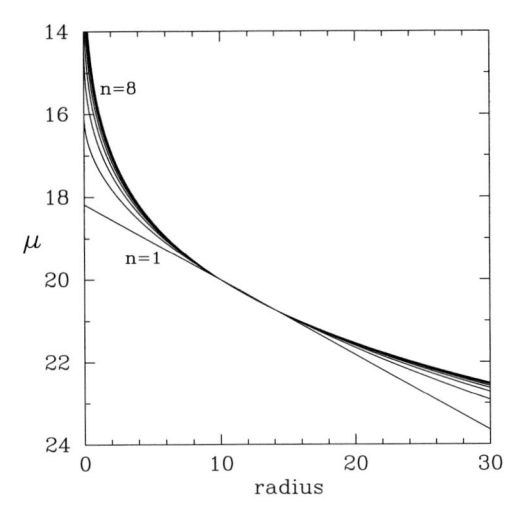

Figure 14.1 Different $r^{1/n}$ Sersic profiles normalized at $r_e = 10''$ and $\mu_e = 20\,\text{mag arcsec}^{-2}$, from [431]. Reproduced by permission of the AAS.

where I' is defined as

$$I' = I_b 2^{-\gamma/\alpha} \exp\left[b_n \left(2^{1/\alpha} \frac{r_b}{r_e}\right)^{1/n}\right] \tag{14.6}$$

This model is characterized by a total of six free parameters (three more than the simple Sersic model), where I_b is the break intensity measured at the break radius r_b. An alternative profile is the double-Sersic model [433]:

$$\begin{aligned} I_{dS}(r) &= I_{Sc}(r) + I_{Sg}(r) \\ &= I(r_{e,c}) \exp\left\{-b_{n,c}\left[\left(\frac{r}{r_{e,c}}\right)^{1/n_c} - 1\right]\right\} \\ &\quad + I(r_{e,g}) \exp\left\{-b_{n,g}\left[\left(\frac{r}{r_{e,g}}\right)^{1/n_g} - 1\right]\right\} \end{aligned} \tag{14.7}$$

where one component corresponds to the galaxy profile (g) and the other to the central light excess (c). Once again, this model is characterized by six free parameters. The simplified core-Sersic model proposed by [436] corresponds to a restriction on the core-Sersic model for $\alpha \to \infty$ and thus only has five free parameters. These models have been defined to quantify the departure in the inner arcseconds from a traditional Sersic model.

Alternative models, whose relative properties are extensively discussed in [437], only adapted to fit the inner regions, are the King model, originally used to fit the profile of globular clusters, or the Nuker profile proposed by [438],

$$I_N(r) = I_b 2^{(\beta-\gamma)/\alpha} \left(\frac{r}{r_b}\right)^\gamma \left[1 + \left(\frac{r}{r_b}\right)^\alpha\right]^{(\gamma-\beta)/\alpha} \tag{14.8}$$

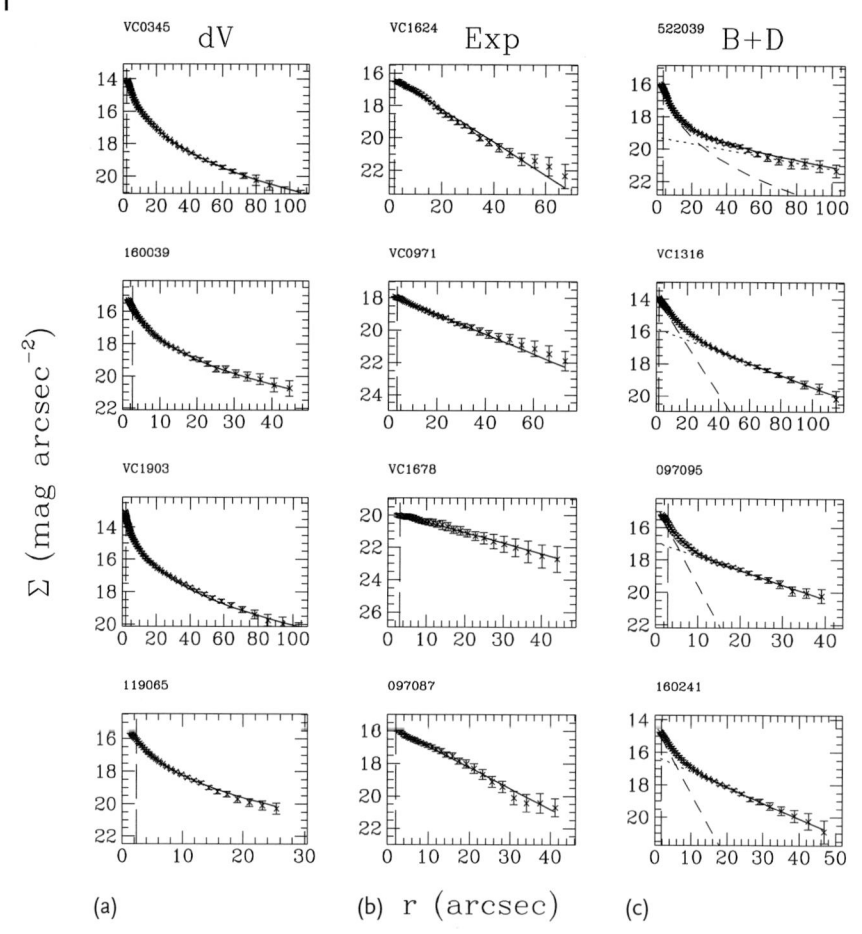

Figure 14.2 Examples of profile decompositions of 12 well known galaxies. From left to right, (a) four elliptical galaxies with a pure de Vaucouleurs decomposition: VCC 345 = NGC 4261, 160 039 = NGC 4839, VCC 1903 = NGC 4621, and 119 065 = NGC 2563; (b) four late-type galaxies with a pure exponential profile: VCC 1624 = NGC 4544, VCC 971 = NGC 4423, VCC 1678 = IC 3576, and 97 087 = UGC 6697; (c) four galaxies with a mixed bulge plus disk profile: 522 039 = NGC 708, VCC 1316 = M87, 97 095 = NGC 3842, and 160 241 = NGC 4889. These are the brightest elliptical galaxies of the clusters A262, Virgo, A1367, and Coma, respectively. The dotted line shows the best fit exponential profile, the short dashed line shows the de Vaucouleurs profile, and the vertical dashed line indicates the radius of the seeing, from [432]. Reproduced with permission (c) ESO.

which has five free parameters and is a blend of power laws with slopes γ, α, and β in the inner, at the break radius r_b, and in the outer regions, respectively. All these models must be fitted to the observed inner profiles after convolution with the instrumental PSF. Figure 14.3 shows some representative g and z band surface brightness profiles of early-type galaxies in the Virgo cluster observed with the ACS camera on HST (other light profiles of early-type galaxies observed with

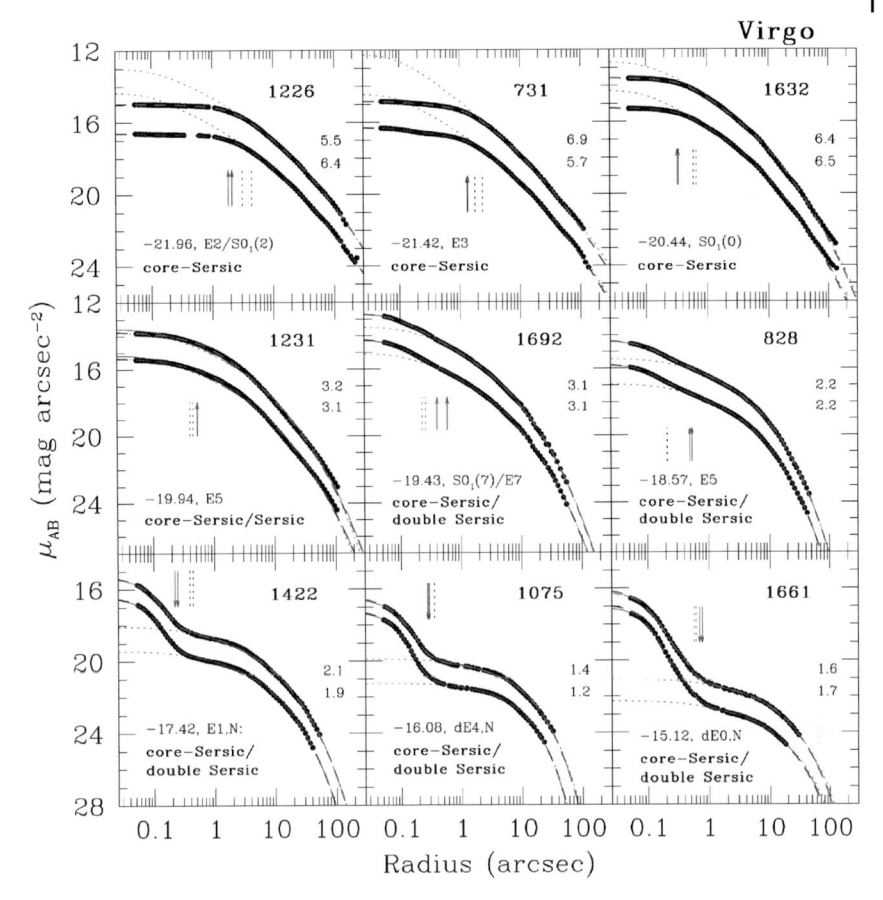

Figure 14.3 Surface brightness profiles of nine representative early-type galaxies in the Virgo cluster (with the VCC name in the upper right corner of each panel) observed with the ACS camera on the HST, adapted from [433]. The profiles are in the g (lower) and z (upper profile) bands. The galaxies are ordered according to decreasing absolute blue magnitude (given in each panel with the morphological type, the best fit Sersic index n in the g (lower) and z (upper) bands). The core-Sersic and the composite profiles (which extend to the outer regions) are shown as short-dashed and long-dashed curves, respectively. The vertical arrows indicate the observed break radius in the two photometric bands. The dotted curves show the inward extrapolation of a simple Sersic component. Reproduced by permission of the AAS.

HST can be found in [437, 439, 440]). Clearly, the simple Sersic model follows the observed data in the inner regions of early-type galaxies with moderate luminosity, while there is a smooth transition from a profile with deficit to excess with respect to a standard Sersic model with decreasing galaxy luminosity. It is worth noting that the reason why both the King and Sersic models can reproduce the observed profiles is that the angular resolution of the available images does not allow the inner regions to be resolved.

14.1.3
The Vertical Light Profile of Late-Type Galaxies

In disk galaxies, the light distribution in the direction perpendicular to the plane of the disk (z) is generally represented by the relation [441]

$$I(r, z) = I(0)e^{-r/h_r} \text{sech}^2(z/h_z) \tag{14.9}$$

where h_z is the characteristic z-scale height. Alternative relations, as well as a detailed description of their properties, can be found in [442]. Color images of edge-on galaxies can be used to reconstruct the vertical light profile of late-type galaxies in different photometric bands. When the observed galaxies are close enough to resolve their stars with the ACS on the HST, color–magnitude diagrams can be used to trace the vertical distribution of the different stellar populations, as done by [443] on six low-mass edge-on spiral galaxies. The characteristic scale height h_z is age-dependent, as shown in Figure 14.4: young main sequence stars are distributed in disks that are thinner than the disks of intermediate-age AGB stars or old RGB stars. This is expected since star formation takes place on the plane of the disk; with time, stars perturbed by the gravitational interactions with other galactic objects migrate in the z-direction, producing thicker disks.

14.2
Structural Parameters

14.2.1
Total Magnitudes, Effective Radii and Surface Brightnesses

The determination of the light profiles is necessary for measuring many parameters used to study the structural properties of galaxies. First of all, the extrapolation to infinity of the fitted light profile is used to estimate total, extrapolated or asymptotic magnitudes. Total magnitudes are generally used to study the integrated properties of statistically complete samples, such as integrated colors or luminosities. Light profiles are necessary for the determination of several structural parameters used in important scaling relations such as the Kormendy relation and the fundamental plane (see Section 17.2). Important structural parameters to remember are the effective or half-light radius r_e (defined as the radius including 50% of the total light of a galaxy), the effective surface brightness μ_e (the surface brightness at r_e), and the mean effective surface brightness $\langle \mu_e \rangle$ (the mean effective surface brightness within r_e). The extrapolation of the fitted light profile to $r = 0$ can be used to estimate the central surface brightness μ_0. Similarly, isophotal diameters and magnitudes can be measured using the fitted light profiles at a given surface brightness.

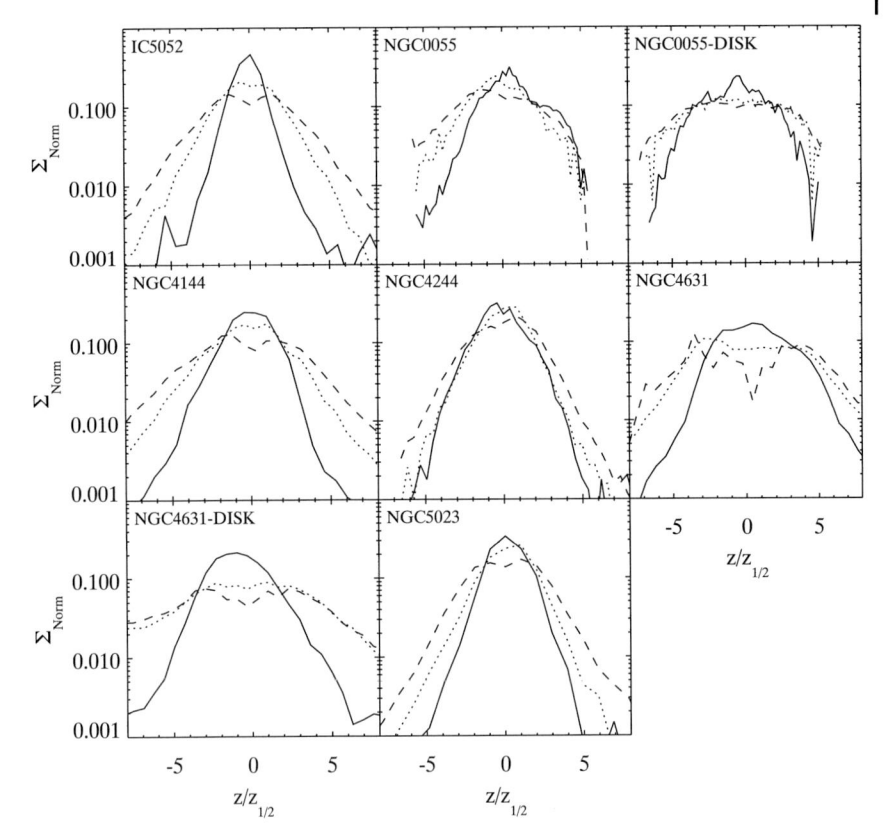

Figure 14.4 The normalized surface density as a function of the scale height for young main sequence stars (solid), intermediate-age AGB (dotted), and old RGB (dashed) stars, from [443]. Each surface density distribution is normalized so that the integrated number is 1. In all the six observed galaxies, the young main sequence distribution is the most peaked, while the old RGB distribution the widest. Reproduced by permission of the AAS.

14.2.2
Bulge to Disk Ratio

Bulge to disk ratios, generally indicated as B/D, or bulge to total ratios (B/T), are used to quantify the relative contribution of the bulge and disk components to the total luminosity of a given galaxy. The contribution to the total luminosity of the bulge and disk components are easily determined by extrapolating the de Vaucouleurs $r^{1/4}$ (Eq. (14.1)) or Sersic $r^{1/n}$ (Eq. (14.4)) radial profiles to infinity, used to fit the bulge component, and the exponential radial profile (Eq. (14.3)) for the disk component.

14.3
Morphological Parameters

Morphology is a qualitative way for classifying galaxies. Many attempts have been presented in the literature for defining alternative quantitative methods. Among these, interesting and widely used examples are the CAS physical morphology classification, proposed by [444], the Gini coefficient and the second-order moment of the brightest 20% of the galaxy's flux [445]. The CAS classification is based on three different parameters, the concentration index C, the asymmetry index A and the clumpiness index S.

14.3.1
Concentration Index

The concentration index is defined as the ratio of the radii that include two different fractions of the total light of a galaxy. Firstly defined by [446], the C_{31} concentration index is given by the simple relation

$$C_{31} = \frac{r_{75}}{r_{25}} \qquad (14.10)$$

where r_{75} and r_{25} are the radii that include 75 and 25% of the total light of a galaxy, respectively, although other definitions have been used in the literature (e.g., [447]). Within the CAS classification system, the C concentration index is defined by [448] as

$$C = 5 \log \left(\frac{r_{80}}{r_{20}} \right) \qquad (14.11)$$

and is graphically represented in Figure 14.5. These concentration indices are tightly related to the shape of the light profile. Exponential radial profiles, for instance, have concentration indices of $C_{31} < 3$ and $C = 2.7$. Bulge-dominated galaxies with a Sersic index of $n > 1$ have $C_{31} \geq 3$ and $C > 2.7$, while $C = 5.2$ for a theoretical de Vaucouleurs $n = 4$ profile [449].

14.3.2
Asymmetry

The asymmetry index A is generally used to quantify the degree of perturbation in galaxies. Although several definitions have been presented in the literature, the most widely used is the asymmetry parameter A of [450, 451], defined as

$$A = \frac{\text{abs}(I - R)}{I} \qquad (14.12)$$

where I is the original galaxy image and R is the image rotated by 180° (see Figure 14.5). In perfectly symmetric galaxies $I = R$, thus $A = 0$; a completely asymmetric galaxy where the absolute value of $I - R$, $|I - R| = I$ has $A = 1$.

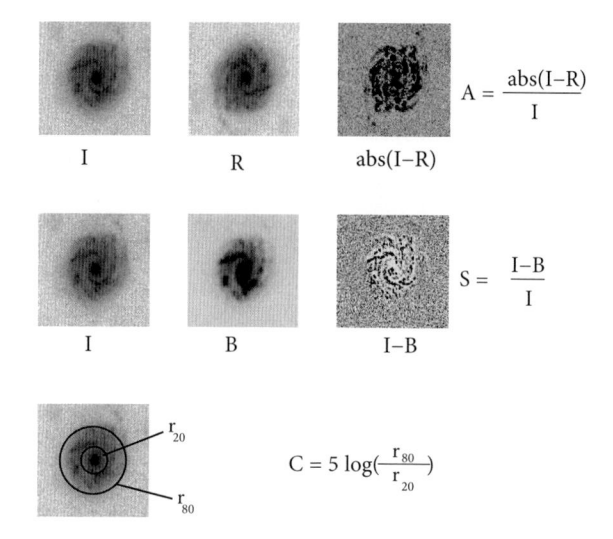

Figure 14.5 Graphical representation of how the three CAS parameters, the concentration index C, the asymmetry A, and the clumpiness S, are measured. I is the original image of the galaxy, R is the image rotated by 180°, and B is the smoothed image, from [444]. Reproduced by permission of the AAS.

14.3.3
Clumpiness

The clumpiness index S has been defined by [444] as

$$S = \frac{I - B}{I} \tag{14.13}$$

where I is the original galaxy image and B is the image smoothed by a given factor (see Figure 14.5). The clumpiness parameter measures the degree of patchiness of a given light distribution, which is generally smooth in ellipticals and clumpy in star-forming systems. In a galaxy with a smooth profile, $I = B$; thus the clumpiness parameter is $S = 0$. In irregular galaxies, where most of the stellar emission comes from compact and bright HII regions and star clusters located in the arm, $I > B$ since in the smoothed image their contribution is completely removed; thus S is large. Table 14.1 gives the mean CAS values for galaxies of different morphological type in the local universe.

14.3.4
The Gini Coefficient G and the Second-Order Moment of the Brightest 20% of the Galaxy's Flux M_{20}

Although the CAS classification system is defined as a nonparametric method because it does not require any model-dependent assumption for the light distribution, as is the case for the determination of the bulge to disk ratio, it does assume

Table 14.1 The average CAS parameters for galaxies of different morphological type, from [444].

Type	C	A	S
Ellipticals	4.4 ± 0.3	0.02 ± 0.02	0.00 ± 0.04
Early-type disks (Sa-Sb)	3.9 ± 0.5	0.07 ± 0.04	0.08 ± 0.08
Late-type disks (Sc-Sd)	3.1 ± 0.4	0.15 ± 0.06	0.29 ± 0.13
Irregulars	2.9 ± 0.3	0.17 ± 0.10	0.40 ± 0.20
Edge-on disks	3.7 ± 0.6	0.17 ± 0.11	0.45 ± 0.20
ULIRGs	3.5 ± 0.7	0.32 ± 0.19	0.50 ± 0.40
Starbursts	2.7 ± 0.2	0.53 ± 0.22	0.74 ± 0.25
Dwarf ellipticals	2.5 ± 0.3	0.02 ± 0.03	0.00 ± 0.06

circular symmetry and requires an accurate knowledge of the central position of the galaxy. At the same time, it might depend on the smoothing length chosen for determining S, a choice that can induce systematic effects dependent on the image's point spread function, the pixel size, the distance to the galaxy, and the angular size of the galaxy [445]. Thus this classification is not ideally defined for irregular galaxies or merging systems. To circumvent these problems, [452] proposed the use of the Gini coefficient G, a statistical model-independent tool generally used to describe the distribution of wealth within a society, to quantify the distribution of flux within the pixels associated with a galaxy. As defined, the Gini coefficient is an alternative way to quantify the concentration of light in a given object: it is indeed tightly related to the concentration index C, but with the advantage that it can detect a peaked light distribution in asymmetric galaxies such as those with perturbed morphologies observed at high redshift [445, 452]. As discussed in [452], however, the two coefficients G and C are not completely equivalent in nearby galaxies where the fraction of peculiar systems is small. The second-order moment of the brightest 20% of the galaxy's flux M_{20}, proposed by [445], is an alternative tracer of the light concentration index since it traces the spacial distribution of any bright structure in a galaxy (nuclei, bars, spiral arms, off-center star clusters) and is thus ideally defined for detecting merging signatures such as double nuclei without assuming any circular symmetry. These coefficients, however, depend on the adopted aperture within which they are measured and, at low signal to noise, on the signal to noise [453].

15
Stellar and Dynamical Masses

Stellar and dynamical masses are key parameters in the study of galaxy evolution. Indeed, current models of galaxy formation predict that dark matter halos within which galaxies are formed are assembled hierarchically, first forming small systems that later merge to give birth to the largest structures observed in the evolved universe (massive galaxies, groups, and clusters). Furthermore, cosmological models predict dark matter halo density profiles [454, 455] that can be directly compared to those determined using kinematical observations of nearby galaxies [456–461], thus providing a unique test bed for comparing the predictions of cosmological simulations with observables. The evolution of the dark matter halo distribution predicted by models can be checked against the distribution of the emitting sources once the mass of the dark matter halos is converted into luminous matter using an appropriate dynamical mass to light ratio. At the same time, the process of star formation transforming the primordial gas into stars along the Hubble time can be understood only once the stellar mass content of galaxies can be correctly determined and compared to that of the gaseous component.

15.1
Stellar Mass Determination Using Population Synthesis Models

Optical and near-IR luminosities of galaxies in various photometric bands ($L(\lambda)$) can be converted into stellar masses by adopting a stellar mass to light ratio $M_{star}/L(\lambda)$. Population synthesis models such as those described in Section 9.1.2 can be used to measure $M_{star}/L(\lambda)$. Indeed, once an IMF is assumed, the ratio between the total stellar mass assembled since its formation and the light emitted at a given epoch depends only on the star formation history of the galaxy. A first attempt to measure how the stellar mass to light ratios (in the B band) changes as a function of the varying star formation history of model galaxies has been done by [196]. Since a varying star formation history implies varying colors, [196] were able to give $M_{star}/L(B)$ values for different colors, and indirectly for different morphological types. This idea has been recently reconsidered by [462, 463] who used different population synthesis models with varying IMF, metallicities, and star formation histories to define standard relations between $M_{star}/L(\lambda)$ and optical

A Panchromatic View of Galaxies, First Edition. Alessandro Boselli.
© 2012 WILEY-VCH Verlag GmbH & Co. KGaA. Published 2012 by WILEY-VCH Verlag GmbH & Co. KGaA.

or near-IR colors as quantitative tracers of the stellar mass of different galaxies. Following these prescriptions, the mass to light ratio $M_{\text{star}}/L(\lambda)$ is given by the relation

$$\log \frac{M_{\text{star}}}{L(\lambda)} \left[\frac{M_\odot}{L_\odot}\right] = a_\lambda + b_\lambda \times \text{Color} \tag{15.1}$$

where a_λ, b_λ, and the different *Color* indices are given in Tables 15.1 (from [462]) and 15.2 (from [463]). These relations have been determined using a grid of population synthesis models of [213] for different star formation histories and metallicities, and adopting a modified Salpeter IMF in the mass range $0.1 M_\odot < m < 125 M_\odot$. An alternative calibration based on more realistic chemo-spectrophotometric models of galaxy evolution including gas infall tuned to reproduce the two-dimensional radial distribution of observed star-forming galaxies [221], computed assuming a [385] IMF, can be found in [387]:

$$\log M_{\text{star}}[M_\odot] = \log L(H)[L(H)_\odot] - 0.84 + 0.69(B - V) \tag{15.2}$$

and

$$\log M_{\text{star}}[M_\odot] = \log L(H)[L(H)_\odot] - 1.08 + 0.21(B - H) \tag{15.3}$$

where magnitudes in the Johnson filters are in the Vega system.

These relations can be easily used to estimate the stellar mass of a given galaxy once its luminosity and colors in a given band are known. The main limitation in this method, however, is that these relations depend on the assumed IMF, metallicity, and the adopted population synthesis model, as illustrated in Figure 15.1.

As stressed by [77], important differences in the mass to light ratio are obtained even within the same population synthesis model whenever different stellar mass losses or metallicities are considered, as shown in Figure 15.2. In particular, in the near-IR bands the modeled stellar mass to light ratio of stellar populations formed after an instantaneous burst with ages in between ~ 0.1 and 2 Gyr strongly depends on the adopted population synthesis model and drastically changes if the thermally pulsing asymptotic giant branch phase (TP-AGB) is considered [77], as depicted in Figure 15.3. These variations are critical whenever near-IR data are used to estimate the stellar mass of giant elliptical galaxies (which have been probably formed after a violent starburst event with a subsequent passive evolution) at high redshifts, when the bulk of their stellar population have ages in this critical range.

Although limited by the choice of the adopted population synthesis model, more precise stellar masses can be directly estimated by fitting the dust-extinction-corrected UV to near-IR spectral energy distribution of galaxies as described in Section 9.1.2. Indeed, stellar masses can be accurately computed using a χ^2 or a Bayesian technique leaving the star formation history, the IMF, and the metallicity as free parameters, provided that photometric and/or spectroscopic data covering the whole wavelength range relative to the stellar emission are available.

Table 15.1 Stellar mass to light ratios as a function of color (from [462]).

Color	a_B	b_B	a_V	b_V	a_R	b_R	a_I	b_I	a_J	b_J	a_H	b_H	a_K	b_K
V-I	−1.919	2.214	−1.476	1.747	−1.314	1.528	−1.204	1.347	−1.040	0.987	−1.030	0.870	−1.027	0.800
V-J	−1.903	1.138	−1.477	0.905	−1.319	0.794	−1.209	0.700	−1.029	0.505	−1.014	0.442	−1.005	0.402
V-H	−2.181	0.978	−1.700	0.779	−1.515	0.684	−1.383	0.603	−1.151	0.434	−1.120	0.379	−1.100	0.345
V-K	−2.156	0.895	−1.683	0.714	−1.501	0.627	−1.370	0.553	−1.139	0.396	−1.108	0.346	−1.087	0.314

Notes: Johnson BVJHK and Kron–Cousins RI magnitudes are in Vega system.

Table 15.2 Stellar mass to light ratio as a function of color (from [463]).

Color	a_g	b_g	a_r	b_r	a_i	b_i	a_z	b_z	a_J	b_J	a_H	b_H	a_K	b_K
u-g	−0.221	0.485	−0.099	0.345	−0.053	0.268	−0.105	0.226	−0.128	0.169	−0.209	0.133	−0.260	0.123
u-r	−0.390	0.417	−0.223	0.299	−0.151	0.233	−0.178	0.192	−0.172	0.138	−0.237	0.104	−0.273	0.091
u-i	−0.375	0.359	−0.212	0.257	−0.144	0.201	−0.171	0.165	−0.169	0.119	−0.233	0.090	−0.267	0.077
u-z	−0.400	0.332	−0.232	0.239	−0.161	0.187	−0.179	0.151	−0.163	0.105	−0.205	0.071	−0.232	0.056
g-r	−0.499	1.519	−0.306	1.097	−0.222	0.864	−0.223	0.689	−0.172	0.444	−0.189	0.266	−0.209	0.197
g-i	−0.379	0.914	−0.220	0.661	−0.152	0.518	−0.175	0.421	−0.153	0.283	−0.186	0.179	−0.211	0.137
g-z	−0.367	0.698	−0.215	0.508	−0.153	0.402	−0.171	0.322	−0.097	0.175	−0.117	0.083	−0.138	0.047
r-i	−0.106	1.982	−0.022	1.431	0.006	1.114	−0.052	0.923	−0.079	0.650	−0.148	0.437	−0.186	0.349
r-z	−0.124	1.067	−0.041	0.780	−0.018	0.623	−0.041	0.463	−0.011	0.224	−0.059	0.076	−0.092	0.019

Color	a_B	b_B	a_V	b_V	a_R	b_R	a_I	$b_.$	a_J	b_J	a_H	b_H	a_K	b_K
B-V	−0.942	1.737	−0.628	1.305	−0.520	1.094	−0.399	0.324	−0.261	0.433	−0.209	0.210	−0.206	0.135
B-R	−0.976	1.111	−0.633	0.816	−0.523	0.683	−0.405	0.518	−0.289	0.297	−0.262	0.180	−0.264	0.138

Notes: SDSS magnitudes are in AB system, Johnson BVR and JHK in Vega system.

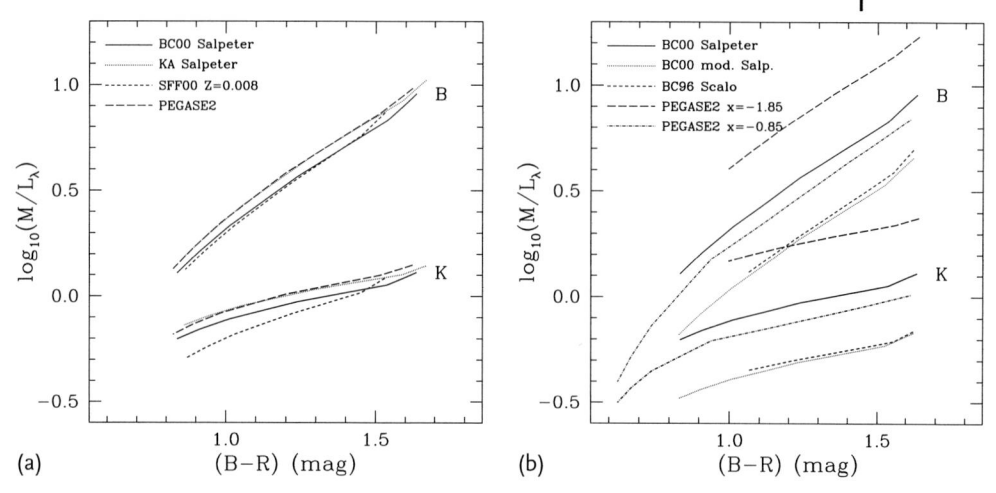

Figure 15.1 Dependence of the $M_{star}/L(\lambda)$–color relation on the (a) adopted population synthesis model and (b) IMF, from [462]. The $M_{star}/L(\lambda)$–color relations (in the Vega system) have been determined for a sequence of exponentially declining star formation rate models 12 Gyr old with fixed solar metallicity. Red colors (B-R $\sim$ 1.5) are for galaxies with strongly declining star formation rates comparable to old starbursts (ellipticals), while blue colors (B-R $\sim$ 1) are for objects with an almost constant star formation rate (spirals). The thin lines are for $M_{star}/L(B)$, the thick ones for $M_{star}/L(K)$. In (a), the different population synthesis models are: solid lines: [213]; dotted lines: [464]; dashed lines: [465]; long dashed lines: [216]. All models have been done with a Salpeter IMF and solar metallicity, except [465] (1/3 solar). In (b), the different population synthesis models are: solid lines: [213] with a Salpeter IMF; dotted lines: [213] with a Salpeter IMF down to $0.6 M_\odot$, and a power slope of $\alpha = -1$ below (modified Salpeter); dashed lines: [213] with a Scalo IMF; long dashed lines: [216] with a steep IMF ($\alpha = -2.85 (x = -1.85)$); dot-dashed lines: [216] with a flat IMF ($\alpha = -1.85 (x = -0.85)$). Reproduced by permission of the AAS.

15.2
Dynamical Mass

Galaxies can be generally considered as systems in dynamical equilibrium, thus their mass can be determined using the virial theorem. This theorem states that any isolated system in statistical equilibrium satisfies the relation

$$2T + U = 0 \tag{15.4}$$

where T and U are the kinetic and the gravitational potential energy of the system, respectively. Under some assumptions, this theorem can be used to measure the mass of different astronomical systems. For instance, for a hot, pressure-supported stellar system, under the simple assumption that the velocity dispersion of the stars (v_i) is independent of their mass (m_i),

$$T = \frac{1}{2} \sum_i m_i v_i^2 = \frac{1}{2} M_{dyn} V^2 \tag{15.5}$$

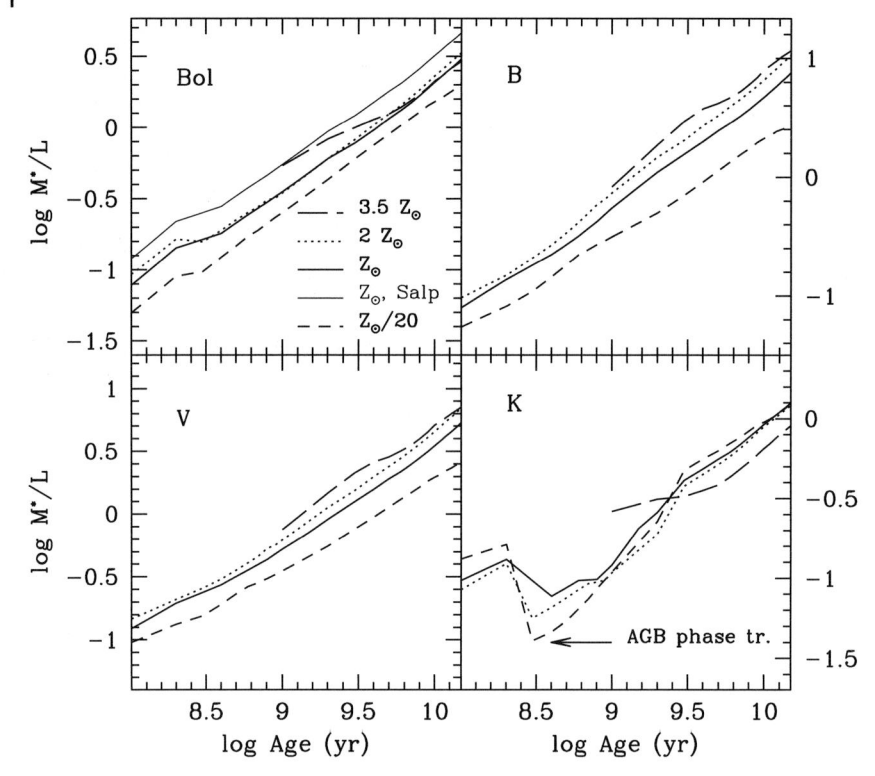

Figure 15.2 The dependence of the $M_{star}/L(\lambda)$ (in solar units) on age for different metallicities after an instantaneous burst of star formation (SSP: single stellar population). Models (thick lines) are computed using a Kroupa [385] IMF and compared to a Salpeter IMF (thin line) for solar metallicity (from [77]). The mass to light ratio is expected to change significantly after $\sim 10^{8.5}$ yr if the TP-AGB phase is considered. Reproduced with the permission of John Wiley & Sons Ltd.

where V is the mean velocity dispersion of the stars and M_{dyn} is the total dynamical mass of the system. The potential energy of the system is

$$U = -\sum_i \sum_{j>i} \frac{G m_i m_j}{|r_i - r_j|} \tag{15.6}$$

where $|r_i - r_j|$ is the distance between the stars i and j ($i > j$ is imposed to count them only once), and r is the weighted mean separation.

In its simplest expression the virial theorem gives the following relation by combining T and U:

$$M_{dyn}(r) = \frac{V(r)^2 r}{\alpha G} \tag{15.7}$$

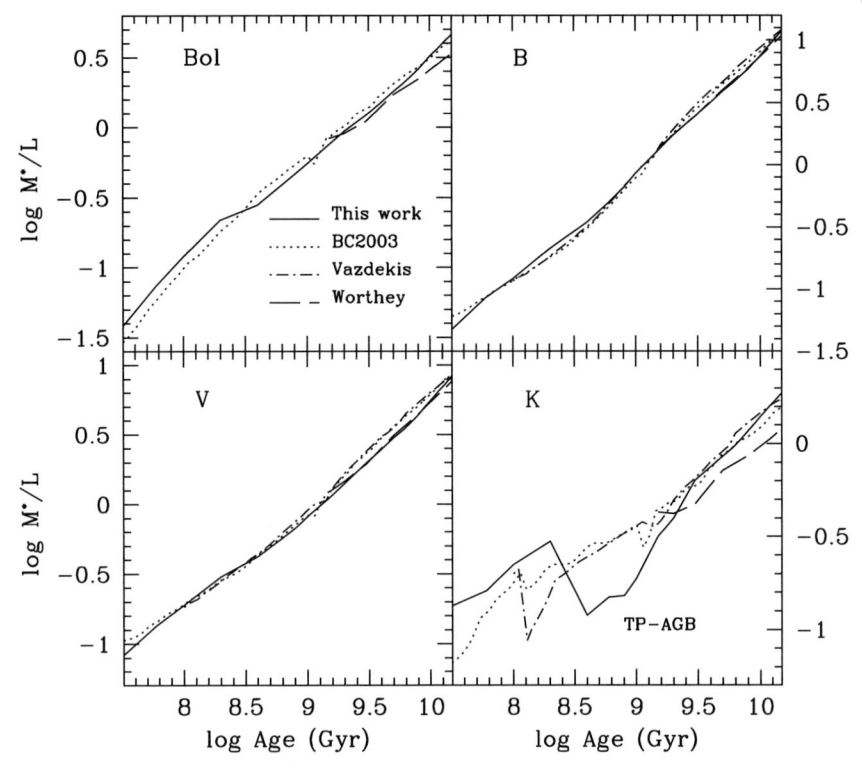

Figure 15.3 The dependence of the $M_{star}/L(\lambda)$ (in solar units) on age after an instantaneous burst of star formation (SSP) for different models computed using a Salpeter IMF (thin line) and solar metallicity: solid line: [77]; dotted line: [213]; dotted-dashed line: [466]; long dashed line: [89] (from [77]). Reproduced with the permission of John Wiley & Sons Ltd.

where α is a parameter which changes according to the mass density distribution within the system. In a spherical system,

$$U = -\frac{\alpha G M_{dyn}^2}{r} \tag{15.8}$$

where $\alpha = 3/2$ for a uniform density and $\alpha > 3/2$ if the central density is higher than the mean density. This theorem, whose derivation can be found in [467, 468], can be applied to derive the mass of any noncollisional system in dynamical equilibrium, from star clusters and giant molecular clouds to spirals and ellipticals, compact groups and clusters of galaxies.

15.2.1
Rotation Curves and the Dark Matter Distribution

Equation (15.7) clearly shows that the mass of an isolated system in dynamical equilibrium can be measured by means of its kinematical properties. In the case of a rotating system such as spiral galaxies, this is done using its observed rotation curve, defined as the variation of the rotational velocity of the galaxy as a function of its radius (see Section 15.2.1.1). The relation between the dynamical mass and the rotational velocity, such as the one given in Eq. (15.7), can also be inverted to trace the expected rotational curve for a known mass density distribution, as generally done for the baryonic components of galaxies. The stellar mass density distribution can be determined using radial light profiles and by assuming constant or variable mass to light ratios, while total gas (HI + H$_2$) density profiles can be inferred directly from HI and CO imaging observations. The relation between the radial mass distribution $M_{dyn}(r)$ and the rotation curve $V(r)$, however, has a much more complex form than the one given in relation (15.7) once the mass density distribution is flat and nonuniform, as is generally the case in disk galaxies, and can be inverted only numerically. Some relations between the dynamical mass and the rotational velocity for the most common geometries often used in the literature are listed below.

Mass point approximation and homogeneous sphere The mass point approximation is the simplest geometrical configuration for a dynamical system, where a crude estimate of the mass of a galaxy within a radius r is given by the relation [469]

$$M_{dyn}(<r) = \frac{rV(r)^2}{G} \tag{15.9}$$

where $V(r)$ is the rotational velocity of the galaxy at r. This relation also holds for a homogeneous sphere. Equation (15.9) can be easily inverted:

$$V(r) = \left(\frac{GM(r)}{r}\right)^{1/2} \tag{15.10}$$

In the Keplerian approximation (mass point approximation), M_{dyn} is a constant, and $V_{rot} \propto r^{-1/2}$ declines with galactocentric distance. It is interesting to note that the mass point approximation (Eq. (15.9)) recovers Kepler's third law of motion, where the orbital period t is equal to $2\pi r/V_{rot} \propto r^{3/2}$.

Homogeneous spheroid of constant density In the case of a homogeneous spheroid of constant density

$$\rho = \frac{3M}{4\pi a^2 b} \tag{15.11}$$

with major and minor axis $2a$ and $2b$, respectively, the mass of the system is given by the relation [469]

$$M_{dyn}(<r) = \frac{aV(r)^2}{\alpha G} \tag{15.12}$$

where

$$\alpha = \frac{3}{2}\frac{a^2}{a^2-b^2}\left[\frac{a}{(a^2-b^2)^{1/2}}\arccos^{-1}\left(\frac{b}{a}\right)-\frac{b}{a}\right] \tag{15.13}$$

15.2 Dynamical Mass

In the interior of the homogeneous spheroid, the mass contained within a similar spheroid (with the same axial ratio b/a as the outer boundary) is

$$M_{\text{dyn}}(<r) = \frac{4\pi \rho c r^3}{3a} \tag{15.14}$$

where r is the distance from the center in the equatorial plane, while the rotational velocity is

$$V(r) = \left(\frac{4\pi G \rho a}{3}\frac{b}{a}\right)^{1/2} r \tag{15.15}$$

and thus $V(r) \propto r$.

Flat-disk approximation Under this approximation, [470] state that the rotational velocity is

$$V(r) = \frac{V_{\max} r}{r_{\max}\left[\frac{1}{3} + \frac{2}{3}\left(\frac{r}{r_{\max}}\right)^n\right]^{3/2n}} \tag{15.16}$$

which implies a total mass $M_{\text{dyn,T}}$ of

$$M_{\text{dyn,T}} = \left(\frac{3}{2}\right)^{3/n} \frac{V_{\max}^2 r_{\max}}{G} \tag{15.17}$$

where $V_{\max}$ is the maximal rotational velocity observed at the radius $r_{\max}$, and n is a parameter changing between 1.5 and 3.

Exponential disk approximation Late-type galaxies generally have exponentially declining surface brightness profiles following the relation (14.3) (see Section 14.1). Surface brightness profiles can be converted into stellar surface density profiles by simply multiplying them by a mass to light ratio

$$\Upsilon_{\text{disc}}^\lambda = \frac{M_{\text{star}}(r)}{L^\lambda(r)} \tag{15.18}$$

whose value can be estimated from population synthesis models. For a stellar disk of zero thickness with an exponential radial surface density profile,

$$\Sigma(r) = \Sigma_0 e^{-r/h_r} \tag{15.19}$$

where h_r is the exponential disk scale length and Σ_0 is the central surface density, the radial variation of the dynamical mass of the stellar disk is

$$M_{\text{dyn}}(r) = 2\pi \Upsilon_{\text{disc}}^\lambda \Sigma_0 h_r^2 \left[1 - \left(1 + \frac{r}{h_r}\right)e^{-r/h_r}\right] \tag{15.20}$$

and the corresponding rotation curve [471] is

$$V(r) = \left\{4\pi G \Upsilon_{\text{disk}}^\lambda \Sigma_0 h_r^2 [I_0(y)K_0(y) - I_1(y)K_1(y)]\right\}^{1/2} \tag{15.21}$$

where I_n and K_n are modified Bessel functions of order n (first and second order, respectively), and $y = \frac{r}{2h_r}$. The exponential disk has a peak velocity at $r = 2.15 h_r$, whose value is given by

$$V_{\max}(r = 2.15 h_r) = 0.88 \sqrt{\pi G \Sigma_0 h_r} \qquad (15.22)$$

Other relations for different mass distributions can be found in [468, 469], while a complete determination of the mass and velocity profile for a real edge-on spiral galaxy is given in [472].

High resolution spectroscopic observations of nearby spiral galaxies revealed that their rotation curves steeply increase with radius in the inner parts and become flat ($V_{\rm rot}$ = const) outwards [471]. Interferometric HI observations of nearby galaxies also indicated that the flat rotation curves extend out to ~ 2 optical radii [473]. While the inner part of the rotation curve can be represented by a solid disk of constant density, the outer disk does not follow Keplerian motion since $M(r) \propto r$, indicating that the total dynamical mass of galaxies increases linearly with distance from the center, even outside the optical radius.

The contribution of the different baryonic galaxy components (star, gas) as a function of the galactic radius to the velocity rotation curve of a galaxy can be directly measured from the observed mass distributions by inverting the dynamical mass vs. rotational velocity relation as previously explained (e.g., Figure 15.4). The radial distribution of the mass of the stellar component is generally determined after decomposing the observed radial light profile in the bulge and disk contribution, and transformed into stellar masses by assuming two different constant stellar mass to light ratios $M_{\rm star}/L$ [472], or whenever possible, by using population synthesis models, as indicated in Section 15.1 [476]. The radial distribution of the total gas component (HI + H_2) can be directly inferred from HI and CO observations by adding the helium contribution (~ 30–40% of hydrogen).

The distribution of the visible matter traced by the stellar light of both the disk and the bulge components is centrally peaked decreasing outwards. This evidence dramatically contrasts the observed distribution of the dynamical mass; the flat rotation curve increases radially, indicating that the dynamical mass to light ratio $M_{\rm dyn}/L$ increases in the outer regions of spiral galaxies [473]. HI interferometric observations allowed us to trace the distribution of the atomic gas and compare it to the distribution of the stellar and dynamical masses. As clearly indicated in Figure 15.4, the sum of the stellar and gaseous masses cannot explain the observed radial variation in the rotation curve of the galaxy, whose values are too high in the outer part. This is thus the clearest evidence of the presence of a dark matter halo in spiral galaxies. The parameters characterizing the dark matter halo, whose presence was theoretically predicted by [477] to stabilize rotating disks, are constrained by fitting a radial density profile which better reproduces the observed rotation curve. These density profiles have different shapes:

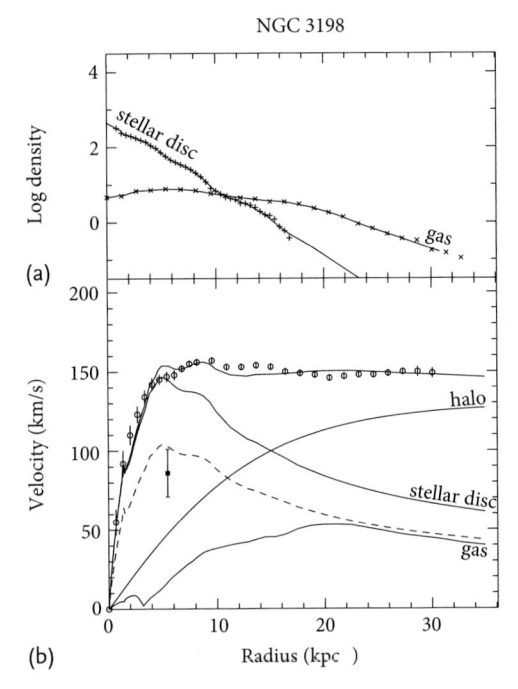

Figure 15.4 (a) the radial distribution of the stellar mass (plus sign) determined from the surface brightness distribution using a constant mass to light ratio and of the HI gas (crosses), in $M_\odot\,pc^{-2}$, of the late-type galaxy NGC 3198, adapted from [474]; (b) the observed HI + He rotation curve (open circles, from [475]) is compared to those deduced for the gas, disk, and halo components and to their sum (solid lines). For the disk component, the solid line gives the maximum disk approximation (corresponding to the maximum contribution that can be provided by the stellar disk to the rotation curve of the galaxy, determined assuming the maximal velocity of the disk that equals at a given r, here at $r \sim 6$ kpc, the observed HI rotation curve), while the dashed line is the rotational velocity of the stellar disk for another analytical solution.

General halo density profile The radial distribution of the halo component can be represented using the general density profiles of [478]:

$$\rho(r) = \frac{\rho_0}{r^\gamma (1 + r^{1/\alpha})^{(\beta-\gamma)\alpha}} \tag{15.23}$$

where ρ_0 is the central density, and α, β, and γ are free parameters.

Navarro, Frenk and White dark matter halo profile Among these radial profiles of differing density, the most used is the Navarro, Frenk, and White (NFW) profile for dark matter halos surrounding formed galaxies, deduced from models in cold dark matter dominated cosmologies by [454, 479]:

$$\rho(r) = \frac{\rho_0}{\frac{r}{r_c}\left(1 + \frac{r}{r_c}\right)^2} \tag{15.24}$$

where r_c is the core radius. This density profile is characterized by a prominent cusp in the core of the galaxy. For this density profile,

$$V(r) = \left\{ 4\pi G \rho_0 r_c^2 \left[-\frac{1}{1+x} + \frac{\ln(1+x)}{x} \right] \right\}^{1/2} \tag{15.25}$$

where $x = r/r_c$.

Pseudo-isothermal sphere profile The pseudo-isothermal sphere density profile, whose analytical form is given by [480],

$$\rho(r) = \frac{\rho_0}{1 + \left(\frac{r}{r_0}\right)^2} \tag{15.26}$$

is generally preferred since it reproduces the flat core density distribution deduced from the rotation curves of spiral and irregular galaxies [461] and low surface brightness systems [456, 481]. Its corresponding rotation curve is [482, 483]:

$$V(r) = \left\{ 4\pi G \rho_0 r_0^2 \left[1 - \frac{r_0}{r} \arctan\left(\frac{r}{r_0}\right) \right] \right\}^{1/2} \tag{15.27}$$

15.2.1.1 Analytical Parameterization of Observed Rotation Curves of Disk Galaxies

The shape of the observed rotation curves, as previously discussed, is tightly related to the distribution of the visible and dark matter in galaxies and is thus a fundamental constraint in models of galaxy formation and evolution. A detailed characterization of the kinematical properties of disk galaxies is thus of key importance for validating and comparing such models with observations. The observed optical and HI rotation curves of disk galaxies have been tentatively fitted with polynomial functions scaled on observed quantities such as optical magnitudes or morphological types. This work has been done with the aim of studying the mean dark matter distributions of different classes of galaxies and providing a standard reference for high-redshift studies. Persic and Salucci [484] first claimed the existence of a universal rotation curve $V_{URC}(r)$. The same authors revisited their analysis using a much larger sample of nearby galaxies with available rotation curves, providing a new analytical expression [485]:

$$V_{URC}\left(\frac{r}{r_{opt}}\right) [\text{km s}^{-1}] = v(r_{opt}) \left\{ \left[0.72 + 0.44 \log\left(\frac{L_B}{L_B^*}\right) \right] \frac{1.97 x^{1.22}}{(x^2 + 0.78^2)^{1.43}} \right.$$

$$\left. + 1.6 \exp\left[-0.4\left(\frac{L_B}{L_B^*}\right)\right] \frac{x^2}{x^2 + 1.5^2 \left(\frac{L_B}{L_B^*}\right)^{0.4}} \right\}^{1/2} \tag{15.28}$$

where $x = r/r_{opt}$, $r_{opt} = 3.2 r_d \sim r_{25}$, r_d is the exponential disk scale length ($r_d = h_r$ in Eq. (14.3)), r_{25} is the isophotal radius at 25 mag arcsec^{-2} which corresponds to the radius encompassing 83% of the total integrated light [486], L_B is the B band luminosity measured assuming $H_0 = 75 \text{ km s}^{-1} \text{ Mpc}^{-2}$, and L_B^* is the characteristic L_B luminosity.

More recently, using an homogeneous sample of ~ 2200 low-redshift galaxies with high quality long-slit optical spectroscopy and I band imaging, [486] fitted the observed rotation curves with the analytical function

$$V_{PE}(r) = V_0(1 - e^{-r/r_{PE}})\left(1 + \frac{ar}{r_{PE}}\right) \tag{15.29}$$

Generally called the Polyex model, this function is characterized by an amplitude V_0, an exponential scale of the inner region r_{PE}, and a slope of the outer part α. Table 15.3 and Figures 15.5 and 15.6 give the mean parameters of the Polyex fitting functions in different classes of I band luminosity, expressed as a function of exponential disk scale lengths, r_d, or optical radii, r_{opt}. Clearly, the shape of the rotation curve steepens in the inner regions and folds closer to the core of the galaxy. The rotation curve also flattens outwards in massive objects, while it has a milder increase that continue in the outer regions in low luminosity objects.

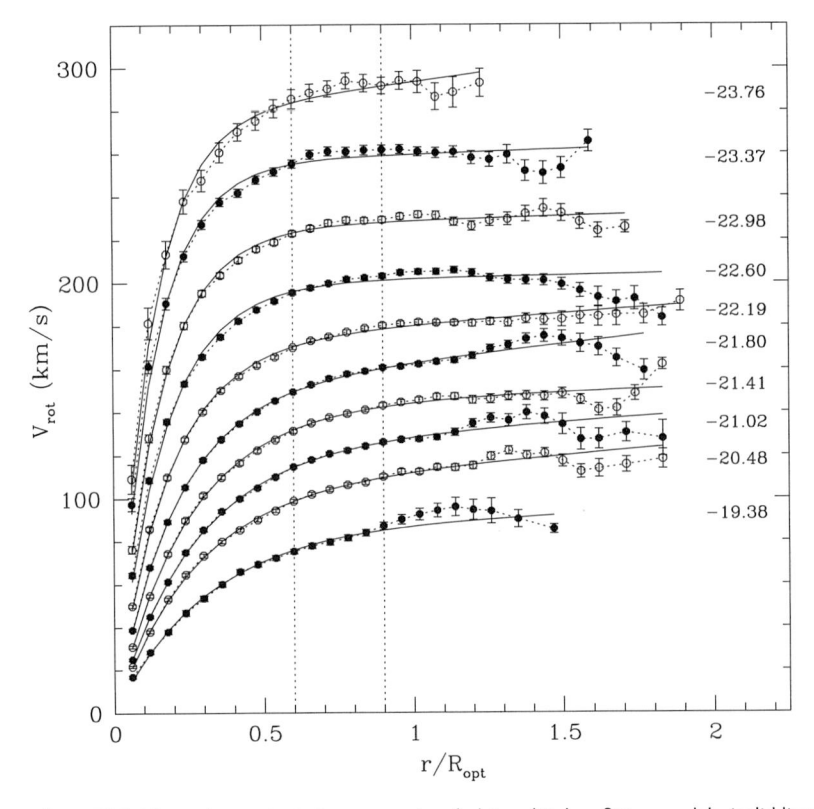

Figure 15.5 Mean observed rotation curves (symbols) and Polyex fitting models (solid lines) as a function of the optical radius of spiral galaxies in different mean I band absolute magnitude intervals ($H_0 = 70 \text{ km s}^{-1} \text{ Mpc}^{-1}$), from [486]. Reproduced by permission of the AAS.

Table 15.3 Mean rotation curves Polyex fitting parameters as a function of exponential disk scale lengths (top) and optical radii (bottom), from [486].

M_I	ΔM_I	$\langle M_I \rangle$	$\langle V_{\Delta r,M} \rangle$	$\langle N_{\Delta r,M} \rangle$	V_0	r_{PE}/r_d	α
−23.80	0.40	−23.76	285.1	43	270 ± 5	0.37 ± 0.02	0.007 ± 0.003
−23.40	0.40	−23.37	258.4	124	248 ± 2	0.40 ± 0.01	0.006 ± 0.001
−23.00	0.40	−22.98	225.2	225	221 ± 1	0.48 ± 0.01	0.005 ± 0.001
−22.60	0.40	−22.60	198.4	324	188 ± 1	0.48 ± 0.01	0.012 ± 0.001
−22.20	0.40	−22.19	175.7	341	161 ± 1	0.52 ± 0.01	0.021 ± 0.001
−21.80	0.40	−21.80	155.1	327	143 ± 1	0.64 ± 0.01	0.028 ± 0.002
−21.40	0.40	−21.41	137.4	263	131 ± 1	0.73 ± 0.02	0.028 ± 0.003
−21.00	0.40	−21.02	120.7	213	116 ± 2	0.81 ± 0.02	0.033 ± 0.005
−20.40	0.80	−20.48	103.7	203	97 ± 2	0.80 ± 0.02	0.042 ± 0.005
−19.00	2.00	−19.37	79.5	92	64 ± 3	0.72 ± 0.05	0.087 ± 0.016

M_I	ΔM_I	$\langle M_I \rangle$	$\langle V_{\Delta r,M} \rangle$	$\langle N_{\Delta r,M} \rangle$	V_0	r_{PE}/r_{opt}	α
−23.80	0.40	−23.76	290.5	44	275 ± 6	0.126 ± 0.007	0.008 ± 0.003
−23.40	0.40	−23.37	260.6	130	255 ± 2	0.132 ± 0.003	0.002 ± 0.001
−23.00	0.40	−22.98	227.8	226	225 ± 1	0.149 ± 0.003	0.003 ± 0.001
−22.60	0.40	−22.60	200.2	328	200 ± 1	0.164 ± 0.002	0.002 ± 0.001
−22.20	0.40	−22.19	176.3	346	170 ± 1	0.178 ± 0.003	0.011 ± 0.001
−21.80	0.40	−21.80	156.5	330	148 ± 2	0.201 ± 0.004	0.022 ± 0.002
−21.40	0.40	−21.41	138.4	267	141 ± 2	0.244 ± 0.005	0.010 ± 0.003
−21.00	0.40	−21.02	121.4	210	122 ± 2	0.261 ± 0.008	0.020 ± 0.005
−20.40	0.80	−20.48	105.1	195	103 ± 2	0.260 ± 0.008	0.029 ± 0.005
−19.00	2.00	−19.38	80.7	93	85 ± 5	0.301 ± 0.022	0.019 ± 0.015

Notes: absolute magnitudes are measured using $H_0 = 70 \, \text{km s}^{-1} \, \text{Mpc}^{-1}$.

15.2.2
The Total Mass of Elliptical Galaxies from Kinematical Measurements

Elliptical galaxies are generally pressure-supported systems where the motion of the stars is comparable to that of a hot gas in thermal equilibrium, characterized by a typical velocity dispersion σ given by (assuming spherical symmetry [469])

$$\sigma^2 = 3\sigma_0^2 \tag{15.30}$$

where σ_0 is the observed velocity dispersion along the line of sight. Thus the mass of elliptical galaxies can be inferred from the relation

$$M(<r) = \frac{3\sigma_0^2 r}{G} \tag{15.31}$$

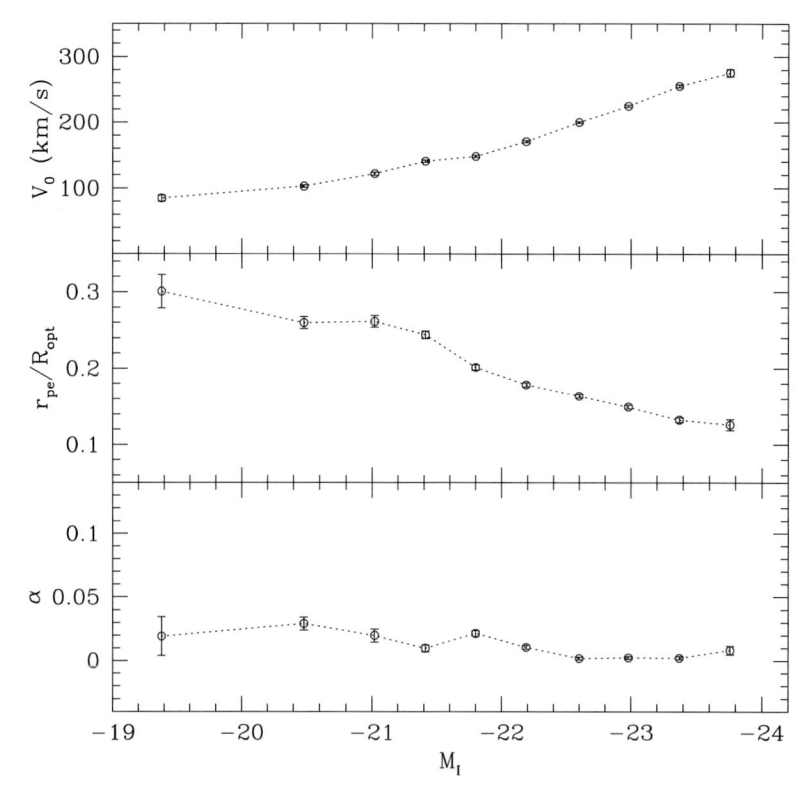

Figure 15.6 Polyex model coefficients of the fits shown in Figure 15.5 plotted as a function of the I band absolute magnitude, from [486]. Reproduced by permission of the AAS.

The velocity dispersion along the line of sight of elliptical galaxies can be determined by measuring the width of absorption line features in their optical spectra and comparing it to that of a set of template stars of different spectral type. This is done in order to disentangle the line broadening due to the limiting resolution of the instrument from the intrinsic line broadening of the galaxy. The spherical symmetry assumption, however, needs to be checked with observational data: indeed, a few bright galaxies [487] and several dwarf ellipticals [488] are rotationally supported systems that do not satisfy the isotropic condition. More recently, the velocity dispersion of elliptical galaxies has been determined using the recessional velocity of globular clusters, enabling the measurement of the dynamical mass out to large radii, as shown in Figure 15.7 [489].

15.2.3
The Total Mass of Elliptical Galaxies from X-ray Measurements

Observations of the hot gas in the halo of elliptical galaxies can be used as an independent measure of the total dynamical mass out to large radii. Accurate mass determinations require, however, high quality X-ray observations of the gas density

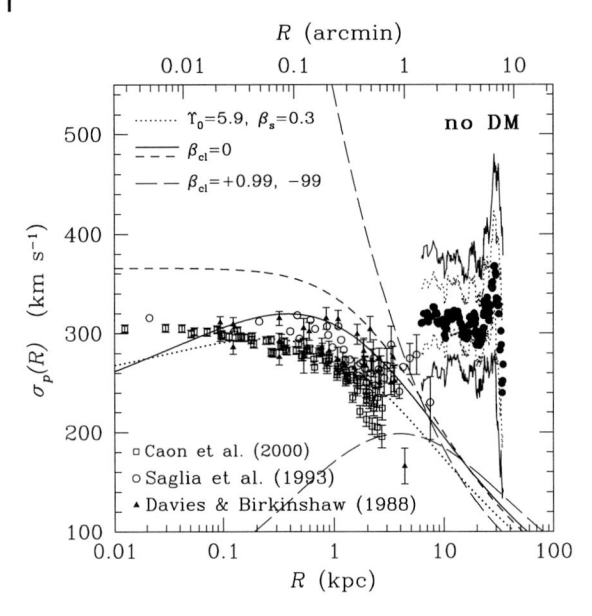

Figure 15.7 The velocity dispersion of the elliptical galaxy M49 is measured from the velocity width of stellar absorption features (empty circles, empty squares, and filled triangles) and from the recessional velocity of globular clusters (filled dots) and compared to different mass model predictions (solid, dotted and dashed lines), from [489]. Reproduced by permission of the AAS.

and temperature profiles, the later being the most critical parameter. This can be achieved only in the most massive ellipticals, where the gas emission dominates over the X-ray emission of compact sources. Furthermore, the galaxy mass can be determined only if the gas is in hydrostatic equilibrium and the gas pressure dominates [53]. Within these conditions, and assuming spherical symmetry, the total mass of stars and dark matter can be estimated using the traditional formulation for hydrostatic equilibrium [53, 490]:

$$M(<r) = -rT(r)\frac{k}{G\mu m_p}\left(\frac{d\log\rho}{d\log r} + \frac{d\log T}{d\log r}\right) \quad (15.32)$$

where r is the radius of the galaxy, k is the Boltzmann constant, $\rho(r)$ is the gas density, $T(r)$ is the gas temperature, m_p is the proton mass, $\mu = 0.61$ is the mean atomic weight of the gas, and G is the universal gravitational constant. An extra value should be added whenever external pressure due to nonthermal turbulence, magnetic fields, or cosmic rays is present [53]. An alternative method is to fit the observed gas densities and temperature radial profiles with standard profiles and compare them to the parameterization of the density and temperature profiles expected for a given mass distribution in hydrostatic equilibrium conditions ([491] and [492]). For instance, assuming the standard parametrizations for the temperature radial profile given in Eq. (6.4) and assuming an isothermal gas (Eq. (6.5)) in hydrostatic equilibrium, the dynamical mass of an elliptical galaxy within a radius r

can be estimated through the relation [74]

$$M(<r)\,[M_\odot] = 3.2 \times 10^{11}(3\beta + \alpha) \left(\frac{T\,[\mathrm{K}]}{10^7}\right)\left(\frac{r\,[\mathrm{kpc}]}{10}\right) \quad (15.33)$$

where T is the temperature expressed in 10^7 K measured at the radius r in units of 10 kpc. Or, equivalently,

$$M(<r)\,[M_\odot] = 1.8 \times 10^{12}(3\beta + \alpha)(T\,[\mathrm{keV}])\left(\frac{r\,[\mathrm{arcsec}]}{10^3}\right)\left(\frac{D\,[\mathrm{Mpc}]}{10}\right) \quad (15.34)$$

where the temperature, expressed in keV, is measured at the angular distance r, expressed in 10^3 arcsec, from the core of a galaxy located at a distance D (in units of 10 Mpc). The difficulty in disentangling the contribution of the cluster intergalactic medium from the X-ray emission of bright ellipticals, combined with the possibility of stellar sources contaminating the total X-ray emission, the presence of inhomogeneities in the gas distribution, and the questionable validity of the hydrostatic equilibrium condition are major sources of uncertainties in the mass determination using this technique.

15.2.4
The Mass of the Supermassive Black Hole

The same analytical formalism used to measure the total dynamical mass of spirals and ellipticals can be adopted, under some assumptions, for measuring the mass of the supermassive black holes hosted in the core of massive galaxies. The presence of central supermassive black holes in galaxies, generally indicated as SBH or SMBH, has already been claimed by [68] to explain the very compact nature of AGNs inferred from their short term flux variability, their nonstellar spectral energy distribution, and their very high bolometric luminosity.

Within the sphere of influence of a supermassive black hole, a Keplerian rotation (or velocity dispersion),

$$V(r) = \left(\frac{GM}{r}\right)^{1/2} \quad (15.35)$$

of stars or gas is a clear sign of the existence of a central mass condensation [493, 494]. The sphere of influence of a SBH is defined as the region of space within which the gravitational potential of the SBH dominates over that of the surrounding stars. That is ([494]),

$$r_h\,[\mathrm{pc}] \simeq \frac{GM_{\mathrm{SBH}}}{\sigma^2} \simeq 11.2 \frac{(M_{\mathrm{SBH}}/10^8\,[M_\odot])}{(\sigma/200\,[\mathrm{km\,s^{-1}}])^2} \quad (15.36)$$

where M_{SBH} is the mass of the supermassive black hole and σ is the velocity dispersion of the surrounding stellar population. As extensively discussed in [494],

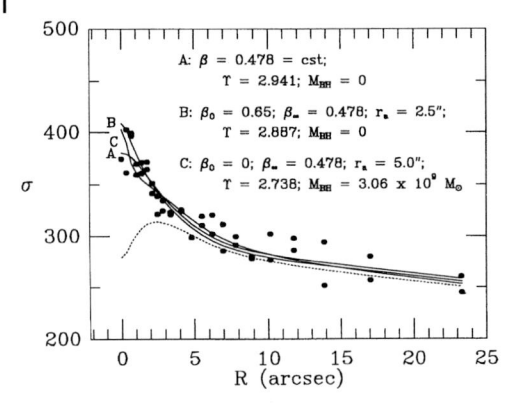

Figure 15.8 The radial variation of the velocity dispersion of M87 obtained using stellar dynamical data compared to different model predictions, from [497]. The Keplerian decrease of σ with r (here indicated with R) is evidence of the presence of a central mass condensation. Models (here represented by solid lines) A and B are radially anisotropic and have no central black hole, while C is isotropic towards the center and requires a central black hole of mass $\leq 5 \times 10^9 M_\odot$ to fit the data (the dotted line shows the same model without the central black hole). Reproduced with the permission of John Wiley & Sons Ltd.

however, the central mass concentration revealed by Keplerian nuclear motion (as shown in Figure 15.8) can be associated with either a SBH or to a dense star cluster. Its very nature can be determined directly only through the detection of relativistic velocities within a few Schwarzschild radii, or by indirectly comparing the measured density within a given radius with the model predictions of dark cluster evaporation and collapse.

Except for the mass in the Milky Way (determined through the direct measurement of the proper motion of individual stars), in external galaxies the mass of the supermassive black hole is generally determined by using either the stellar dynamics obtained from absorption lines of integrated spectra (as shown in Figure 15.8), or the gas dynamics of nuclear gas disks using emission lines (Figure 15.9). Even taking advantage of the subarcsecond resolution of the HST to spacially resolve the sphere of influence of the supermassive black hole, this technique can be used only for very nearby galaxies and can never unambiguously determine whether the mass concentration is due to a singularity or a dense star cluster. An alternative method for measuring the gas dynamics in the core of galaxies is the detection of H_2O maser clouds confined to a thin, regular disk located a few parsecs from the singularity. VLBA observations at 22 GHz of the water maser have an angular resolution a factor of ~ 200 better than the HST. Thus they are ideal for measuring the dynamics of the gas on angular distances much closer to the black hole than those determined from optical spectroscopy, and for removing an ambiguity in the nature of the central mass concentration. This method has been successfully applied to NGC 4258 [495] and has been extended to a few other relatively nearby objects [494].

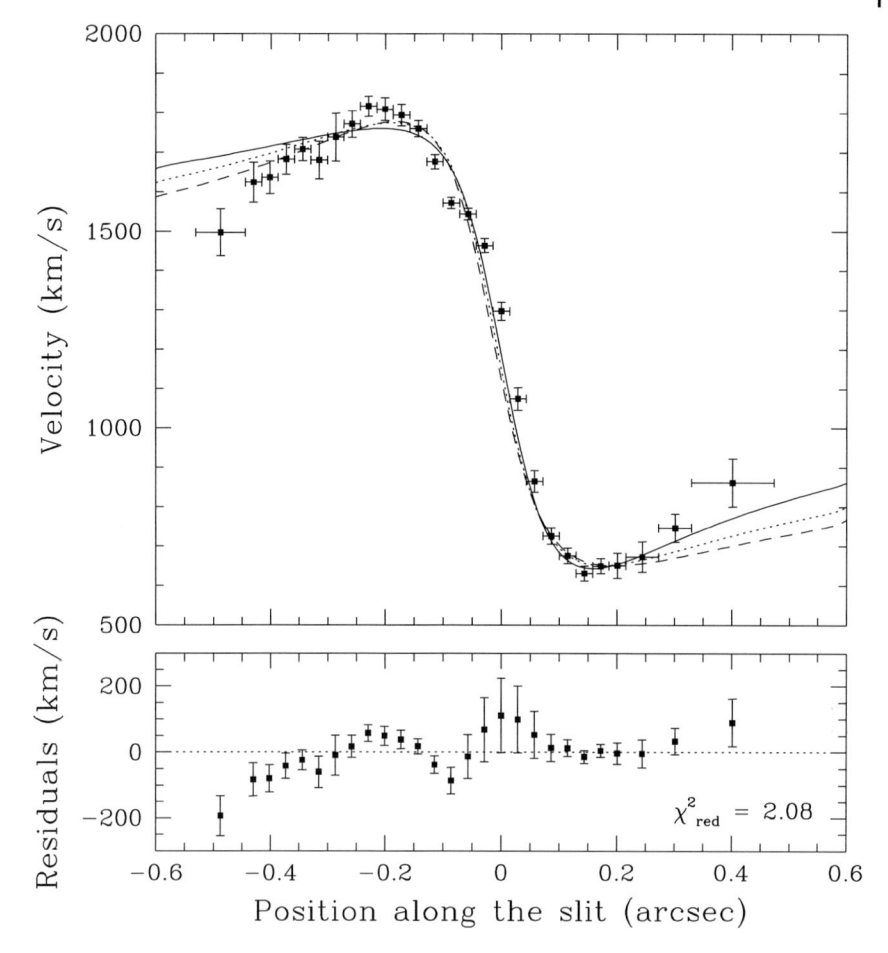

Figure 15.9 The rotation curve of the ionized gas of M87 obtained with HST by [496]. The different lines correspond to thin disk models with different parameters. The estimated central black hole mass is of a few 10^9 $M_\odot$. Reproduced by permission of the AAS.

The determination of the mass of the supermassive black hole from the observed motion of stars or gas in the nuclei of galaxies is not straightforward since observed rotational velocities and velocity dispersions are available only along the line of sight. Some assumptions should be placed on the mass distribution (for instance, a spherical mass distribution should be assumed in pressure-dominated systems, and an axisymmetric one should be assumed in rotationally supported systems) to determine the most appropriate mass models that should be used for transforming observed rotational velocity or velocity dispersion radial profiles into mass radial profiles. Other uncertainties in the mass to light ratio must be assumed for transforming observed stellar surface densities into masses. These are related to the unknown inclination of rotating systems, the presence of turbulent motions in the gas, the uncertain position of the slit of the spectrograph relative to the dynam-

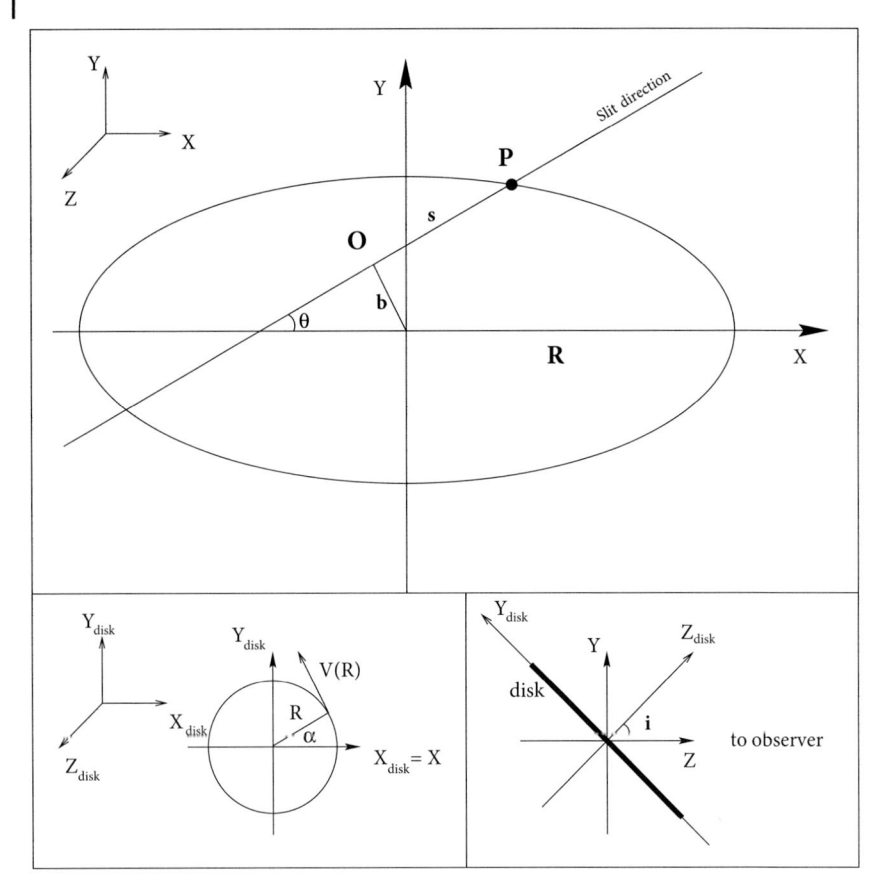

Figure 15.10 The schematic representation of the reference frames used in the determination of the Keplerian rotation curve. XY is the plane of the sky. X and Y are directed along the major and minor axes of the disk, respectively, while Z is directed towards the observer. X_{disk} and Y_{disk} define the reference frame on the disk plane such that $X_{disk} = X$, from [496]. Reproduced by permission of the AAS.

ical center of the galaxy, or others uncertainties introduced by the different setup of the instrumentation (as illustrated in the case of M87 using either gas emission lines [496] or stellar absorption lines [497]).

Macchetto et al. [496] give a simple example on how the inner rotation curve of M87, determined using several optical emission lines using the FOC spectrograph on the HST, can be used to estimate the mass of the supermassive black hole hosted in the core of this powerful radio galaxy. Ignoring the effects due to the finite width of the slit and of the PSF, we follow their simplified treatment and leave the more complete description of the method to this original reference. If X and Y are the reference axes on the plane of the sky in the direction of the major and minor axes of the emitting nuclear disk (see Figure 15.10), then a point with coordinates $P(X, Y)$ is at a radius r, such that

$$r^2 = X^2 + \frac{Y^2}{(\cos i)^2} \tag{15.37}$$

where i is the inclination of the disk with respect to the line of sight. Let O be the center of the slit (the closest point from the kinematical center of the galactic disk, located at a distance b), s be the distance along the slit from O, and θ be the angle of the slit with respect to the major axis of the disk (see Figure 15.10). The coordinates of the points located along the slit are given by

$$X = -b \sin \theta + s \cos \theta \tag{15.38}$$

and

$$Y = b \cos \theta + s \sin \theta \tag{15.39}$$

If $V(r)$ is the circular velocity tangential to the disk at a distance r, then its projection along the line of sight is $-V(r) \cos \alpha \sin i$, with

$$\tan \alpha = \frac{Y}{X \cos i} \tag{15.40}$$

Thus, we can derive that the velocity along the slit is

$$V = V_{\text{sys}} - (G M_{\text{SBH}})^{1/2} \frac{X(\sin i)}{(X^2 + [Y^2/(\cos i)^2])^{3/4}} \tag{15.41}$$

where V_{sys} is the systemic velocity of the galaxy. The mass of the supermassive black hole, M_{SBH}, can be determined by fitting the observed rotation curve and leaving V_{sys}, M_{SBH}, i, θ, and b as free parameters. Since smearing effects due to the PSF are important, and should be considered [496], this simple formalism only gives a first-order approximation.

Part Three   Constraining Galaxy Evolution

16
Statistical Tools

Galaxy number counts, luminosity, and mass functions are statistical tools defined to deal with the large samples of local or high redshift galaxies now available to the astronomical community. They are often compared to model predictions to constrain models of galaxy evolution as well as to measure different cosmological parameters.

16.1
Galaxy Number Counts

The galaxy number counts give the number of galaxies per bin of magnitude per unit area. Number counts are generally measured in logarithmic scale, and are represented by plotting the logarithm of the number of galaxies, log N, as a function of the observed magnitude, m, down to a given magnitude limit. At frequencies other than the UV, optical, or near-infrared, number counts are represented by a log N vs. log S relation, where S is the flux density at which the number of sources is measured.

First, the observed counts should be corrected to remove the contamination of stars. These can be recognized by looking at their radial light profiles, extended in galaxies and peaked in stars, or by using the color–magnitude relation. Galaxies populate a different region of the diagram since they have colors that are different than those of high Galactic latitude stars, which are mainly evolved systems [498]. Fields at high Galactic latitudes are also required to minimize the effect of the Galactic extinction, which reduces the flux and reddens the color of the detected sources.

Number counts have been proposed as a possible tool for measuring cosmological parameters. In a hypothetical scenario with Euclidian cosmology where there is no galactic evolution, the density distribution of galaxies, n_0, is uniform, and all galaxies have the same luminosity, L_i, the number of objects with an apparent flux density brighter than S is proportional to the volume within which these galaxies can be observed:

$$N(>S) \propto n_0 D_i^3 \propto n_0 \left(\frac{L_i}{4\pi S}\right)^{3/2} \propto S^{-3/2} \qquad (16.1)$$

A Panchromatic View of Galaxies, First Edition. Alessandro Boselli.
© 2012 WILEY-VCH Verlag GmbH & Co. KGaA. Published 2012 by WILEY-VCH Verlag GmbH & Co. KGaA.

Or, equivalently:

$$N(< m) \propto 10^{0.6m} \tag{16.2}$$

Here, m is the limiting magnitude and D_i is the distance within which a galaxy of luminosity L_i can be observed:

$$D_i = \left(\frac{L_i}{4\pi F}\right)^{1/2} \tag{16.3}$$

For a Schechter luminosity function (see Section 16.2), $N(< m)$ is given by [500]:

$$N(< m) \propto D^{*3}(m) \int dL\phi(l) \left(\frac{L}{L*}\right)^{3/2} \tag{16.4}$$

where $D^{*3}(m)$ is the limiting depth of the survey for a L^* galaxy. Deviations from this expected Euclidian log N vs. log S relation (16.1), sometimes represented in its differential form,

$$\frac{dN}{dS} \propto S^{-5/2} \tag{16.5}$$

as in Figure 16.1, can be imputed to a different cosmological model and thus used to constrain cosmological parameters. However, accurate K-corrections are required to account for the fact that the rest frame wavebands of the detected galaxies change as the surveys get to low flux densities. This ideal configuration for the use of source counts as a probe of cosmological models is however based on the unrealistic assumption of no galaxy evolution. Recent studies of the star formation history have indeed shown that the different stellar populations inhabiting galaxies might have significantly evolved in the last billions of years, which suggests a variation of the measured flux density S with redshift. At the same time, current models of galaxy evolution predict that galaxies have been formed by the merging of small structures, an evolutionary process that should induce a variation of the source number density with z. Thus, number counts do not only depend on the cosmological model but also on the evolutionary path that gave birth to galaxies. For this reason, they are used to simultaneously constrain cosmological and evolutionary models, as first done by [501, 502] in the optical bands and by [503] in the radio continuum. It has been later shown that UV and optical number counts might also be affected by the extinction induced by the intergalactic medium which becomes important when the detected sources are at very high redshift [504]. Furthermore, at longer wavelengths (in the infrared and submillimeter domain), deep surveys are soon limited by confusion. All these uncertainties and required corrections have finally made galaxy number counts a quite limited tool for constraining galaxy evolution unless they are combined with other statistical tools for a more complete and coherent vision on the evolutionary process that gave birth to the present galaxy population.

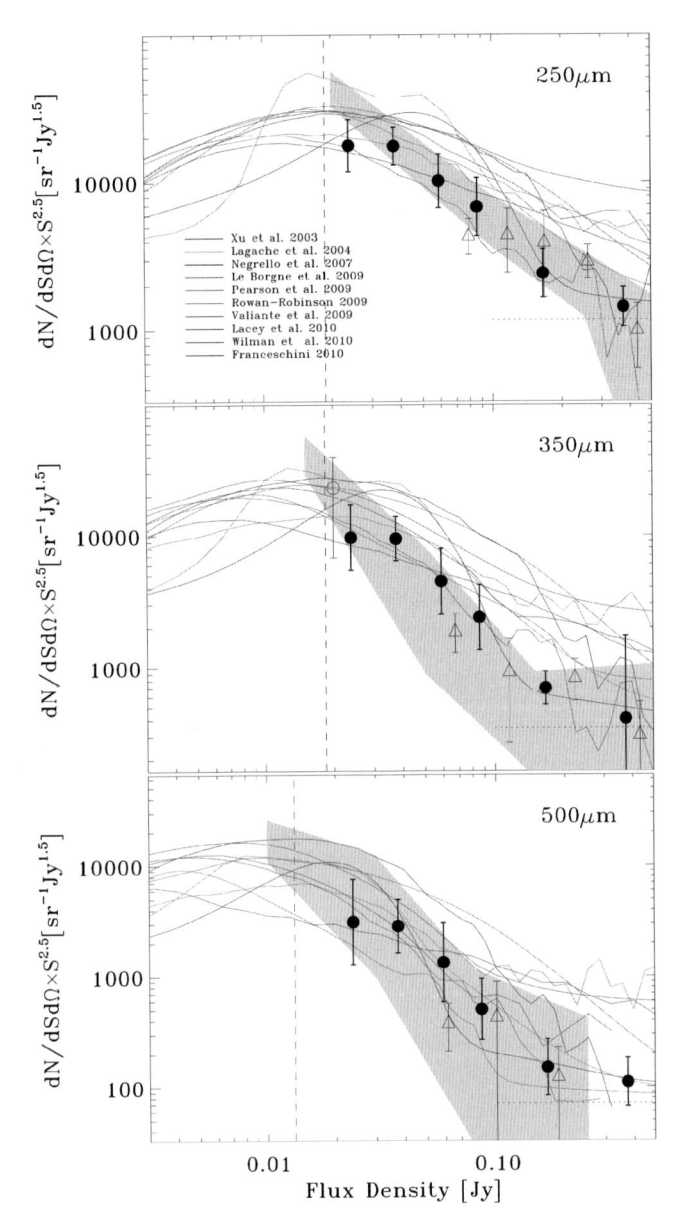

Figure 16.1 The Herschel 250, 350, and 500 μm galaxy number counts compared to different galaxy evolution model predictions, from [507]. Reproduced with permission (c) ESO.

16.1.1
Observed Number Counts

Number counts have been measured at almost all frequencies, from X-rays to centimeter radio wavelengths. AGNs dominate the number counts in the soft (0.5–

2.0 keV) and hard (2–8 keV) Chandra X-ray bands, with a contribution of galaxies (star-forming, starbursts, and ellipticals) increasing at low flux densities [505], as depicted in Figure 16.2. XMM number counts in other X-ray spectral bands have been presented by [506].

Galaxy counts in the UV bands can be found in [508], in the optical bands in [499] (see Figure 16.3), and in the near-infrared in [509]. Deep UV, optical, and near-infrared counts are all consistent with a decrease of the mean star formation activity of galaxies since $z \sim 1$ to the present epoch. Indeed, the contribution of blue galaxies increases at faint magnitudes (e.g., [500]). A stronger evolution than that observed in optical and near-infrared bands is generally observed in the far-infrared or in the submillimeter domain (e.g., [507] and references therein), where the emission is dominated by starburst and star-forming galaxies (see Figure 16.1). A pure evolution in density seems to be excluded by the 24 μm Spitzer number counts, which rather indicate a combined increase of density and luminosity [510].

A flattening of the number counts is observed in the radio continuum at 20 cm (1.4 GHz) for flux densities smaller than a few mJy [511], as shown in Figure 16.4. It has also been shown that the radio population above $\sim$ 50 μJy is composed principally of AGNs (50–60%), star-forming galaxies (30–40%), and QSOs ($\sim$ 10%) [512] (Figure 16.5). Galaxy counts at 1.4 GHz and at other radio frequencies can be found

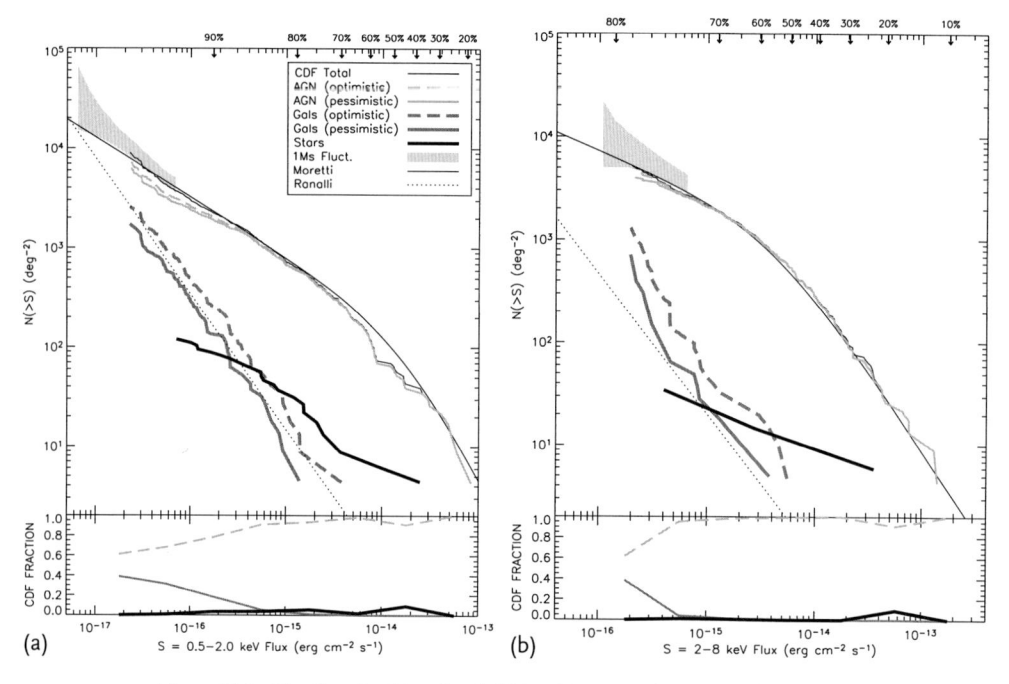

Figure 16.2 The Chandra (a) soft and (b) hard X-ray number counts for AGNs (solid and dashed thin gray curves), galaxies (solid and dashed dark gray curves) and stars (thick solid black curves), from [505]. Reproduced by permission of the AAS.

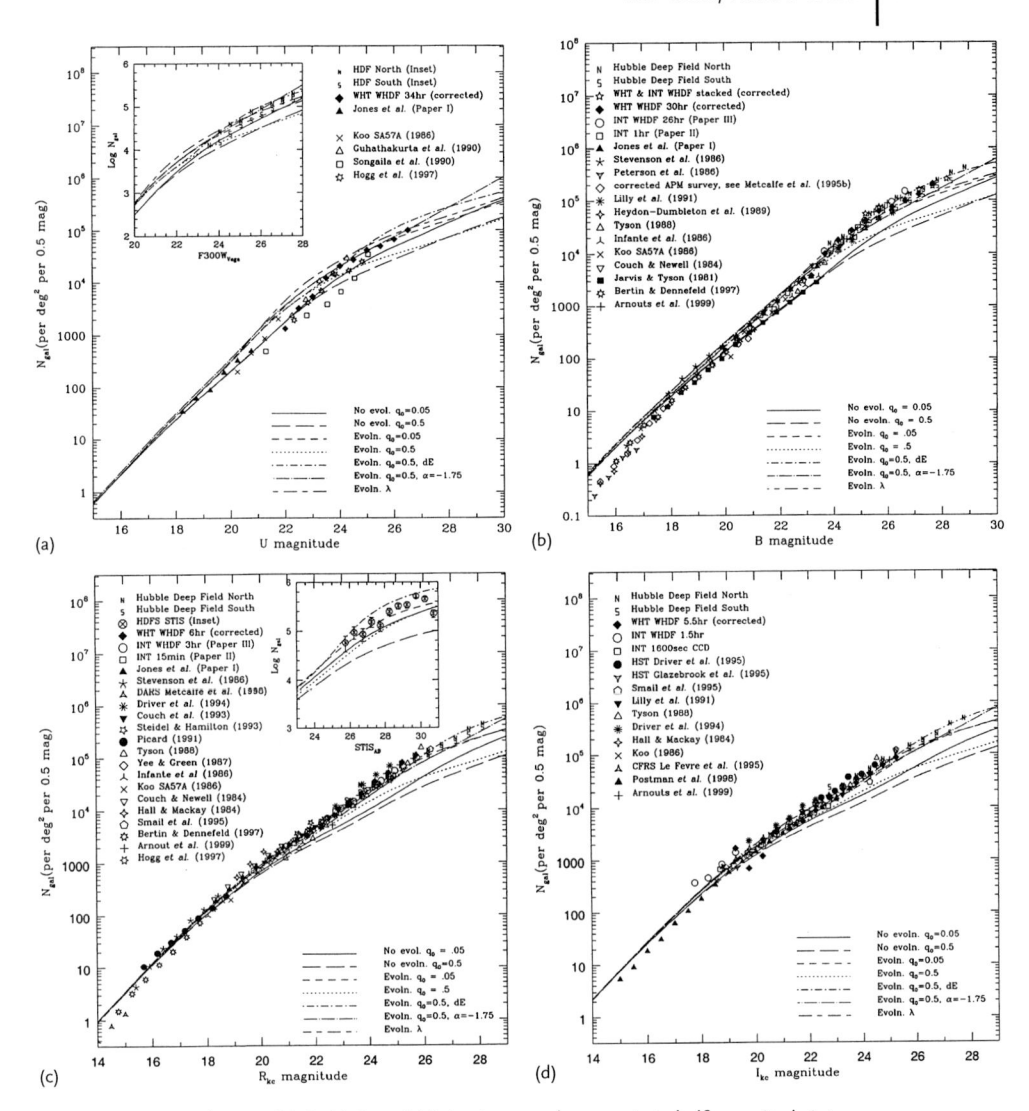

Figure 16.3 Optical (a) U; (b) B; (c) R; and (d) I galaxy number counts in half-magnitude intervals from different sources in the literature (symbols) compared to the predictions of different models (lines), from [499]. Reproduced with the permission of John Wiley & Sons Ltd.

in [513] (15 GHz), [514, 515] (8.5 GHz), [516] (5 GHz), [517] (1.4 GHz), and references therein, or in the recent review of [518].

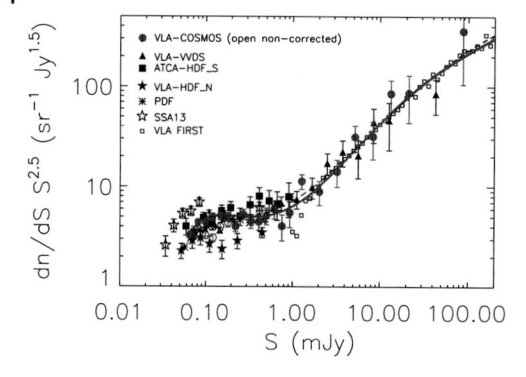

Figure 16.4 The radio source counts at 1.4 GHz from the VLA-COSMOS survey (filled dots) compared to that of other surveys, from [519]. A flattening in the number counts is observed for flux densities $S \leq 2$ mJy. Reproduced by permission of the AAS.

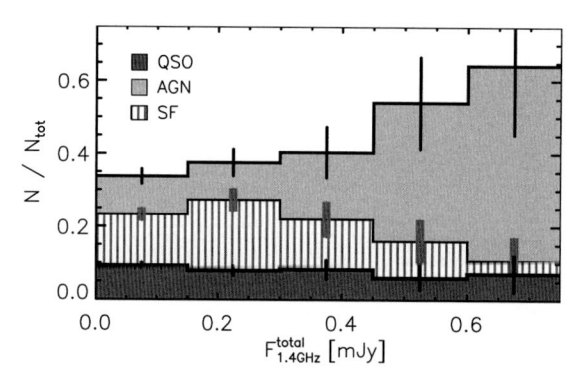

Figure 16.5 The relative contribution of star-forming (SF), AGNs, and QSOs to the radio number counts at 20 cm (1.4 GHz), from [512]. Reproduced by permission of the AAS.

16.2
Luminosity Function

The luminosity function is a statistical function that gives the number of objects per unit of comoving volume within a given interval of luminosity ($[L, L + dL]$) or absolute magnitude ($[M, M + dM]$). When dark matter halo, stellar, gas, or dust masses are considered, the luminosity function is called a mass function. The mass of the different galaxy components are determined as indicated in the previous sections. In particular, the stellar mass is generally computed after fitting the spectral energy distribution of the detected galaxies and considering the possible observational biases for the different galaxy populations (i.e., star-forming vs. quiescent, [520–524]). Given that the formalism for determining luminosity and mass functions is the same, they are hereafter treated without distinction.

Luminosity and mass functions play a major role in extragalactic astronomy and observational cosmology since they are generally used to compare, in a statistical sense, the observed and predicted galaxy distributions within the universe [525,

526]. They are also of fundamental importance for interpreting source number counts (see Section 16.1) and galaxy clustering (Chapter 19), to test theories of galaxy formation (e.g., [527]), or to measure the variation of the cosmic star formation density as a function of redshift (see Section 16.3.1).

The determination of the luminosity function requires complete catalogues down to a given flux density limit with measured distances for all the detected sources. Optical spectroscopy and photometric redshifts (see Section A.1) are generally used for measuring the distance of all the extragalactic sources detected in ultraviolet, optical, and near-infrared blind surveys. In the X-ray, the far-infrared, the submillimeter, and the centimeter radio domains the identification of the detected sources and the measure of their distance is far more complicated, not only by confusion problems (because of their extended emission, detected sources might have several optical counterparts) but also because of the very heterogeneous nature of the emitting sources. Typical examples of hardly identifiable sources are the Ultra Luminous Infrared Galaxies and submillimeter galaxies (SMG) whose high dust content combined with a very active star formation activity makes them extremely bright objects in the infrared but highly extinguished sources in the optical or ultraviolet bands, or radio-loud quasars frequent in deep radio surveys and characterized by a point-like optical aspect similar to that of stars. This difficulty is often overcome by constructing the spectral energy distribution of all the possible counterparts and comparing it to that of different representative objects. For measuring a possible galaxy evolution with z, luminosity functions must be determined in well defined rest frame bands. Observed magnitudes or flux densities must thus first be transformed into rest frame values using appropriate K-corrections which might strongly depend, at least in particular bands, on the nature of the emitting source.

Luminosity functions can be well determined only when the statistics are significant, with more than ≥ 50 objects. To have a sufficient number of sources per bin of luminosity, volumes of important depth are often used. Given the obvious strong relationship between the limiting absolute magnitude detectable in a flux limited survey and the redshift shown in Figure 16.6, corrections for completeness are required. Indeed, a faint galaxy of absolute magnitude M drops out of the survey for distances $z > z_{max}$.

To correct for this bias, several recipes have been proposed in the literature. The most widely used is the $1/V_{max}$ method first presented by [528] to measure the luminosity function of radio quasars. This method is based on the idea that a galaxy of a given luminosity can be observed only within a fraction of the sampled volume if $\min(z_l, z_{min,i}) < \max(z_{max,i}, z_u)$, where z_l and z_u are the lower and upper cut-offs for the redshift range where the luminosity function is determined, and $z_{min,i}$ and $z_{max,i}$ the lower and upper redshift limits within which a galaxy of absolute magnitude M_i can be detected in the survey [529][1]. The contribution of this galaxy to the corrected luminosity function is then calculated assuming a statisti-

---

1) A minimum redshift $z_{min,i}$ must be considered whenever the survey samples only galaxies of apparent magnitude $m_i > m_l$, as is generally the case in deep surveys not appropriate to deal with nearby, extended sources.

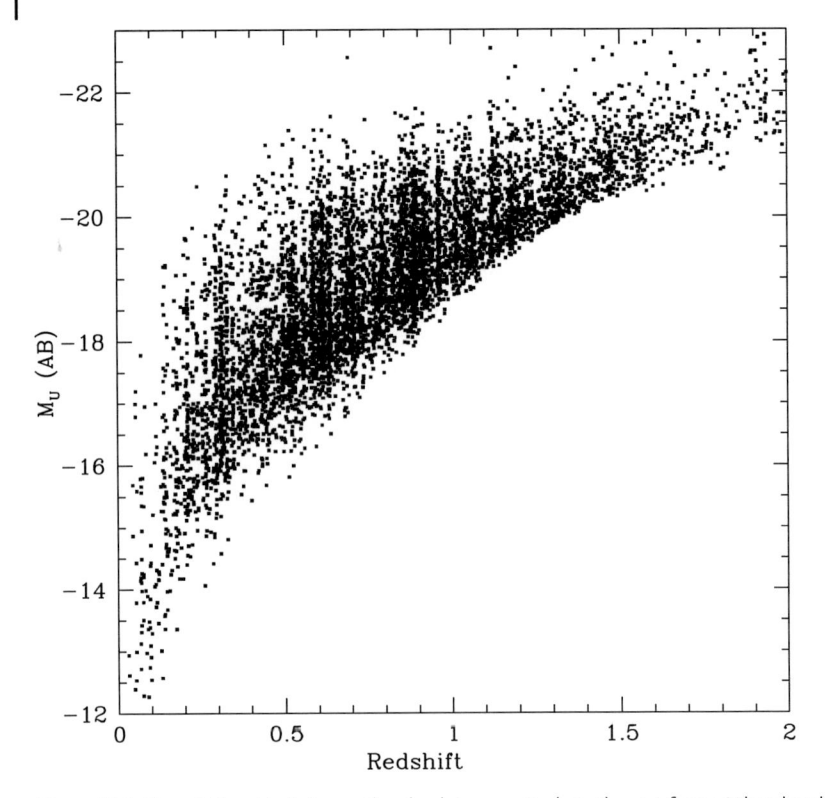

Figure 16.6 The relationship between the absolute magnitude in the rest frame U band and the redshift for galaxies in the VVDS-F02 field, kindly made available by O. Ilbert.

cal weight $V_{tot}/V_{max,i}$, where V_{tot} is the total volume of the survey and $V_{max,i}$ is the volume within which the galaxy can be observed. This correction can be important whenever the depth of the sampled volume is not infinitesimal, as is generally the case [528, 530]. This method is valid only with the assumption that galaxies are homogeneously distributed within the sampled volume, and does not require a parametric form of the luminosity function. Alternative techniques used to measure the luminosity function of galaxies can be found in [529]. Among these, we remind the C^+ method proposed by [531] and [532], a non parametric estimator of the luminosity function which has the advantage of not requiring a uniform distribution of galaxies, but with the main limit that it can only measure the shape of the luminosity function but not its absolute normalization. The stepwise maximum likelihood (SWML, [533]) and the STY [534] estimators are both based on the maximum likelihood method which assumes a given form for the luminosity function and estimates the probability that a galaxy can be observed in the absolute magnitude range $M_{faint} > M > M_{bright}$. The estimated luminosity function is the one that maximises this probability. As for the C^+ method, these two estimators, which differ from each other in the parametric form of the assumed luminosity functions, cannot be used for determining the absolute normalization but have the advantage of not requiring any assumption on the distribution of galaxies within the sampled volume. Different weights can be used for galaxies with spectroscopic and photometric redshifts [535]. Corrections should also be adopted to take into account the fact that, because of their different spectral energy distributions and thus adopted K-corrections, not all galaxy types are visible in the same absolute magnitude range, particularly at high redshifts when K-corrections become important [536].

Different techniques are used for determining the galaxy luminosity function in clusters, where the difficulty is that of selecting cluster members in the absence of spectroscopic redshift, while no corrections for completeness are required, since all galaxies are at the same distance [537]. Photometric redshifts generally do not have a sufficient resolution (~ 0.03–$0.05z$) to unambiguously identify cluster objects, because the typical velocity dispersion of a cluster is on the order of 1000 km s^{-1}. They are thus used to preselect potential candidates up to redshift $z \geq 1$ [538]. From these, cluster members are identified according to their position on the color–magnitude relation (cluster early-type galaxies, the dominant population in high-density environments, form a very well defined color–apparent magnitude relation given their similar distance). The cluster luminosity function can also be determined by applying statistical corrections when the possible contribution of background and foreground objects in the sampled volume is estimated using the field luminosity function or observed number counts in an adjacent field of similar depth [539].

In the nearby universe, where the foreground contamination is negligible, the background contribution to the number of sources per bin of luminosity detected in the field of the cluster can be statistically subtracted using number counts [540]. This, however, requires the assumption that the cosmic variance is not affecting the cluster background. A different technique is the completeness-corrected method, initially proposed by [541], based on the assumption that the spectroscopic sample, composed of spectroscopically identified members, is representative of the entire cluster. That is, that the fraction of cluster members is the same in the incomplete spectroscopic sample as in the complete photometric sample. This method is valid with the (quite strong) assumption that spectroscopic targets have been selected randomly and not according to well defined criteria (color, surface brightness, etc.). In the very local universe (velocity $\leq 3000 \text{ km s}^{-1}$), surface brightness criteria can be used to identify cluster members, as successfully done by Sandage and collaborators in the Virgo cluster [542].

16.2.1
Parametrization of the Luminosity Function

Luminosity functions in the ultraviolet, optical, and near-infrared bands are generally parametrized using the Schechter luminosity function, an analytical function composed of a Gaussian to represent the luminosity distribution of massive galaxies and an exponential for the dwarf galaxy population. The Schechter luminosity function is defined as [543]

$$\phi(L)dL = \phi^* \left(\frac{L}{L^*}\right)^\alpha \exp\left(-\frac{L}{L^*}\right) d\left(\frac{L}{L^*}\right) \qquad (16.6)$$

where the parameters ϕ^*, L^*, and α are determined by fitting the data. They are the number of galaxies per unit volume, the characteristic luminosity at which the luminosity function exhibits a rapid change in the slope, and the slope of the exponential distribution at its faint end, respectively. The luminosity function, expressed as a function of the absolute magnitude, is given by the relation [529]

$$\phi(M)dM = 0.4\ln 10\phi^* 10^{-0.4(\alpha+1)(M-M^*)} \exp[-10^{-0.4(M-M^*)}]dM \qquad (16.7)$$

where M^* is the characteristic absolute magnitude. It is worth mentioning that α and L^* or M^* are not independent parameters. Alternative parametrizations of the luminosity function, sometimes more appropriate for representing the distribution of galaxies in particular bands, can be found in the literature. In the infrared domain, for instance, the Schechter luminosity function is not adapted to account for the observed excess of bright galaxies. For this reason, in this spectral domain the luminosity function is fitted with a double exponential profile parametrized as follows [510, 544, 545]:

$$\phi(L) = \phi^* \left(\frac{L}{L^*}\right)^{1-\alpha} \exp\left\{-\frac{1}{2\sigma^2} \log_{10}^2\left[1 + \left(\frac{L}{L^*}\right)\right]\right\} \qquad (16.8)$$

16.2.2
Luminosity Distributions and Bivariate Luminosity Functions

The luminosity function in a given band can be determined only by considering all the sources extracted from a blind survey in that band down to a limiting magnitude or flux density. Counting the number of sources in different luminosity bins in a given band but selected in another band only provides a simple luminosity distribution unless the different biases induced by the various selection criteria are correctly considered using complex statistical tools (bivariate luminosity function, [546, 547]). This is, for instance, the case when optically selected galaxies are counted in different bins of radio continuum or X-ray luminosities. Bivariate luminosity functions are used to consider, with a correct weight, both detected and undetected sources [548, 549]. Following the prescription of [548], originally applied to derive the fractional X-ray distribution function of optically selected galaxies, the differential probability distribution $f(P)$ that galaxies develop a source of a given luminosity P, taking into account the number of detected objects $n_d(P_k)$ in each bin of luminosity P_k, as well as the upper limits, is given by the relation

$$f(P_k) = \frac{n_d(P_k) \times \left(1 - \sum_{j=1}^{k-1} f(P_j)\right)}{n_u(P_l < P_k) + n_d(P \leq P_k)} \qquad (16.9)$$

where $n_d(P \leq P_k)$ is the number of detected objects with $P \leq P_k$, while $n_u(P_l < P_k)$ is the number of undetected objects with $P_l < P_k$. The uncertainty on $f(P_k)$ is given by

$$\sigma f(P_k) = \frac{f(P_k)}{\sqrt{n_d(P \geq P_k)}} \tag{16.10}$$

while the cumulative distribution is given by

$$F(\geq P_k) = \sum_{j=1}^{k} f(P_j) \tag{16.11}$$

Total or normalized luminosities P must be used in the determination of the differential and cumulative probability distributions. An example of an application of this method can be found in [550] for the determination of the fractional radio distribution function of optically selected galaxies in the Virgo cluster.

16.2.3
The Observed Luminosity Functions

Luminosity functions at different wavelengths and in different intervals of z, from the local to the high redshift universe, have been extensively presented and discussed in the literature. Due to lack of space, only some of the most representative results are listed here. The determination of the hard (2–10 keV) X-ray luminosity function of AGNs has been presented in [551], and that of normal star-forming galaxies is in [552–554]. All authors find a mild evolution with redshift up to $z \sim 3$ for the AGNs and $z \sim 0.8$ for the star-forming galaxies, witnessing a luminosity-dependent density evolution consistent with a picture where AGNs dominate the X-ray background. At the same time, they show the decrease in star formation activity also observed at optical and infrared wavelengths.

The luminosity function in the GALEX UV bands (1529 and 2316 Å) has been presented in [555] (see Table 16.1) for the local universe and in [558] for redshift up to $z = 1.2$. The largest number of luminosity functions has been determined in the optical bands. Among these, it is certainly worth mentioning the luminosity function in the five u, g, r, i, z photometric bands determined using SDSS data for $\sim 150\,000$ galaxies [20] at $z = 0.1$, shown in Figure 16.7, whose Schechter parameters are given in Table 16.1. Although these values are generally taken as reference, they have been significantly revised by [557] to account for the presence of very low surface brightness galaxies detected in the very local universe ($z < 0.1$), whose contribution becomes dominant at very low luminosities. If fitted using a Schechter luminosity function with two exponential components, the presence of dwarf, low surface brightness galaxies increases the slope of the low luminosity exponential component from $\alpha \sim -1.0$ to $\alpha \sim -1.5$ [557], thus getting closer to the value expected from cosmological models for the dark matter halo mass function ($\alpha \sim -2$). They have also shown that the faint end of the luminosity

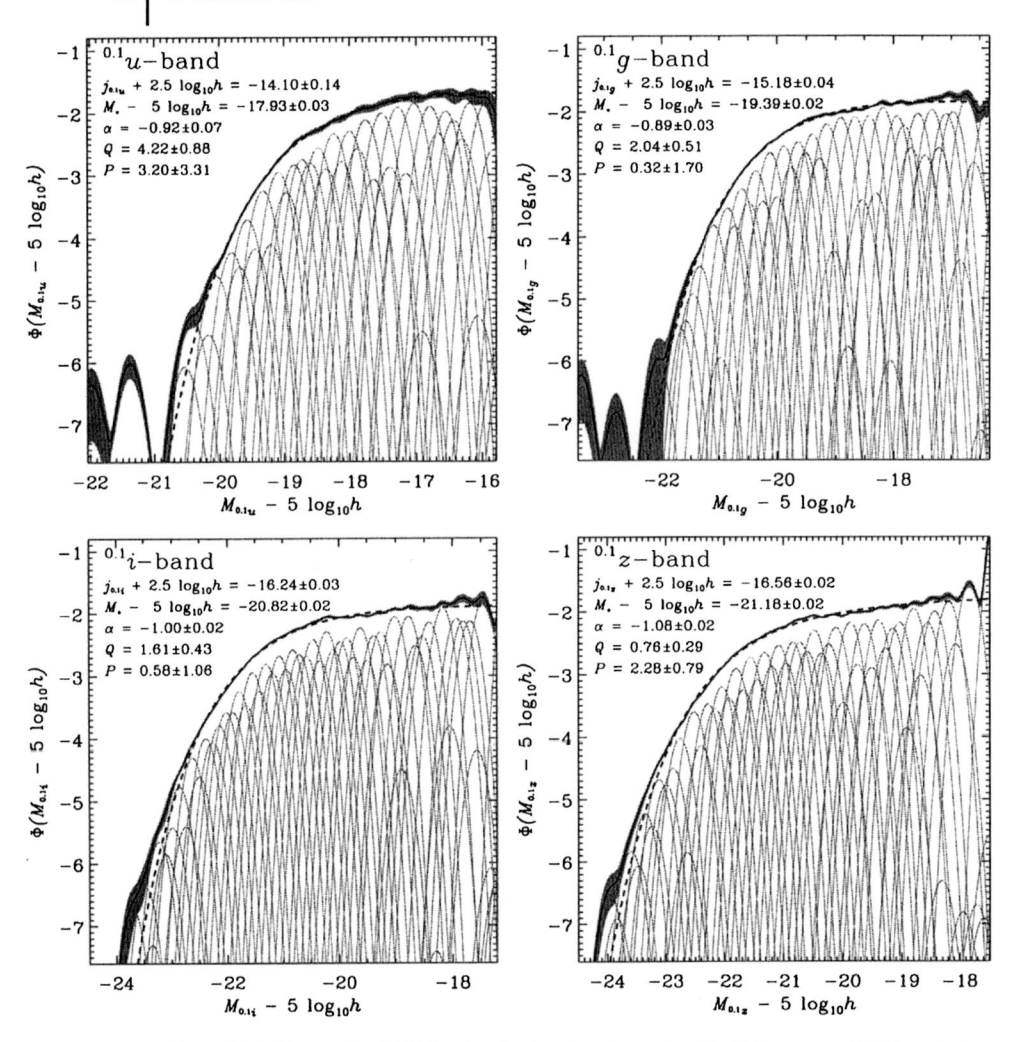

Figure 16.7 The u, g, i, z SDSS luminosity function determined by [20] using $\sim$ 150 000 galaxies at $z = 0.1$, whose Schechter parameters are given in Table 16.1. Reproduced by permission of the AAS.

function of field galaxies is dominated by blue, star-forming objects while red, quiescent objects are the main population at high luminosities, as depicted in Figure 16.8. This color cut only roughly corresponds to a morphological separation into spirals and irregulars (blue), and early-type ellipticals and lenticulars (red). Optical spectroscopic luminosity functions at higher redshift have been presented by [535, 559, 560]. Concerning active galaxies, the SDSS spectroscopic survey has shown that, in the local universe, Seyfert 1 and 2 are equally frequent, while Seyfert 1 outnumber Seyfert 2 at high redshift [241].

In the near-infrared, local luminosity functions obtained with the 2MASS survey have been presented by [556, 561], and at large redshifts by [522, 562]. In the

Table 16.1 The UV, optical, and near-infrared Schechter luminosity functions for galaxies in the nearby universe.

Band	ϕ^* ($\times 10^{-2}\,h^3\,\mathrm{Mpc}^{-3}$)	M_*	α	$\Omega_0, \Omega_\Lambda, H_0$ (km s^{-1} Mpc^{-1})	Ref.	Notes
$FUV(1529\,\text{Å})$	0.43 ± 0.06	-18.04 ± 0.11	-1.22 ± 0.07	0.3, 0.7, 70	[555]	
$NUV(2316\,\text{Å})$	0.55 ± 0.06	-18.23 ± 0.11	-1.16 ± 0.07	0.3, 0.7, 70	[555]	
u	3.05 ± 0.33	-17.93 ± 0.03	-0.92 ± 0.07	0.3, 0.7, 100	[20]	a
g	2.18 ± 0.08	-19.39 ± 0.02	-0.89 ± 0.03	0.3, 0.7, 100	[20]	a
r	1.49 ± 0.04	-20.44 ± 0.01	-1.05 ± 0.01	0.3, 0.7, 100	[20]	a
i	1.47 ± 0.04	-20.82 ± 0.02	-1.00 ± 0.02	0.3, 0.7, 100	[20]	a
z	1.35 ± 0.04	-21.18 ± 0.02	-1.08 ± 0.02	0.3, 0.7, 100	[20]	a
J	1.04 ± 0.16	-22.36 ± 0.02	-0.93 ± 0.04	0.3, 0.7, 70	[556]	
K_S	1.02 ± 0.15	-22.38 ± 0.03	-0.95 ± 0.04	0.3, 0.7, 70	[556]	

^{a} $\alpha \sim -1.5$ are determined when considering the low surface brightness galaxy population [557].

near-infrared, the local luminosity function corresponding to the deepness of the 2MASS survey has a relatively flat slope, as indicated in Table 16.1.

All these ultraviolet, visible, and near-infrared luminosity functions show a stronger evolution with z in the blue than in the red bands, consistent with a rapid decrease of the mean star formation activity of galaxies in the last few billions of years ($z \leq 1$). The interpretation of these results in terms of galaxy evolution and their connections with the formation of the red sequence through merging sequences are extensively discussed in [524, 560]. A slope $\alpha \sim -2.0$ is the value expected in a λCDM hierarchical structure formation for the dark matter halos mass functions [525, 526]. Luminosity functions can be converted into stellar mass functions or directly compared to those deduced from the dark matter halo mass functions using semianalytical models of galaxy formation, as shown in Figure 16.9. Clearly, the observed knee in the mass function can only be reproduced by introducing the effect of feedback from AGNs and supernovae discussed in Chapter 18, both required to suppress the star formation activity in high- and low-mass objects, respectively.

The UV, optical, and near-infrared luminosity functions of galaxies in nearby clusters have generally revealed a relatively steeper slope than in the field, as extensively discussed in [10]. At these wavelengths, the steepening at low luminosities is principally due to dwarf ellipticals whose contribution becomes dominant for absolute magnitudes $M_B \geq -17$, as elegantly shown in the beautiful work of Sandage and collaborators [537, 542] (see Figure 16.10). The evidence of a steeper slope in the optical luminosity function of cluster galaxies with respect to the field, however, has been recently questioned by [563], who remarked that once low surface brightness galaxies are considered, as generally done in nearby cluster surveys, the slope of the fitted Schechter functions are $\alpha \sim -1.5$ in both high and low density regions [557].

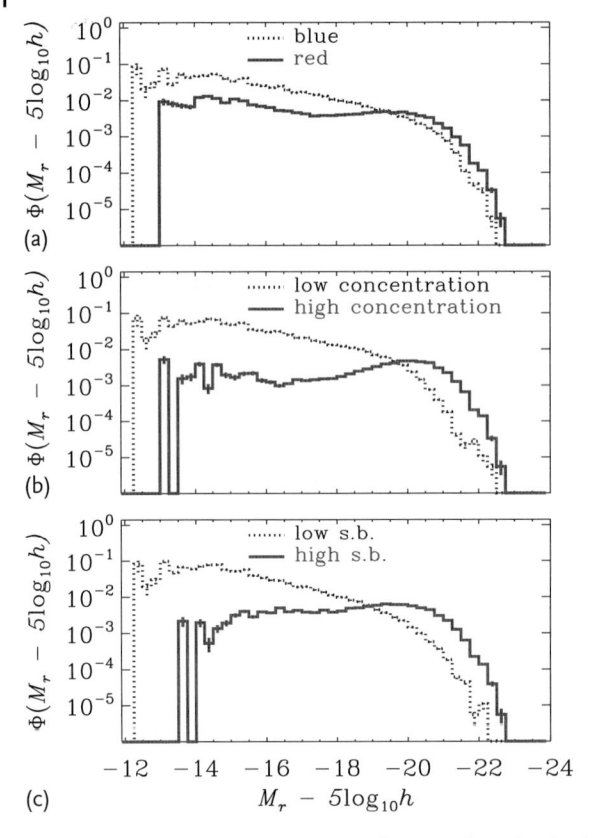

Figure 16.8 The r band SDSS luminosity function of very local galaxies of different (a) color; (b) light concentration, and (c) surface brightness, determined by [557]. Reproduced by permission of the AAS.

The local infrared luminosity function has been determined using IRAS data by [544, 564, 565] and later revisited by [545]. At high redshifts ($z \sim 1$), the infrared luminosity function has been determined using Spitzer data at 24 µm by [510, 566]. The comparison with the local universe revealed a strong evolution, consistent with a decrease of the star formation activity of infrared galaxies after $z \sim 1$. Indeed, the high redshift universe ($0.5 < z < 1$) is dominated by luminous infrared galaxies with luminosities $10^{11} L_\odot \leq L_{IR} \leq 10^{12} L_\odot$, the probable progenitors of local massive spirals.

The local radio continuum luminosity function at 1.4 GHz (20 cm) has been first determined by [567] and later revised by [173, 568–571], and has been determined at other frequencies by [572] (20 GHz) and [573] (95 GHz). As clearly depicted in Figure 16.11, the radio continuum luminosity function at 1.4 GHz of extragalactic radio selected sources has two different components: a flat component due to AGNs which dominates at radio luminosities $L_{radio}(1.4\,\text{GHz}) \geq 10^{23.2}\,\text{W Hz}^{-1}$, and a steep component due to which normal star-forming galaxies dominate at lower

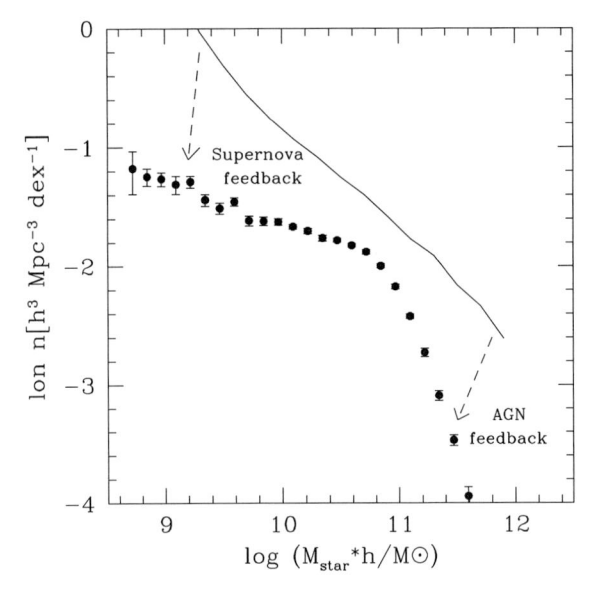

Figure 16.9 The dark matter halo mass function deduced from λCDM cosmological simulations scaled to the baryonic mass (solid line) is compared to the observed stellar mass function of [463] (filled dots). To reproduce the observed knee in the stellar mass function, star formation must be suppressed at high and low luminosities. In semianalytical models, the suppression of the star formation is regulated by the feedback of AGNs (for high masses) and supernovae (low masses), as described in Chapter 18 (courtesy of A. Cattaneo).

radio luminosities [518, 570]. A mixture of AGNs and star-forming galaxies should also dominate the radio continuum luminosity function at high redshifts.

16.3
Luminosity Density

The luminosity density, generally indicated as ρ_L, is the integrated luminosity of galaxies and can be directly measured by integrating the luminosity function down to a limiting magnitude L_{lim}. This value corresponds to the observational limit or to the extrapolated value of the survey when the non parametric or parametric luminosity functions are used, respectively. The luminosity density is quite an important parameter in the study of the global evolution of the galaxy population with cosmic time since it can be easily transformed into star formation densities whenever the observed bands are directly related to the emission of the young stellar components (using the recipes given in Chapter 13) or in stellar mass densities when the adopted filters are sensitive to the bulk of the stellar component. In the

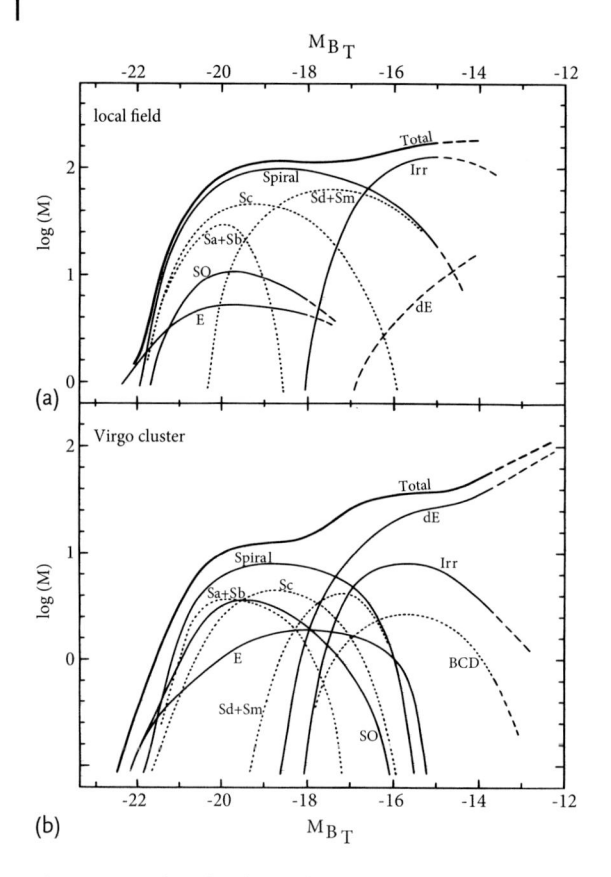

Figure 16.10 The B band optical luminosity function of (a) field and (b) Virgo cluster galaxies, adapted from [537]. The slope of the faint end of the luminosity function is steeper in the Virgo cluster than in the field (however, see text). The optical luminosity function is dominated by late-type spirals and irregulars in the field and by early-type ellipticals and lenticulars as well as dE and dS0 in the cluster. Reprinted with permission from the Annual Review of Astronomy & Astrophysics, Volume 26 (c) 1988 by Annual Reviews, http://www.anualreviews.org.

nonparametric case, ρ_L is given by the relation

$$\rho_L = \sum_{i=1}^{Nb} L_i \phi_i dL_i \tag{16.12}$$

where Nb is the number of bins of luminosity of range dL_i, L_i is the mean luminosity of each bin, and ϕ_i is the estimated value of the nonparametric luminosity function within the bin i. In the case of a Schechter luminosity function, the luminosity density is given by the relation [543]

$$\rho_L = \int_{L_{\lim}}^{\infty} L\phi(L)dL = \phi^* L^* \Gamma\left(\alpha + 2, \frac{L_{\lim}}{L^*}\right) \tag{16.13}$$

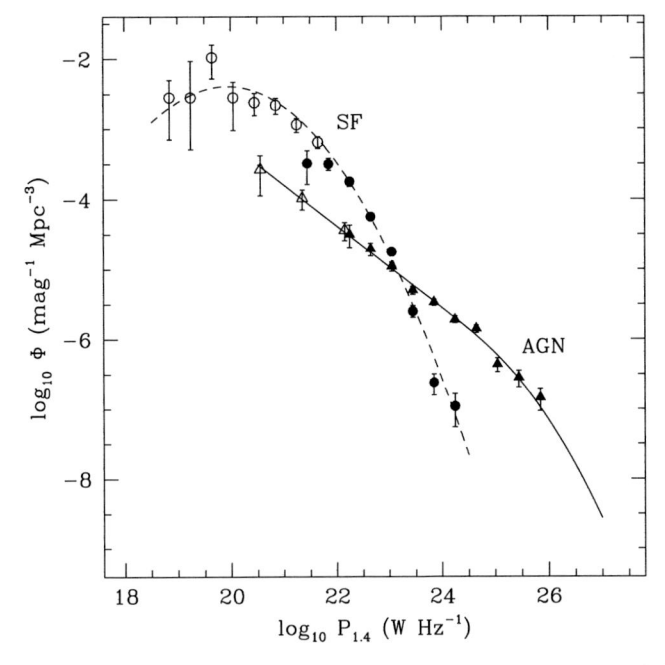

Figure 16.11 The local 1.4 GHz radio continuum luminosity function for AGNs (triangles) and star-forming galaxies (circles) obtained by [570] (filled symbols and solid lines) and [568] (empty symbols and dotted lines), from [570]. Reproduced with the permission of John Wiley & Sons Ltd.

where Γ is the gamma function. Obviously, since surveys generally do not completely sample the faint end of the luminosity function, the luminosity density determined with nonparametric functions only gives a lower limit of the real luminosity density.

16.3.1
The Cosmic Star Formation History and Build Up of the Stellar Mass

Luminosity densities can be converted into physical densities using the recipes given in the Part Two. Measured for different ranges of redshift, luminosity functions can thus be used to trace the variation of some important physical parameters per unit of comoving volume through the cosmic time. Among these, the most widely determined is the variation of the star formation density, ρ_{SFR}, with look-back time, also referred to as the cosmic star formation history (e.g., [574–576]). As discussed in Chapter 13, an accurate determination of the star formation rate of galaxies can be obtained by combining multifrequency data necessary, for instance, for the correction of dust extinction (rest frame UV and H_α data). At the same time, multifrequency data can provide independent estimates of the star formation density at a given redshift, and make the determination of ρ_{SFR} possible at some redshift where other indicators are not accessible for technical reasons (atmospheric transmission,

(a)

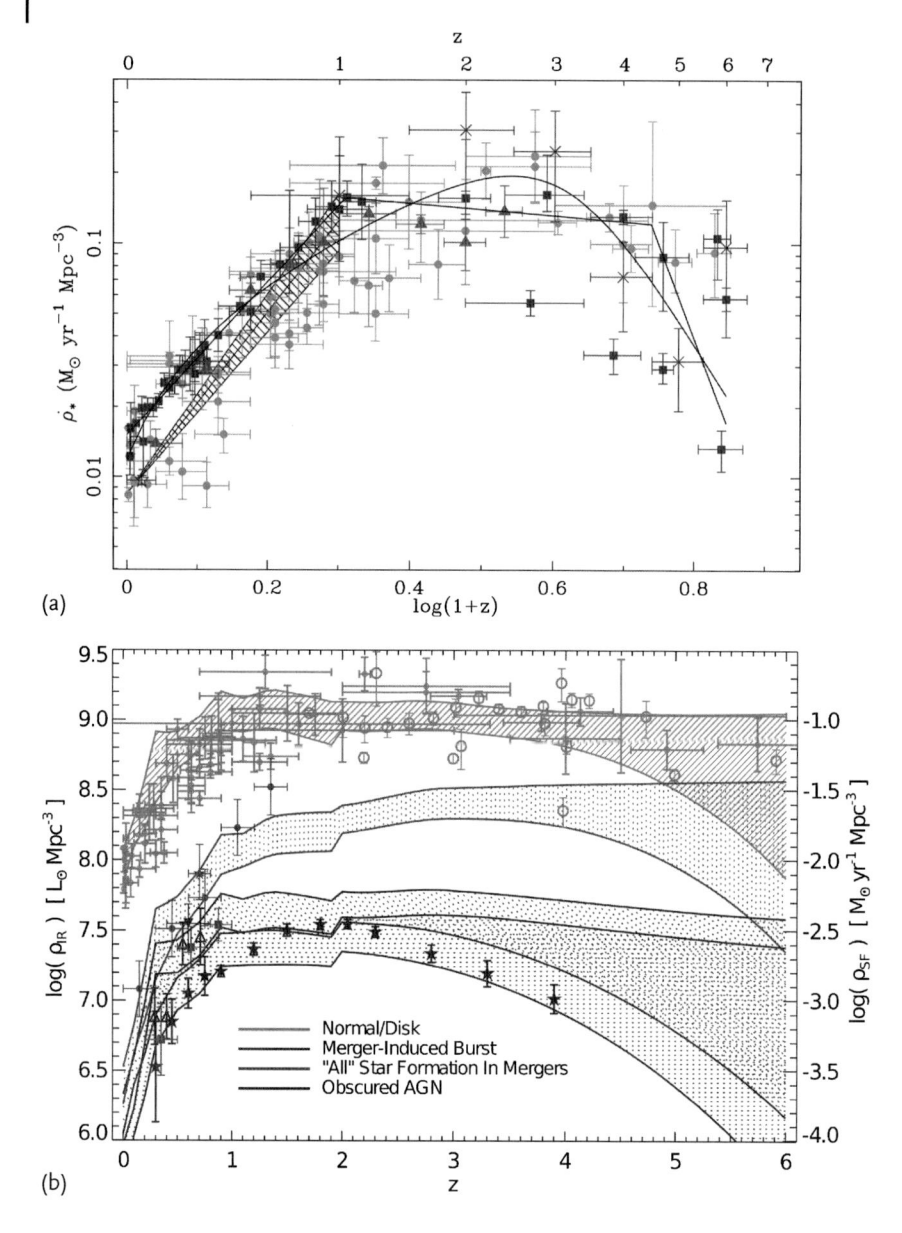

(b)

sensitivity of the detectors). A recent example of the reconstruction of the star formation density of the universe with look-back time, up to $z \sim 6$, can be found in [577, 578] (see Figure 16.12). Star formation densities are determined using a compilation of multifrequency photometric and spectroscopic data, including UV rest frame imaging, H_α and [OII] emission lines, infrared 24 μm, and centimeter radio 1.4 GHz data. For a correct comparison, luminosity densities must be converted into star formation rate densities using the same assumption on the IMF.

◀ **Figure 16.12** (a) Evolution of the star formation rate density, measured in $M_\odot\ \text{yr}^{-1}\ \text{Mpc}^{-3}$, with redshift, from [577]. Star formation rates are determined assuming a Salpeter IMF. Different symbols are used to indicate star formation rates determined using different tracers (1.4 GHz, H_α, far-infrared (24 μm), UV rest frame). Reproduced by permission of the AAS; (b) evolution of the star formation rate density, measured using infrared data, with redshift for different classes of objects, from [579]. Plotted in order of decreasing importance (from top to bottom), are normal disk galaxies, galaxies undergoing a merger-induced bursts (global star formation within the galaxy), and the contribution of the merger to the star formation activity of galaxies undergoing a merging event and obscured AGNs. At all redshifts, the contribution of AGNs and of mergers to the total star formation history of the universe is relatively small. Reproduced with the permission of John Wiley & Sons Ltd.

Using distinct luminosity functions for different classes of objects, [579] have recently tried to quantify the contribution of AGNs, star-forming disks and mergers to the cosmic star formation history of the universe (Figure 16.12).

As for the cosmic star formation rate, stellar mass functions determined at different redshift and luminosity distributions of other entities have been used to trace the evolution with cosmic time of the stellar mass density (e.g., [524] and references therein), of the metal mass density [577] and of the gas density [392]. Among these, the most widely studied is the evolution of the stellar mass density distinctly for quiescent and star-forming galaxies (see Figure 16.13). Determining the cosmic variation of these parameters with time is of particular importance for constraining models of galaxy evolution.

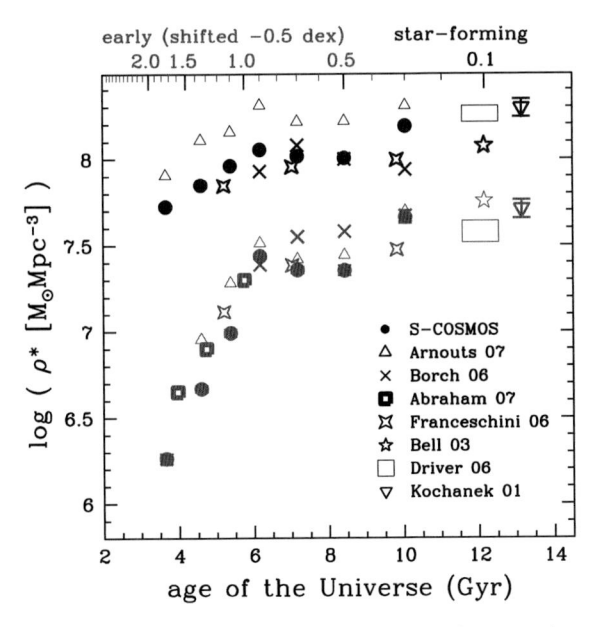

Figure 16.13 Evolution of the stellar mass density, measured in $M_\odot\ \text{Mpc}^{-3}$, with the age of the universe for early-types (lower symbols, shifted vertically by -0.5 dex) and star-forming galaxies (upper symbols) separately, from the compilation of [524]. Reproduced by permission of the AAS.

17
Scaling Relations

Galaxies span a large range in luminosity, from $M_B \sim -23$ in the most massive cD in the center of rich clusters [580] down to $M_B \sim -8$ in the dwarf spheroidals of the local group [581], a range corresponding to $\sim$ a factor of 10^6 in mass. Despite this large dynamic range, galaxies follow several well-defined scaling relations linking a physical, structural, spectrophotometric, or kinematical quantity with a measure of their size, luminosity, or mass. These scaling relations are often interpreted as a clear indication for a common origin of galaxies of different luminosity and thus are generally compared to simulations to constrain models of galaxy formation and evolution [525, 526, 582]. At the same time, they give accurate information on the different relationships between various physical (gas content, star formation), structural (light distribution), spectrophotometric (stellar populations, metallicity), and kinematical (velocity dispersion in ellipticals and rotational velocity in spirals) parameters, making them a powerful tool to study the physical processes acting in galaxies of different luminosity and morphological type [198, 392]. The very low dispersion observed in some of these scaling relations, such as the Tully–Fisher [583] and the fundamental plane [584], make them powerful distance indicators, easily accessible for large samples of galaxies at different redshift. Thus, they are ideal tools for tracing the three-dimensional distribution of the baryonic matter in the Universe.

The most famous scaling relations have been extensively sampled and analyzed in the nearby universe, but some of them are still poorly constrained for galaxies at high redshift. Sometime in the past their use in the study of the intrinsic properties of galaxies has been proposed in an incorrect way, when luminosity-luminosity (or mass) relations have been claimed to prove the physical relationship between two independent variables. Indeed, as ironically shown by [585], it should always be remembered that these luminosity–luminosity or luminosity–mass relationships come from the fact that "bigger galaxies have more of everything."

Scaling relations can be divided into different categories according to whether or not they deal with spectrophotometric, structural, physical, or kinematical entities, although some of them use mixed quantities. Here we describe some of the most famous scaling relations often used in the literature to characterize the properties of different populations of objects. We should however remember that a much larger number of important scaling relations exist in the literature. Some of them

A Panchromatic View of Galaxies, First Edition. Alessandro Boselli.
© 2012 WILEY-VCH Verlag GmbH & Co. KGaA. Published 2012 by WILEY-VCH Verlag GmbH & Co. KGaA.

are an extension or a projection of those listed here. We are forced to omit them for lack of space.

17.1
Spectrophotometric Relations

Spectrophotometric scaling relations link variables directly connected to the stellar emission – such as UV-optical-near-infrared colors, the metallicity, and the present and past star formation activity – to the total stellar or dynamical mass of galaxies. Because the gas is the principal feeder of star formation, here we also describe the gas mass to stellar mass scaling relation. All these scaling relations consistently show that mass is the main parameter that drives galaxy evolution [198, 392, 586]. With the exception of the atomic gas, which will be detectable for high redshift sources once SKA is operational, the data required for determining these spectrophotometric scaling relations are now accessible for large samples of galaxies at different redshifts.

17.1.1
The Color–Magnitude and Color–Color Relations

Galaxies follow well-defined color–magnitude (CMR) and color–color relations. Color–magnitude relations between the integrated colors and the absolute magnitude show that galaxies become redder with increasing luminosity, while color–color relations consistently show that this reddening occurs simultaneously in almost all bands[1]. Given the tight relationship between the color, age, and metallicity of the underlying stellar population, the color–magnitude and color–color relations are strongly related to the evolutionary path that gave birth to galaxies. First discovered for early-type galaxies in the optical bands [187, 587–591], it has been later shown that color–magnitude relations extend to late-type systems [592, 593]. They are present for both galaxy populations in the UV, optical, and near-infrared spectral domain and are valid for objects belonging to high and low density environments [594]. After the advent of all-sky surveys such as SDSS, 2MASS, and GALEX, color–magnitude relations are generally called "sequences," with early-types following a tight "red sequence" and late-type galaxies a dispersed "blue sequence" (or "blue cloud") [37, 595]. Examples of UV, optical, and near-infrared extinction-corrected color–stellar mass relations (where the stellar mass is obtained from the absolute H band magnitude as described in Section 15.1) determined for a homogeneous sample of nearby galaxies are given in Figure 17.1.

Several attempts have been made using population synthesis models to understand the origin of these relations and see whether the observed reddening of colors with increasing luminosity is primarily related to an increase of the metallicity or of the mean age of the underlying stellar populations [464, 596]. In early-type galax-

---

1) The UV upturn observed in early-type galaxies is an exception, see below.

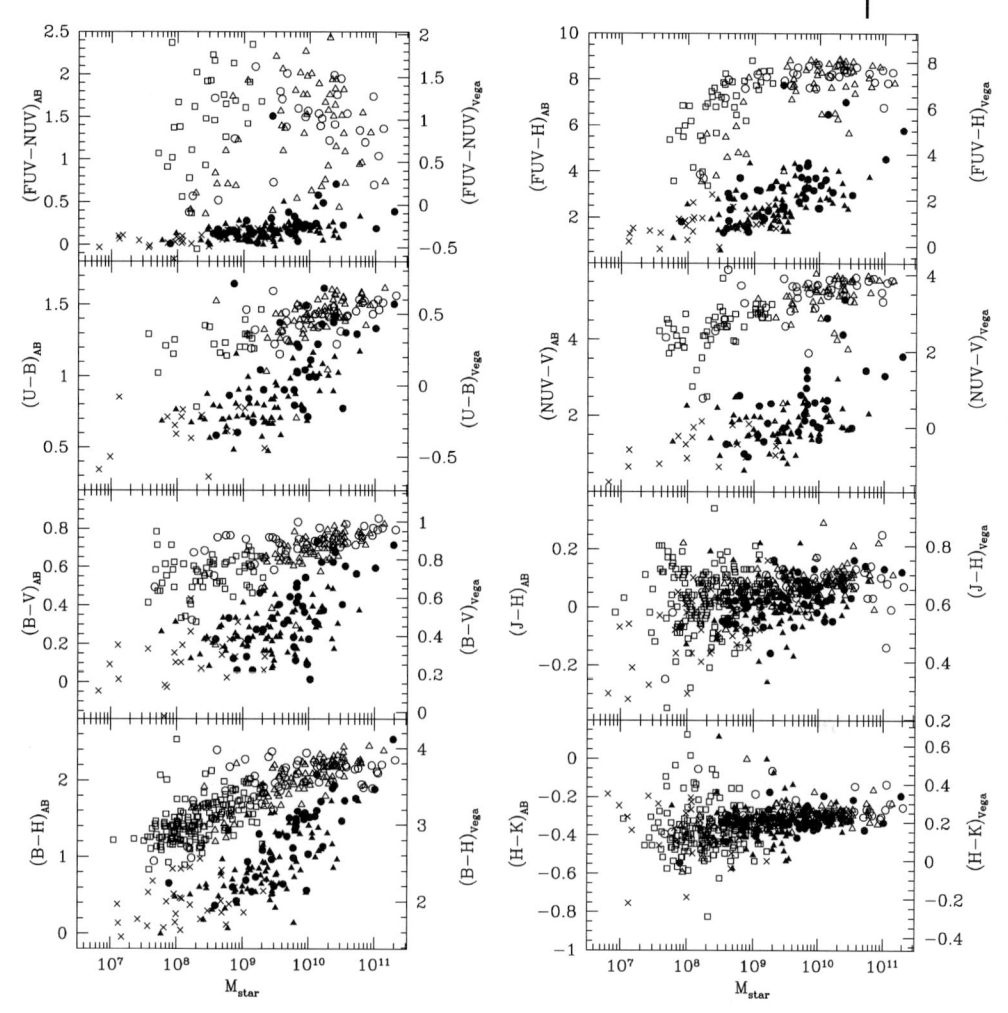

Figure 17.1 The mean extinction-corrected UV, optical, and near-infrared color–stellar mass (in AB and Vega systems) relations of nearby galaxies of different morphological classes (open circles for ellipticals, open triangles for S0-S0a, open squares for dE-dS0, filled circles for Sa-Sbc, filled triangles for Sc-Sd, crosses for Sm-Im-BCD). These relations are equivalent to the color–magnitude relations (CMR), given the tight relation between the H band near-infrared absolute magnitude and the stellar mass, measured here using the recipes given in Section 15.1. *UBVJHK* are in the Johnson photometric system, and *FUV* and *NUV* are in the GALEX ultraviolet bands.

ies, if the most recent results agree in indicating that both metallicity and age are responsible for the observed color–magnitude relations (see [16] for a review), then the much larger dynamic range in the colors observed in late-type galaxies must result from their different star formation histories; i.e., massive galaxies must have formed most of their stars in the past, while dwarf systems are still active at the present epoch [189, 221, 593]. A clear exception to the reddening of colors with increasing luminosity is the inverted color–magnitude relation of elliptical galaxies observed in the far-UV band, where the UV emission is dominated by evolved stellar populations (UV upturn, [78, 181, 597]).

It is worth mentioning that the tightness of the color–magnitude relation observed in early-type galaxies, whose scatter is close to the observational errors, suggests that it is a possible distance indicator. For this reason, CMRs are often used to infer the membership of early-type galaxies to rich clusters (see Section 16.2). Color–magnitude relations are also used as strong observational constraints for semi-analytical models of galaxy evolution [598, 599] or to reconstruct the different star formation histories of cluster and field galaxies at different redshifts [9, 563, 600–603]. Indeed, CMRs are the clearest evidence that in both early- [189, 294] and late-type [198, 392, 393, 586, 604] populations, mass is the principal driver of galaxy evolution. While massive galaxies have formed most of their stars in the past (and thus have red colors at the present epoch), low-mass systems are still forming stars at the present epoch and thus have blue colors[2]. This observational evidence is one of the different manifestations of the downsizing effect. The scatter in the color–magnitude or color–color relations has been interpreted as evidence of recent bursts that perturbed the relatively smooth star formation activity of galaxies along their life. Significant variations of the optical colors, however, are only possible when the starburst episode is a major event in the star formation history of the galaxy, as clearly shown in the seminal work of [195]. The UV bands, where the emission is dominated by young stellar populations, are only moderately more sensitive to recent variations of the star formation activity of galaxies [605]. Abrupt variations of the star formation activity of galaxies are probably present in low luminosity, star-forming systems, where the activity is dominated by a few, giant HII regions, or in major gas-rich mergers, when most of the gas is rapidly transformed into stars. For this reason, the larger scatter observed in the color–magnitude relations of isolated early-type galaxies compared to that of cluster objects has often been used to infer an older formation of elliptical galaxies in high density environments [16].

17.1.2
The Mass–Metallicity Relation

Galaxies have a metal content frozen into the gaseous component that increases with increasing luminosity or stellar mass. This relation is of paramount importance to the study of galaxy evolution because it relates two fundamental parameters: the stellar mass, which traces the amount of gas locked up in stars, and the metallicity, which reflects the fraction of gas that has been reprocessed during stellar evolution. The metal content is particularly sensitive to internal (AGN and supernovae feedback) and external (gas inflow) processes, often proposed by cosmologists to explain the difference between the observed and the expected statistical distribution of galaxies. A typical example is the relative lack of bright and

---

2) There is, however, for the lowest mass systems, some degeneracy due to a lower average metal content of the stars and thus a lower line blanketing in the blue.

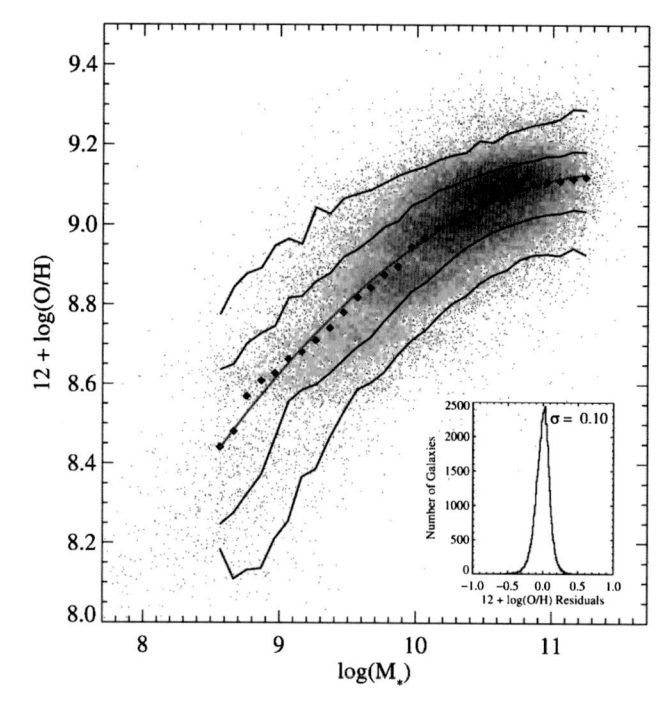

Figure 17.2 The stellar mass–metallicity relation determined using SDSS data for ∼ 53 000 galaxies by [281]. The large black filled diamonds represent the median value in bins of 0.1 dex in stellar mass, while the solid lines represent the contours that enclose 60 and 95% of the data. The solid line on the black diamonds shows the polynomial fit to the data. Reproduced by permission of the AAS.

dwarf galaxies observed in the local luminosity function with respect to those determined from the distribution of dark matter halos and predicted by semianalitycal models of galaxy formation [598, 606–608].

The relationship between the gas metallicity and the optical luminosity of galaxies was first discovered by [609] in dwarf galaxies (see also [610, 611]). This relationship was later extended to normal, star-forming objects in the nearby universe by [279, 612], and at intermediate [613, 614] and high [615, 616] redshifts. This relation has been confirmed on a large statistical sample of ∼ 53 000 galaxies with SDSS data by [281] (see Figure 17.2). Using their own recipes for determining stellar masses and oxygen abundances, [281] have shown that the gas metallicity index $12 + \log(O/H)$ is related to the absolute B band magnitude and to the stellar mass by the following relations:

$$12 + \log\left(\frac{O}{H}\right) = -0.185(\pm 0.001) M_B + 5.238(\pm 0.018) \quad (17.1)$$

$$12 + \log\left(\frac{O}{H}\right) = -1.492 + 1.847(\log M_{star}) - 0.08026(\log M_{star})^2 \quad (17.2)$$

In early-type galaxies devoid of gas, the average metallicity of the stars can be qualitatively traced by several optical absorption lines, and quantitatively measured by combining spectral indices with high resolution population synthesis models. Several relations linking metallicity sensitive Lick/IDS indices, such as the Mg_2, to the total luminosity or the velocity dispersion σ of elliptical galaxies empirically confirm that a metallicity–mass relation is present in quiescent galaxies devoid of gas [187, 617, 618]. Using stellar metallicities determined using population synthesis models, [294, 596, 619] have shown the existence of a stellar metallicity–mass relation in early-type galaxies that is equivalent to that observed in star-forming systems. This relation extends to the bulges of spiral galaxies [620] and to dwarf spheroidals in the local group [621].

The observed mass–metallicity relation is expected if galaxies evolve as closed boxes, that is, if they do not exchange matter with their surrounding environment and follow an exponentially delayed star formation history (according to Sandage, as described in Section 9.1.2), where massive galaxies have formed most of their stars in the past and dwarf systems are still producing stars at an almost constant rate [612]. As analytically deduced in [87], under these assumptions the metallicity Z of a closed box is given by the relation [612, 622, 623]

$$Z = y_Z \ln \frac{M_{gas} + M_{star}}{M_{gas}} = y_Z \ln(\mu^{-1}) \qquad (17.3)$$

where M_{gas} and M_{star} are the total mass of the gas and stars, μ is the gas fraction, and y_Z is the true yield of primary elements produced by massive stars (for example, oxygen). The true yield is defined as the mass of primary elements freshly produced and ejected by a generation of stars in units of the mass that remains locked in long-lived stars and compact remnants [87]. The true yield y_Z can be replaced by the effective yield y_{eff} when Z and μ in Eq. (17.3) are replaced by the observed metallicities (determined using the $12 + \log(O/H)$ index), and gas fractions (where the total gas content is measured considering hydrogen in both the atomic and molecular phases, and helium) [612]. The observed metallicity–mass relation naturally follows from Eq. (17.3). Departures from this relation have been interpreted as evidence of supernovae driven gas outflows in dwarf systems [281, 612], an interpretation which has been recently questioned by [624, 625]. The low effective yields observed in low luminosity galaxies could indeed result from gas dilution produced by freshly infalling gas still present in these slowly evolving systems but totally absent in massive objects, where the bulk of the stellar mass has been formed at early epochs. The mass–metallicity relation and its evolution with z has thus become a major observational constraint to both chemical models of galaxy evolution [625, 626] and cosmological simulations [627–631].

17.1.3
The Mass–Gas Relation

Although only marginally based on stellar emission data, the stellar mass–gas relation can be included in the spectrophotometric scaling relations given the major role played by the gaseous component, the principal feeder of the star formation process, in building up the stellar mass. Variations of the atomic, molecular and total gas content along the Hubble sequences have been found since the first systematic surveys of nearby samples, as reported in [632]. The variation of the normalized HI gas content as a function of the H band luminosity, proxy of the total stellar mass, was initially reported by [198] and later confirmed on a larger statistical basis by [315, 392]. Boselli et al. [392] determined the mass–gas relation using, for the first time, total gas masses determined as the sum of the HI and H_2 components and including 30% of the contribution from helium. This relation has been recently extended to include low luminosity systems by [633].

Figure 17.3 shows the relationship between the total gas mass (determined using the relation (11.1)) and the normalized total gas mass with the stellar mass in a nearby sample of late-type galaxies. There is an obvious linear relationship between the total gas mass and stellar mass ("bigger galaxies have more of everything,"

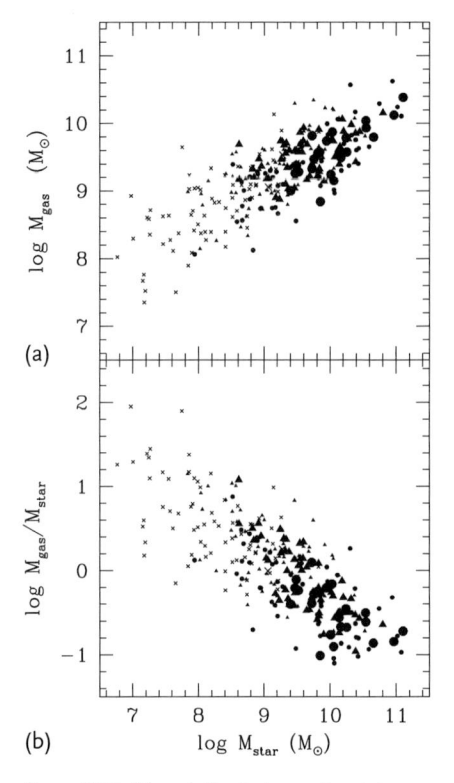

Figure 17.3 The relation between the stellar mass and total gas mass, determined using a nearby sample of resolved galaxies. The total gas mass M_{gas} is defined as in Section 11.1, $M_{gas} = 1.3(M(HI) + M(H_2))$ when the molecular gas content is measured using CO data (big symbols), while $M_{gas} = 1.3(M(HI) + 0.15M(H_1))$ when CO data are not available [315]. The multiplicative factor 1.3 accounts for the contribution of helium. Filled circles represent Sa-Sbc galaxies, filled triangles represent Sc-Sd, and crosses represent Sm-Im-BCDs.

Figure 17.3a), but the total amount of gas per unit stellar mass strongly decreases with the increasing mass which indicates that dwarfs are gas-rich while massive objects are gas-poor.

17.1.4
The Mass–Star Formation Rate Relation

The present day star formation activity and the past star formation history of galaxies are tightly related to their morphological type [196, 220] or their total luminosity [198, 392, 393]. These relations are depicted in Figure 17.4, where the star formation rate, determined as described in Chapter 13 using extinction-corrected H_α and UV data, and the birthrate parameter b are plotted as a function of the stellar mass for a nearby sample of isolated, late-type galaxies. Dwarf systems have, on average, birthrate parameters close to unity, indicating that, although probably bursty on short time scales, their mean star formation activity has been $\sim$ constant since their formation. On the contrary, massive galaxies produced most of their stars in

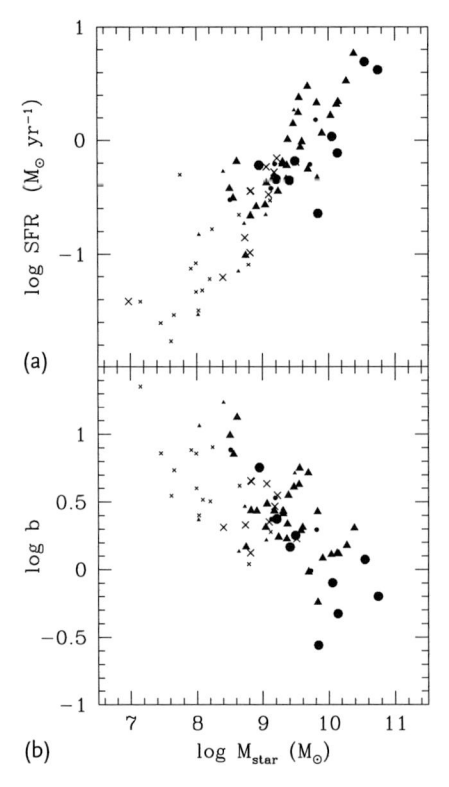

Figure 17.4 (a) The present day star formation rate and (b) the birthrate parameter (defined in Chapter 13) of a sample of nearby, isolated galaxies are plotted as a function of the total stellar mass. Filled circles represent Sa-Sbc galaxies, filled triangles represent Sc-Sd galaxies, and crosses represent Sm-Im-BCDs.

the past and presently have only a minor star formation activity. These relations, shown here for nearby galaxies, are also valid for high redshift objects [634, 635].

17.2 Structural Relations

Structural scaling relations, such as the surface brightness–absolute magnitude relation or the Kormendy relation, (which links the effective surface brightness to the effective radius), have been mainly used to see if early-type galaxies, dwarf ellipticals, and spheroidals form a unique sequence of objects. These relations can be easily determined for large samples of galaxies at different redshifts since they only require that galaxies are resolved, as is generally the case for images in the optical and near-infrared bands. They also require that the automatic pipelines for flux extraction provide a rough determination of the light profile of the detected sources, as is generally the case for the most common surveys available to the community (SDSS, 2MASS, COSMOS...). These scaling relations are generally valid whether the surface brightnesses are measured at or within the effective radius, or are central surface brightnesses. Seeing corrections are necessary in this latter case.

17.2.1 The Surface Brightness–Absolute Magnitude Relation

The importance of the surface brightness–absolute magnitude relation (see Figure 17.5) in the study of the origin of the red sequence became evident after the work of Sandage and collaborators on the Virgo cluster [437, 636–638]. This relation has been often used to study any possible difference in the origin of normal ellipticals with respect to that of dwarf systems. The most recent works on complete samples of nearby early-type galaxies all agree in showing a continuous sequence in the surface brightness–absolute magnitude relation down to dwarf early-type systems, with the exception of the brightest ellipticals ($M_B \leq -20.5$) [433, 434, 437, 639–641]. These objects, which are generally defined as "core" galaxies, lack a distinct stellar nucleus and have a central surface brightness and an inner profile slope lower than that which is expected by extrapolating the scaling relation defined by fainter systems [434, 437]. A continuity in the structural properties of normal and dwarf ellipticals has been claimed by [431], who have shown that the observed surface brightness–absolute magnitude relation is naturally produced by the fact that early-type galaxies have a Sersic light profile, with a Sersic index n increasing with galaxy luminosity (see Section 14.1).

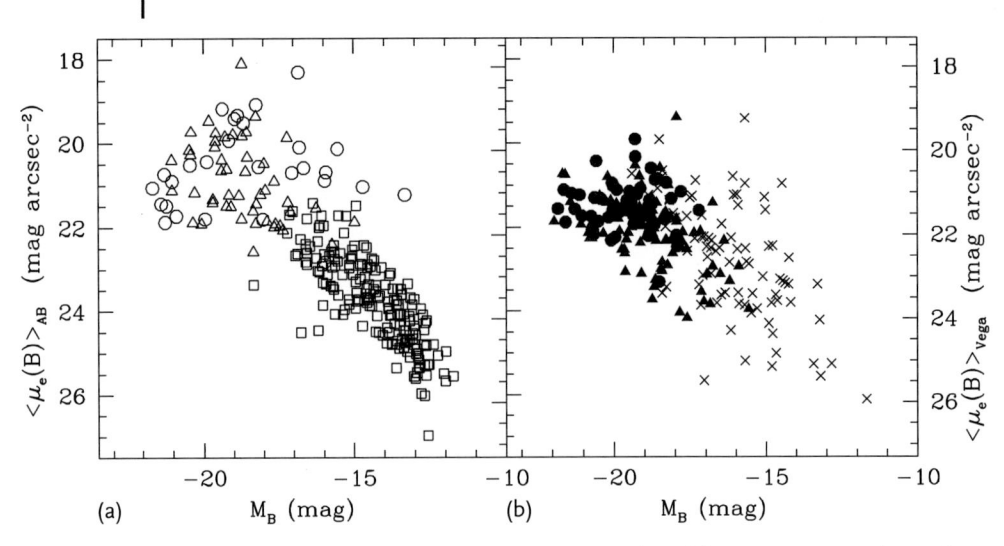

Figure 17.5 The B band effective surface brightness (defined as the mean surface brightness within the effective radius R_e, the radius including half of the total light) vs. absolute magnitude relation for a sample of nearby (a) early-type and (b) late-type galaxies, adapted from [642]. Open circles are for ellipticals, open triangles for S0-S0a, open squares for dwarf ellipticals, filled circles for Sa-Sbc, filled triangles for Sc-Sd, and crosses for Sm-Im-BCDs.

17.2.2
The Kormendy Relation

The Kormendy relation, [637] links the effective surface brightness to the effective radius of quiescent and star-forming systems. As shown in Figure 17.6, in elliptical galaxies the effective surface brightness decreases with the effective radius. That is, compact galaxies similar to M32 have the highest surface brightness observed among nearby extragalactic objects. In early-type spirals (Sa-Sc), the effective surface brightness does not change significantly with the effective radius, while it seems to decrease with R_e in late-type systems (Scd-Im-BCD) [642].

17.3
Kinematical Relations

High resolution ($R \geq 5000$) optical or radio spectroscopy allows the kinematical data to be included in the determination of several major scaling relations. Velocity dispersions measured using absorption lines in the optical spectra of dynamically hot systems (early-type galaxies) or rotational velocities measured using optical emission lines or the 21 cm HI line in rotationally supported objects (late-type galaxies) are of paramount importance, since they are directly related to the total dynamical mass of galaxies (see Section 15.2). For this reason, kinematical scaling

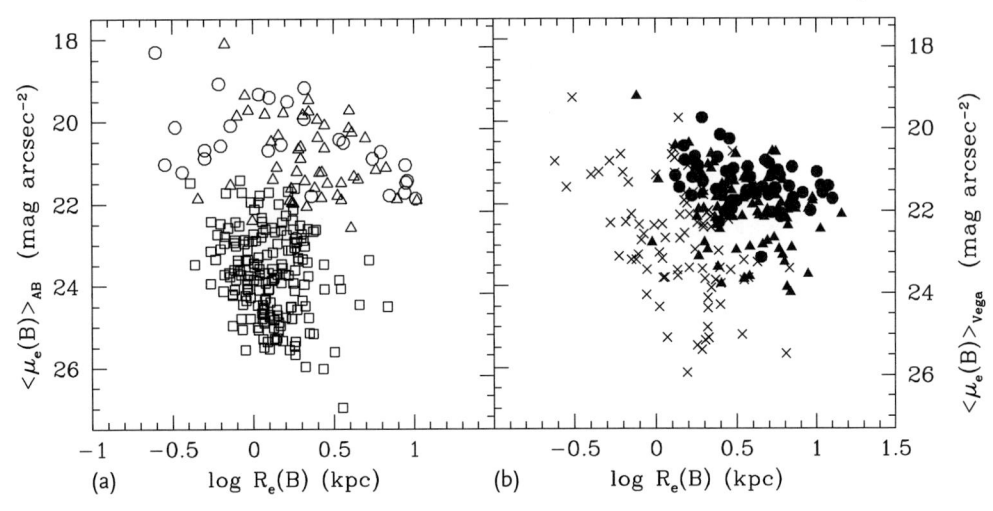

Figure 17.6 The B band effective surface brightness vs. effective radius relation (Kormendy relation) for a sample of nearby (a) early-type and (b) late-type galaxies, adapted from [642]. Open circles are for ellipticals, open triangles for S0-S0a, open squares for dwarf ellipticals, filled circles for Sa-Sbc, filled triangles for Sc-Sd, and crosses for Sm-Im-BCDs.

relations are tightly connected to the dark matter content of the different kinds of galaxies and are thus one of the most important direct observational constraints for models of galaxy formation and evolution. Indeed, N-body and hydrodynamical simulations predict the formation and evolution of dark matter halos: the evolution of the different baryonic components (gas, stars), whose mass is determined from the assumed cosmological model, is deduced by assuming different recipes for cooling the hot gas and transforming it into stars through the Schmidt law, and by considering the energy re-injected into the interstellar medium from AGNs and supernovae (feedback process; see Chapter 18), as done in semianalytical models. The relative small scatter in the Tully–Fisher relation for spirals and in the fundamental plane for ellipticals makes them ideal tools for measuring distances. These relations are often used to study peculiar motions in the nearby universe [643, 644] or for determining the three-dimensional structure of nearby clusters such as Virgo [645–647]. Thanks to the combination of high performance, high resolution spectrographs with 8 m class telescopes, these kinematical scaling relations can now be measured up to redshifts > 1 [648].

17.3.1
The Tully–Fisher Relation

In spiral galaxies, the luminosity is tightly related to the maximal rotational velocity V_{max} by the relation

$$L \propto V_{max}^a \qquad (17.4)$$

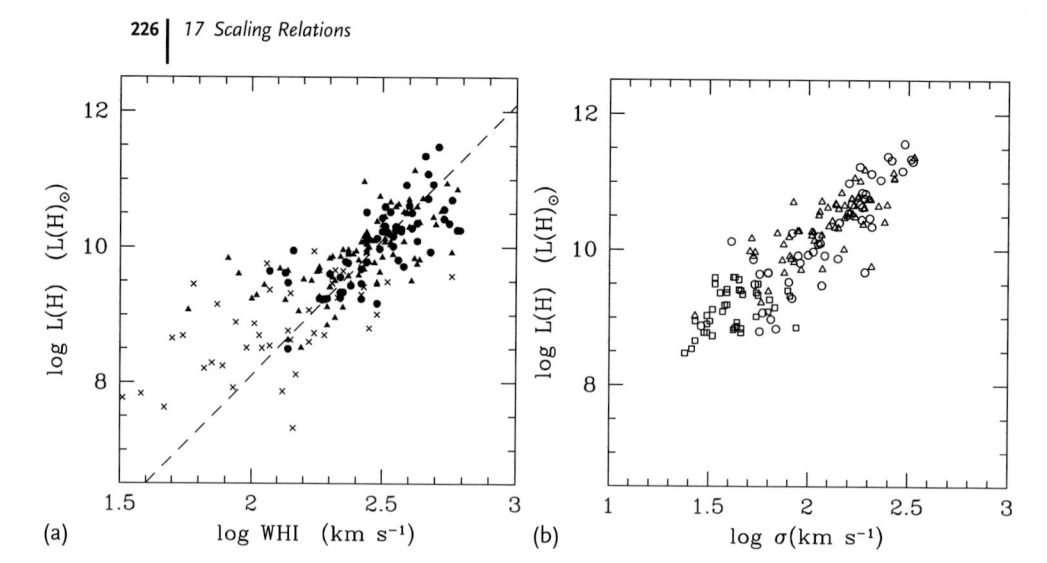

Figure 17.7 The H band (a) Tully–Fisher and (b) Faber–Jackson relations for a sample of nearby galaxies. Open circles are for ellipticals, open triangles for S0-S0a, open squares for dwarf ellipticals, filled circles for Sa-Sbc, filled triangles for Sc-Sd, and crosses for Sm-Im-BCDs. The dashed line is the expected relation for exponential disks with the constant mass to light ratio given in Eq. (17.12).

as shown in Figure 17.7a. Given the distribution of the atomic hydrogen outside the optical disk, the width of the HI line at 21 cm, corrected for inclination and turbulent motions as indicated in Section 11.1.1, is the most direct measure of the maximal rotational velocity of spiral galaxies in the local universe, while optical emission lines are generally used for high redshift objects. The slope of this scaling relation, generally called Tully–Fisher relation [583], changes with the observed band and gradually increases from $\alpha \sim 2.5$ in the B band to $\alpha \sim 4$ in the near-infrared K band (e.g., [649–651]). The most recent calibrations determined using 2MASS near-infrared data are [652]:

$$M_K - 5 \log h = -22.030 - 10.017(\log W_{HI} - 2.5) \tag{17.5}$$

$$M_H - 5 \log h = -21.833 - 9.016(\log W_{HI} - 2.5) \tag{17.6}$$

and

$$M_J - 5 \log h = -20.999 - 9.070(\log W_{HI} - 2.5) \tag{17.7}$$

where W_{HI} is the HI maximal rotational velocity (see Section 11.1.1). The coefficients of these 2MASS near-infrared Tully–Fisher relations slightly change with morphological type and luminosity. These relations can be used for measuring distances since they relate a distance-independent variable (W) to a distance-dependent (M or L) variable.

The Tully–Fisher relation can be easily interpreted for exponential disks [467]. In the case where the mass follows an exponential light distribution given by the relation (14.3), $I_{\exp}(r) = I(0)\exp[-r/h_r]$, the total mass of the disk is given by:

$$M = \int_0^\infty 2\pi I(0) e^{-\frac{r}{h_r}} dr = 2\pi I(0) h_r^2 \int_0^\infty x e^{-x} dx = 2\pi I(0) h_r^2 \quad (17.8)$$

Most of the mass of the disk lies within $r \sim h_r$. Therefore, the maximum of the rotation curve roughly corresponds to the Keplerian velocity at a distance of $r \sim h_r$ from the center. Placing all the mass at the center of the disk and equating the centripetal and gravitational accelerations, we obtain that the maximal rotational velocity $V_{\max}$ can be determined from the relation

$$\frac{V_{\max}^2}{h_r} \simeq \frac{2\pi G I(0) h_r^2}{h_r^2} \quad (17.9)$$

from which we can derive

$$V_{\max} \simeq (2\pi G I(0) h_r)^{1/2} \quad (17.10)$$

or

$$h_r \simeq \frac{V_{\max}^2}{2\pi G I(0)} \quad (17.11)$$

Substituting h_r in Eq. (17.8), we obtain that

$$M \propto \frac{V_{\max}^4}{I(0)} \quad (17.12)$$

which, for a constant mass to light ratio, and with the adoption of the Freeman's law that the central surface brightness of bright spiral galaxies is roughly constant, corresponds to Eq. (17.4). The slope of the Tully–Fisher relation measured in the near-infrared K band is $\alpha \sim 4$ ($\alpha \sim 10$ when the luminosity is expressed in absolute magnitudes as in Eq. (17.5)), indicating that the near-infrared bands are good tracers of the total dynamical mass of galaxies [198]. Near-infrared magnitudes are generally preferred in the determination of the Tully–Fisher relations also because they are less affected by internal attenuation and recent episodes of star formation than the optical bands. Notice that this derivation of the Tully–Fisher relation has been recently questioned by [651], who, following different logical arguments, obtain a slope of $\alpha = 3$ (see Eq. (17.4)) in agreement with their own data.

In recent years, two main efforts have been made to extend and improve the study of the Tully–Fisher scaling relation. The first one considers its baryonic form, where the luminosity L in Eq. (17.4) is substituted by the total baryonic mass of the galaxy [653–655]:

$$M_{\text{bar}} \propto V_{\max}^{\alpha_{\text{bar}}} \quad (17.13)$$

where the baryonic mass M_{bar} is defined as the sum of the stellar and gas masses

$$M_{\text{bar}} = M_{\text{star}} + M_{\text{gas}} \quad (17.14)$$

In this form, the change of slope in the classical Tully–Fisher relation observed at low luminosities is removed, and a linear relation in the velocity range 20 km s^{-1} < V_{max} < 300 km s^{-1}, corresponding to five orders of magnitude in mass, is observed [654]. The second development has been that of extending the Tully–Fisher relation, originally defined for disk galaxies, to also consider the contribution of bulges and include dynamically hot systems [656, 657]. This approach lead to the determination of a two-dimensional surface within the space defined by the effective radius, the effective surface brightness, and the internal velocity of these systems. The later was determined to consider the contribution of the rotational velocity and the velocity dispersion, an analogue to the fundamental plane of elliptical galaxies [657].

17.3.2
The Faber–Jackson Relation and the Fundamental Plane

In pressure supported, hot systems such as elliptical galaxies, the Tully–Fisher relation is replaced by the Faber–Jackson relation [658]

$$L \propto \sigma^x \tag{17.15}$$

where σ is the central velocity dispersion of the galaxy, and x is the exponent whose value ranges between three and five (Figure 17.7b). Elliptical galaxies are located along a plane, generally called the fundamental plane (see Figure 17.8), formed by the total luminosity L, the effective surface brightness I_e (with I_e defined as $10^{-0.4(\mu_e - 27)}$), and the velocity dispersion σ [584, 659],

$$L \propto \sigma^{8/3} I_e^{-3/5} \tag{17.16}$$

whose dispersion is significantly smaller than that observed for the Faber–Jackson relation (the later is a one-dimensional projection of the fundamental plane). The fundamental plane is now generally represented in the form

$$\log R_e = A \log \sigma + B I_e + \gamma = A \log \sigma + b \mu_e + \gamma \tag{17.17}$$

where R_e is the effective radius, and I_e and μ_e are the effective surface brightness. The former is defined as

$$I_e = \frac{L}{2\pi R_e^2} \tag{17.18}$$

and the latter is expressed in mag arcsec^{-2} (with $b = B/ - 2.5$ in Eq. (17.17)). A compilation of A and B (or b) coefficients of Eq. (17.17) can be found in Table 17.1, while the intercept γ is a distance-dependent coefficient. Among these coefficients, those most widely used are those of [660] in the r Gunn band and those of [661] in the K band. With the assumption that all elliptical galaxies form a homologous family (that is, that dwarfs have the same kinematical and structural properties of massive ellipticals scaled to low luminosities), then the virial theorem predicts

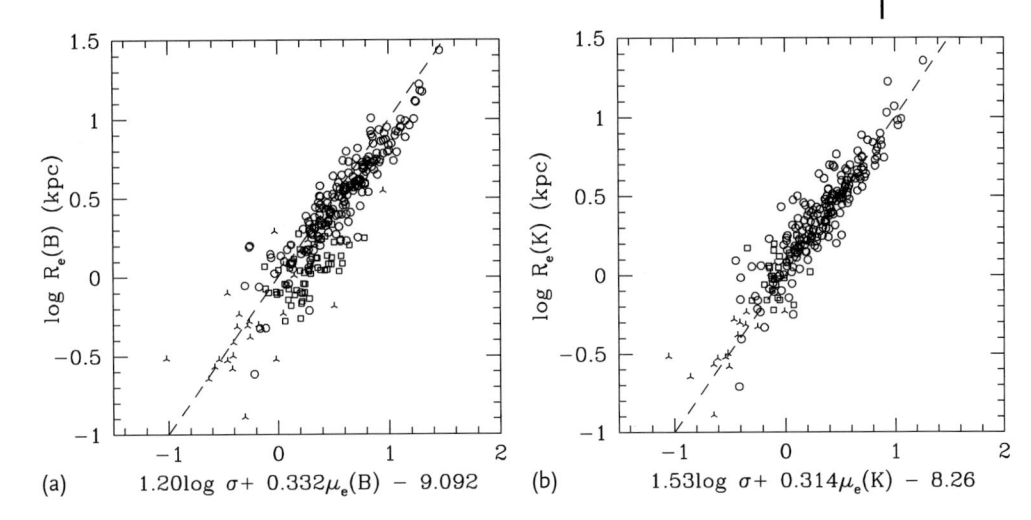

Figure 17.8 The (a) B and (b) K band fundamental plane relations for a sample of nearby early-type galaxies. Open circles are for ellipticals, open squares are for dwarf ellipticals, and asterisks are for spiral bulges (done with data kindly made available by E. Toloba). The dashed lines are the fitted relations given in Table 17.1.

the existence of an infinitely thin fundamental plane whose form can be easily determined by remembering that [662]

$$M(r) = c_2 \sigma^2 r = L(r)\frac{M}{L} = c_1 r^2 I_e \frac{M}{L} \quad (17.19)$$

where c_1 and c_2 are structure constants, thus inverting the relation (17.19)

$$r = \left(c_2 c_1^{-1}\right) \sigma^2 I_e^{-1} \left(\frac{M}{L}\right)^{-1} \quad (17.20)$$

In the case of a constant mass to light ratio, the fundamental plane takes the form of

$$\log R_e \propto 2\log \sigma + 0.4\mu_e \quad (17.21)$$

where R_e is the effective radius (in kpc) and μ_e is the effective surface brightness (in mag arcsec^{-2}). The A and b coefficients of Eq. (17.17), however, lie in the range 1.2–1.6 and 0.30–0.35, respectively (see Table 17.1). The difference between the expected (2 and 0.4) and observed coefficients is generally called the tilt of the fundamental plane. This tilt has been first imputed as a systematic variation of the mass to light ratio, parametrized as

$$M/L \sim L^\beta \quad (17.22)$$

where β is related to the coefficient A of Eq. (17.17) as [663, 664]

$$\beta = \frac{2 - A}{2 + A} \quad (17.23)$$

Table 17.1 The A, b and γ fitted parameters of the fundamental plane in different photometric bands.

Band	A	b	Ref.
U-Johnson	1.08	0.340	[660]
B-Johnson	1.20	0.332	[660]
g-Gunn	1.16	0.304	[660]
r-Gunn	1.24	0.328	[660]
K-Johnson	1.53	0.314	[661]
Virial	2.00	0.400	

Among other possible explanations of the tilt, we should remember variations in the stellar population along the fundamental plane due to a varying age, metallicity, or IMF [663, 665], changes in the dark matter content [665], or deviation from homology both in kinematical structure [666] and in matter (light) distribution [667–669]. The most recent results based on SDSS data of ∼ 16 000 early-type galaxies indicate that the variation of the stellar populations are not enough to explain either the tilt or the thickness of the fundamental plane. Variations of the dynamical mass to light ratio or possibly of the IMF are thus required. These works have also shown that the thickness of the fundamental plane is tightly related to the star formation history of galaxies – those objects which lie below the plane are characterized by a truncated star formation that produced less stars, while those above the plane with a relatively high number of stars produced by a longer star formation history [670–672].

It is worth mentioning that [673] have recently tried to extend the fundamental plane to incorporate any spheroid-dominated stellar system, from dwarf ellipticals to the intracluster stellar population of galaxy clusters. This exercise led to the definition of the fundamental manifold of spheroids.

17.3.3
The k-Space

The k-space is an orthogonal rotation of the global parameter space defined by the velocity dispersion σ, the effective radius R_e and $\langle\mu_e\rangle$, the mean effective surface brightness within R_e [662, 674]. This new coordinate system, which provides face-on and edge-on views of the fundamental plane, has been defined to have coordinates directly related to variables with a physical meaning; that is, galaxy mass (k_1), mass to light ratio (k_3) and a third quantity that depends primarily on surface brightness (k_2) [661, 662]:

$$k_1 = \frac{\log \sigma^2 + \log R_e}{\sqrt{2}} \propto \log\left(\frac{M}{c_2}\right) \tag{17.24}$$

$$k_2 = \frac{\log \sigma^2 + 0.8\langle\mu_e\rangle - \log R_e}{\sqrt{6}} \propto \log\left(\frac{c_1}{c_2}\right)\left(\frac{M}{L}\right)\langle\mu_e\rangle^3 \tag{17.25}$$

and

$$k_3 = \frac{\log \sigma^2 + 0.4\langle\mu_e\rangle - \log R_e}{\sqrt{3}} \propto \log\left(\frac{c_1}{c_2}\right)\left(\frac{M}{L}\right) \quad (17.26)$$

The k-space, intialy measured for early-type systems in the optical bands, was later extended to all stellar systems – including globular clusters, disk galaxies, groups and clusters of galaxies [674] – and measured in the near-infrared bands [661, 675].

17.4
Supermassive Black Hole Scaling Relations

Since the theoretical prediction of their existence and their first observation, it is now well established that bulge-dominated massive spiral and elliptical galaxies host a supermassive black hole (generally indicated with SBH or SMBH) in their center. The importance of supermassive black holes resides in the fact that they play a major role in galaxy evolution. They are associated with the central AGN and its accretion disk, which in turn regulate the star formation history of massive galaxies and the pollution of the intergalactic medium through feedback. The study of the relationship between the physical properties of the supermassive black hole and those of the host galaxy soon revealed that they are tightly related. These scaling relations are of paramount importance for constraining models of galaxy evolution since the feedback processes of AGNs that are associated with these supermassive black holes are one of the key parameters that regulate the matter assembly in all semianalytical models of galaxy formation [607, 676–681].

The most important scaling relations for this class of object are those linking the total mass of the supermassive black hole, M_{SBH}, to the absolute magnitude [493, 682, 683] of the host galaxy, often called the Magorrian relation, or to its central velocity dispersion [683, 684]. The scatter in the relations decreases when only ellipticals are considered, and when the mass of the central black hole is measured in those objects where the sphere of influence is resolved (see Section 15.2.4; [494]). Once bulges are included, the relation is tighter when only the luminosity of the bulge is considered or when luminosities are measured in the near-infrared bands [494, 685]. The slope and the intercept of these scaling relations significantly change from author to author. A recent fit to the data gives [686] (see Figure 17.9)

$$\log M_{SBH}\,[M_\odot] = (8.95 \pm 0.11) + (1.11 \pm 0.18)\log\left(\frac{L_V}{10^{11}}\right)\;[L_{\odot,V}] \quad (17.27)$$

for elliptical and lenticular galaxies, and

$$\log M_{SBH}\,[M_\odot] = (8.12 \pm 0.08) + (4.24 \pm 0.41)\log\left(\frac{\sigma_e}{200\,\mathrm{km\,s^{-1}}}\right) \quad (17.28)$$

including spiral bulges, where σ_e is the effective velocity dispersion defined as

$$\sigma_e^2 = \frac{\int_0^{R_e}(\sigma^2 + V^2)I(r)dr}{\int_0^{R_e} I(r)dr} \quad (17.29)$$

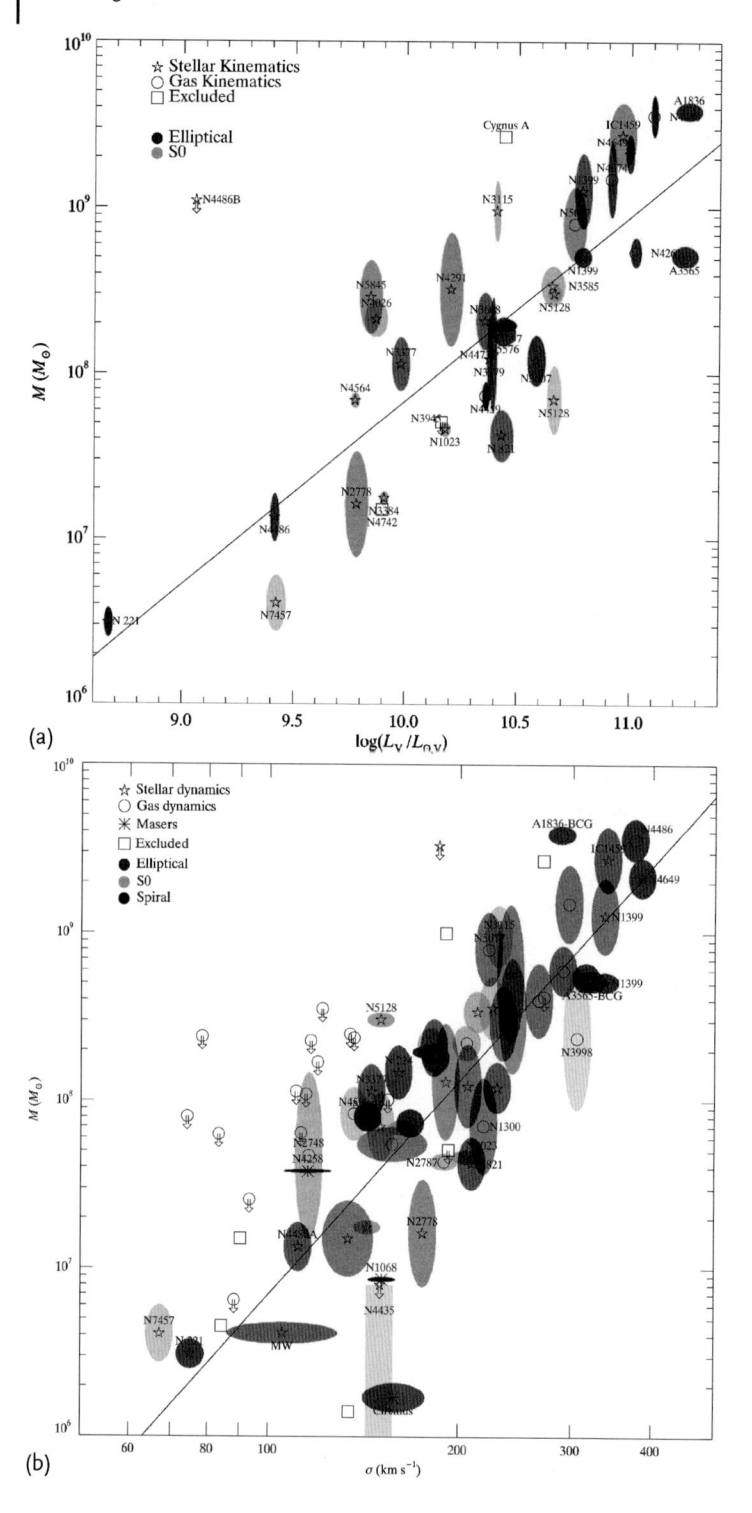

17 Scaling Relations

(a)

(b)

Figure 17.9 The scaling relations between (a) the mass of supermassive black hole and the V-band luminosity and (b) the velocity dispersion σ for elliptical galaxies and spiral bulges, adapted from [686]. The symbols indicate the method used for determining M_{SBH}: stellar dynamics (stars), gas dynamics (circles), masers (asterisks); arrows are for upper limits to M_{SBH}. The shaded areas indicate the error on the data. Spiral bulges are excluded from the M_{SBH} vs. L_V relation. The solid lines shows the best fit to the data (determined excluding uncertain values indicated with boxes), whose values are given in Eqs. (17.27) and (17.28). Reproduced by permission of the AAS.

where R_e is the effective radius of the galaxy, V is the rotational component of the spheroid, and $I(r)$ the light profile.

Models of galaxy formation predict that the total dynamical mass of galaxies (dark matter mass) plays a fundamental role in the formation of the supermassive black holes [687–690]. It is thus expected that the mass of the supermassive black hole is strongly related to the total dynamical mass of galaxies, as indeed shown by [691, 692] using total dynamical masses measured using rotation curves, or [693] using halo masses determined using gravitational lensing. It has been also shown that the mass of the supermassive black hole is tightly correlated with the concentration index parameter [694] or the Sersic index n [695]. These last relations are quite obvious given the tight relation between the luminosity of the bulge and the shape of the radial light profile of galaxies.

In optically selected, radio-quiet AGNs[3], the mass of the supermassive black hole is also related to the luminosity of the nuclear radio source. This correlation, however, is not shared by powerful AGNs [697]. Recent works have also shown that galaxies hosting a supermassive black hole are located on a fundamental plane that is formed by the nuclear radio and X-ray luminosities, and M_{SBH} [698]. This scaling relation probably originates from the physical link between the properties of the central black hole, the accretion rate traced by the X-ray luminosity, and the jet production, whose emission is observable at radio wavelengths [699].

---

3) AGNs are divided into radio-quiet and radio-loud according to their radio properties using different definitions, see [696].

18
Matter Cycle in Galaxies

The baryonic component of the universe, originally issued from the primordial nucleosynthesis and principally composed of atomic hydrogen ($\sim 75\%$ in mass), is processed and transformed during the evolution of galaxies. The atomic hydrogen, generally located on an extended disk whose size overcomes the optical size of galaxies, has to cool down and collapse into giant molecular clouds to reach the critical density necessary for the formation of new stars. Only a fraction of the available gas is transformed into stars of different size, whose mass distribution follows an initial mass function (see Section 13.1). These newly formed stars, in particular those with the largest mass, produce and inject metals, recycled gas and kinetic energy into the interstellar medium through stellar winds. This phenomenon, generally called feedback, is very important to the regulation of the star formation process that shapes galaxy evolution [525, 526, 700, 701] (see also Section 16.2.3). Metals aggregate to form dust grains, which are the principal coolers of the interstellar medium (through the absorption of the stellar heating radiation) and important catalysts for the formation of molecular hydrogen. The recycled gas can reach 30–40% of the total mass locked into stars [196] and is thus an important fraction of the gaseous component that can be reprocessed in future episodes of star formation. The kinetic energy injected into the interstellar medium, particularly that which is produced during supernova explosions where stellar winds can reach $20\,000\,\text{km s}^{-1}$, can expel the gaseous component from galaxies [608, 702] and create turbulent motions within the interstellar medium, with a subsequent compression of the gas that can trigger star formation [703]. In galaxies hosting a supermassive black hole, energy can be injected into the interstellar medium by the central AGN [607].

The matter cycle in galaxies is probably the best example of a physical process that can be studied only through multifrequency analysis. The gaseous component that feeds the star formation process can be observed through the HI and CO radio lines, and the activity of star formation can be inferred using optical emission lines or the UV emission once the data are corrected for dust extinction; the later is quantifiable through the infrared emission. The metal production can be determined using optical emission lines, the cooling of the gas can be determined through molecular or atomic emission lines in the infrared domain, and the different stellar populations formed during this matter cycle over the cosmic time are determined using UV to near-infrared spectral energy distributions.

A Panchromatic View of Galaxies, First Edition. Alessandro Boselli.
© 2012 WILEY-VCH Verlag GmbH & Co. KGaA. Published 2012 by WILEY-VCH Verlag GmbH & Co. KGaA.

18.1
The Star Formation Process

The process of star formation is one of the most important physical processes in astrophysics. As previously mentioned, this process is responsible for the transformation of the primordial gas into stars and is thus a principal factor in the formation and evolution of galaxies. The star formation process is tightly connected to the reionization of the universe and it is responsible for the production of most of the heavy elements present in the intergalactic medium.

For an accurate description of the star formation process, which is beyond the scope of the present book, we refer the reader to the recent review of [704]. It is worth mentioning that star formation is regulated by three key nonlinear and multidimensional dynamical processes: turbulence, magnetic field, and self-gravity. The turbulence of the interstellar medium is generally driven by local events such as supernova explosions, by protostellar, O star, and Wolf–Rayet winds or expanding HII regions, by large-scale motions induced by galactic rotations such as spiral density waves or bars, and by external perturbations due to the interaction of galaxies with their surrounding environment [703]. Turbulence creates over-dense regions where the gravitational collapse can take place, and can also counterbalance gravitational forces. The relation between turbulence and self-gravity is modulated by the presence of magnetic fields, important in the interstellar medium [704]. In this section we briefly describe and discuss the process of star formation on galactic scales.

18.1.1
The Schmidt Law

Despite the physical complexity of the relationship between the gaseous phase of the interstellar medium and the star formation process, observations of galaxies on global scales revealed the existence of a surprisingly tight correlation between the average star formation rate per unit area, Σ_{SFR}, and the mean surface density of the cold atomic and molecular gas, Σ_{gas}, extending over several orders of magnitude in gas surface density [705]. This relation, often parametrized with a power law, is generally called the Schmidt or Kennicutt–Schmidt law [706], and has the form

$$\Sigma_{SFR} = A \Sigma_{gas}^N \qquad (18.1)$$

When star formation rates and gas surface densities are measured in units of $M_\odot \, \text{yr}^{-1} \, \text{kpc}^{-2}$ and $M_\odot \, \text{kpc}^{-2}$ using disk-averaged data, $A = (2.5 \pm 0.7) \times 10^{-4} \, \text{yr}^{-1}$ and $N = 1.4 \pm 0.15$ [705]. The available multifrequency data at high angular resolution for nearby galaxies enabled us to extend these relations to galactic scales ($\sim 1 \, \text{kpc}$) [708–713]. The first studies of the Schmidt law showed the presence of a threshold in the gas column density at $\sim 5 M_\odot \, \text{kpc}^{-2}$, below which star formation does not occur [708, 709]. The presence of this threshold, however, has been recently questioned since it is not observed when star formation radial profiles are determined using ultraviolet data [712, 714]. The most recent and exhaustive study of the relationship between the gas surface density and the activity of star formation on sub-kpc scales is based on the HI and CO interferometric data of 18 nearby galaxies [707]. This work has shown that the star formation rate and the molecular hydrogen H_2 surface densities are linearly related ($N = 1.0 \pm 0.2$) when Σ_{gas} is replaced by Σ_{H_2} in Eq. (18.1). Thus, H_2 forms stars at a constant efficiency in spiral disks with gas densities in the range $3\,M_\odot\,pc^{-2} \leq \Sigma_{H_2} \leq 50\,M_\odot\,pc^{-2}$, with a gas depletion time scale on the order of $\sim 2 \times 10^9$ yr [707, 715]. The relationship between the star formation rate and total gas (HI + H_2) surface densities dramatically varies among and within spiral galaxies, with N ranging between 1.1 and 2.7, while the correlation is almost absent between Σ_{SFR} and Σ_{HI}. A change in slope in the $\Sigma_{SFR} - \Sigma_{gas}$ relation is observed for gas surface densities of $\sim 9\,M_\odot\,kpc^{-2}$, with $N \sim 1$ above this break where the gas content is dominated by the molecular phase, and $N > 1$ below (see Figure 18.1). The declining efficiency in transforming the atomic gas dominating the outer disks into new stars is probably related to the physical process controlling the formation and destruction of molecular hydrogen [715–717].

As clearly summarized in [705], however, the Schmidt law offers little insight into the underlying physics governing the star formation process. If the star formation rate is driven by gravitational instabilities in the disk, then the time scale for collapse is related to the local density as [718]

$$t_{SFR} \propto \Sigma_{gas}^{-1/2} \tag{18.2}$$

If ϵ is the efficiency of star formation (see Section 13.4), then the rate of star formation would be inversely proportional to t_{SFR}, and thus

$$SFR \propto \epsilon \Sigma_{gas} \propto \frac{\Sigma_{gas}}{(G\Sigma_{gas})^{-0.5}} \propto \frac{\Sigma_{gas}}{t_{SFR}} \propto \Sigma_{gas}^{1.5} \tag{18.3}$$

and the Schmidt law would have an exponent $N \sim 1.5$, close to the observed value [718, 719]. In practice, however, the time scale for gas to collapse might strongly depend on other local factors which depend on the physical properties of the interstellar medium (such as gas heating and cooling, turbulence, the presence of magnetic fields, shocks, and so on), making the theoretical explication of the Schmidt law much more complex. A relation similar to Eq. (18.1), but with the gas surface density modulated by the rotation is given by

$$\Sigma_{SFR} = A' \frac{\Sigma_{gas}}{t_{dyn}} = A' \Sigma_{gas} \Omega_g \tag{18.4}$$

where t_{dyn} is the dynamical time scale, corresponding to the orbital time scale of the disk. Ω_g is the average angular frequency of the rotating gas disk. Relation (18.4) is expected if the star formation rate per unit gas mass in clouds is constant, and if the formation of clouds decreases locally with the orbital time, or increases with the frequency of the crossing of spiral density waves [713]. This relation (18.4) is

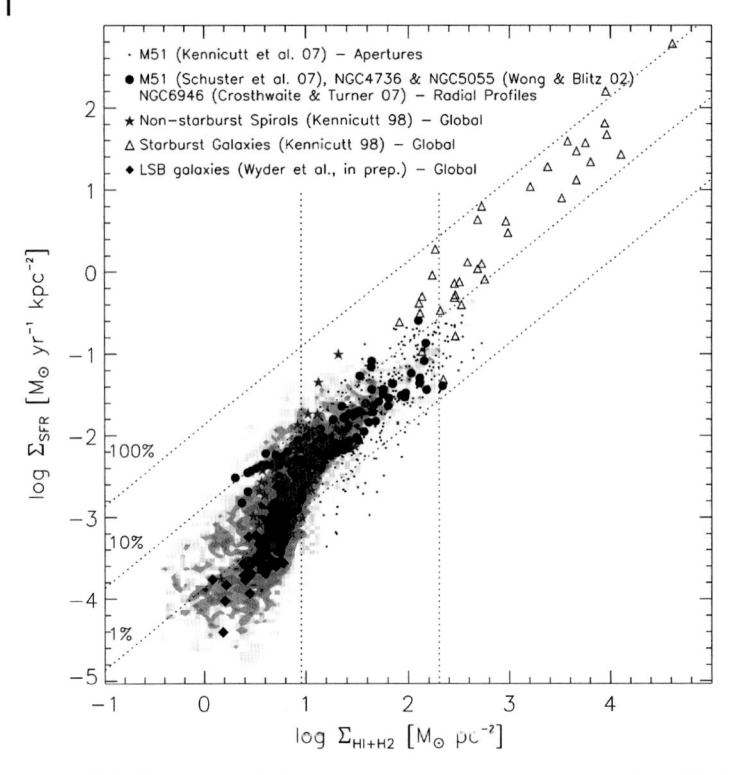

Figure 18.1 The relationship between the star formation rate and the total (HI + H2) gas surface density of resolved spiral galaxies, adapted from [707]. The diagonal dotted lines indicate a constant star formation efficiency able to consume 1, 10, and 100% of the total gas (including helium) in 10^8 yr. The vertical dotted lines indicate three different regimes for the star formation law (see [707] for details). Reproduced by permission of the AAS.

indeed observed in the disks of resolved galaxies, where $\Omega_g(R) \sim V(R)/R$ in the flat part of the rotation curve [705, 708, 711].

Besides any explication for the possible physical origin of the observed relation, the Schmidt law is one of the most important empirical relations in extragalactic astronomy. It is used in galaxy formation and evolution models as a key recipe for transforming the primordial gas into stars. In this context, it is worth remembering that its most updated empirical determination in resolved galaxies has been obtained for gas surface densities in the range $3\,M_\odot\,pc^{-2} \leq \Sigma_{H_2} \leq 50\,M_\odot\,pc^{-2}$ [707], densities significantly lower than those expected in starburst galaxies dominating the high redshift universe. A relationship between the gas surface density and the star formation rate is present in ULIRGs, and it seems to be more efficient than in normal, star-forming disks [720–722]. The efficiency might indeed be related to the presence of metals and dust, which in turn can regulate the time scale for cooling the gas. Furthermore, the A and N coefficients in Eq. (18.1) might slightly change with the assumed IMF, the adopted population synthesis model used to estimate star formation rates, and the assumed X conversion factor to measure molecular hydrogen surface densities.

18.2
Feedback

The feedback process is any action that the different stellar populations, as well as the presence of a nuclear massive black hole, make on the interstellar medium of the host galaxy, that can in turn affect the star formation process.

18.2.1
The Feedback of AGNs

Elliptical galaxies and the bulges of spiral galaxies contain a supermassive black hole. Its mass, M_{SBH}, is proportional to the stellar mass of the bulge or of the whole elliptical galaxy, indicated here without distinction by M_{bulge}, and is roughly given by the relation

$$M_{SBH} \sim 0.001 M_{bulge} \tag{18.5}$$

deduced from Eq. (17.27). The central black hole is fed by the accretion disk, where the infalling material spirals around the deep potential well, heating up by viscous friction. The gravitational potential energy of a particle of mass m located at a distance R from the supermassive black hole is

$$E_{grav} = \frac{-G m M_{SBH}}{R} \tag{18.6}$$

where G is the gravitational constant. If the particle of mass m is rotating around the black hole with a velocity V, then its kinetic energy is

$$E_{kin} = \frac{1}{2} m V^2 \tag{18.7}$$

For a particle on a circular orbit at a distance R from the black hole, where the gravitational acceleration $G M_{SBH}/R^2$ equals the centrifugal acceleration V^2/R, the total energy is given by

$$E(R) = E_{kin} + E_{grav} = \frac{1}{2} m V^2 - \frac{G m M_{SBH}}{R} = -\frac{1}{2} \frac{G m M_{SBH}}{R} \tag{18.8}$$

Decelerated by viscous drag, the energy of the particle is transformed into heat and radiated. The maximal radiated energy is obtained when all the energy of the particle is released at the Schwarzschild radius

$$R_S = \frac{2 G M_{SBH}}{c^2} \tag{18.9}$$

Here, the velocity of the particle equals the speed of light c:

$$\Delta E = E(\infty) - E(R_S) = \frac{1}{2}\frac{Gm M_{SBH}}{R_S} = \frac{1}{4}mc^2 \qquad (18.10)$$

In practice, because of the nature of the accreting disk, only a fraction of this maximal energy, ~ 0.06–$0.4\, mc^2$, is radiated, a fraction significantly larger than that produced by the nuclear process acting inside stars ($\sim 0.007\, mc^2$).

The infalling material is simultaneously attracted by the gravitational forces and repelled by the radiation pressure. The latter is equal to the momentum transfer of the emitted photons to the absorbing gas, per unit time. Close to the black hole, where the gas is ionized, the radiation pressure is principally due to the pressure exerted by the photons on the electrons (Thomson scattering). A photon of energy $E = h\nu$ has a momentum E/c, and the momentum transfer exerted by the radiation emitted by a source of luminosity L per unit time and unit area is $L/(4\pi R^2 c)$. Assuming that electrons are not shielded by those further in (that is, that the plasma is Compton thin), the force exerted by the radiation on each electron, whose cross-section is σ_T (Thomson cross-section), is

$$F_{rad} = \frac{\sigma_T L}{4\pi R^2 c} \qquad (18.11)$$

Although the radiation pressure on protons is negligible, they are dragged away with the electrons since they are attracted electrostatically. If the radiation force exceeds the gravitational forces, then

$$F_{grav} = \frac{G M_{SBH} m_p}{R^2} \qquad (18.12)$$

where m_p is the mass of the proton. That is, if the luminosity exceeds the Eddington luminosity [494]

$$L_{Edd} = \frac{4\pi G M_{SBH} m_p c}{\sigma_T} \sim 3 \times 10^4 \frac{M_{SMH}}{M_\odot} L_\odot \sim 1.3 \times 10^{46} \frac{M_{SMH}}{10^8\, M_\odot}\, \text{erg s}^{-1} \qquad (18.13)$$

then the radiation pressure blows out the surrounding gas and stops the infall of fresh material. The Eddington luminosity is thus the maximal luminosity that a black hole of mass M_{SBH} can produce[1,2].

Photons and radio jets of charged particles produced by the AGN can release energy to the interstellar medium and blow away the entire gas content, thus stopping the star formation activity of the host elliptical galaxy or of the bulge. The gas can be either heated by the thermal energy-driven winds and expand to escape the galaxy, or be pushed out by the pressure of momentum-driven winds [723]. The interstellar

1) In practice, super-Eddington luminosities of values close to those given in Eq. (18.13) are possible if more realistic nonspherical configurations for the accretion disk are assumed.
2) Notice that the formalism used to describe the accretion of material in supermassive black holes is the same as the one used to describe the accretion of matter in X-ray binaries (see Section 2.1.1.1).

medium can be heated through the photoionization of metals such as iron, which retain their inner electrons even at high temperatures, and by Compton scattering, where part of the energy of the photons is transferred to the electrons. Jets can heat the gas through shocks. The radiative pressure exerted on the electrons in the interstellar medium by the photons emitted by the AGNs, described above, is mainly important to the surroundings of the nuclear black hole. However, the radiation pressure from AGNs can also be important to dust grains at large distances if the dust content is significant, given their large cross-section. Jets transfer momentum to the interstellar medium through ram pressure. Several attempts have been made to quantify these effects despite the difficulty of estimating their efficiency within the complex interstellar medium of galaxies. Indeed, metals are present in the gas surrounding the central black hole at a very low level, while photoionization is efficient only in a narrow frequency band. Energy is transferred to electrons through Compton scattering at a very low level, while the radiation pressure exerted on dust grains is efficient only if dust covers a large solid angle. Jets are bipolar and generally cover a small solid angle; thus, they can escape the galaxy with a minor impact on its interstellar medium. Furthermore, these physical processes can have a different impact depending on whether or not the host galaxy is isolated or embedded in a hot intracluster medium [723].

Despite these difficulties, several works succeeded in modelling these physical processes and reproducing the observed scaling relations for supermassive black holes described in Section 17.4, in particular the relation between the mass of the black hole and the velocity dispersion of the bulge (Eq. (17.28)) which can be generalized in the form

$$M_{SBH} = k\sigma^\alpha \qquad (18.14)$$

with $\alpha \sim 4$ [607, 679, 724–726]. The gaseous phase of the interstellar medium is removed from the galaxy when the velocity of the wind exceeds the typical velocity dispersion of the bulge, σ, whose value can be estimated by considering the galaxy or the bulge as an isothermal sphere with a mass–density profile of $\rho \propto R^{-2}$ and of mass M_{bulge} [726]:

$$M_{bulge} = \frac{2\sigma^2 R}{G} \qquad (18.15)$$

If the dependence on the velocity dispersion of the bulge σ is stronger for M_{SBH} (Eq. (18.14)) than for M_{bulge} (Eq. (18.15)), then the feedback from AGNs is efficient at removing the interstellar medium of the host galaxy and quenching the star formation process only for black hole masses exceeding $\sim 10^7\ M_\odot$, or equivalently for galaxies of stellar masses above $\sim 10^{10}\ M_\odot$.

These evidences have major consequences in the study of galaxy evolution through the cosmic time [723]. Included in semianalytical models of galaxy formation [727, 728], the feedback of AGNs is now considered to be primarily responsible for preventing the gas from cooling and, as a consequence, for shutting down the star formation process in the most massive objects. This abrupt decrease of the star formation activity in massive objects is necessary to reproduce the lack of bright galaxies present in the observed luminosity function with respect to the dark matter halo mass function predicted by models as depicted in Figure 16.9, and the formation of the red sequence [687, 701, 728–733].

18.2.2
The Feedback of Massive Stars

Massive stars lose an important fraction of their mass into the interstellar medium through stellar winds. The energy injected into the interstellar medium can either trigger star formation through the compression of the surrounding gas, or heat the interstellar medium and expel it through galactic fountains. A large fraction of the energy released into the interstellar medium comes from supernovae. Type II supernovae result from the explosion of massive stars, and type Ib probably result from the explosion of massive stars in late Wolf–Rayet stages. The expelled material, which is on the order of $\sim 0.25\ M_\odot$, interacts with the circumstellar material released by the massive star through stellar winds and with the surrounding interstellar medium. In these objects, the matter that was not expelled collapses to form a neutron star or a black hole. Neutron stars emit high-energy particles that supply extra energy to the expanding shell. In type Ia supernovae, which result from the explosion of a white dwarf after it accretes material from a nearby companion, the expanding remnant only interacts with the interstellar medium because of the lack of circumstellar envelopes. Since a white dwarf is completely destroyed during the explosion, there is no formation of a compact remnant [75, 97]. Despite their difference in luminosity (type Ia are more luminous than type II and Ib), the total kinetic energy in the ejected envelope is $\sim 4 \times 10^{50}$ erg for all types.

The evolution of supernovae remnants follows three different phases, as extensively described in [75, 97]. During the first phase, the density of the ejected material is larger than that of the surrounding interstellar medium (free expansion). The material expelled by the stars, which initially has a velocity on the order of $\sim 20\,000$ kms^{-1}, forms, in the presence of magnetic fields, a magnetohydrodynamic shock whose velocity decreases to conserve the momentum. This phase ends when the mass swept up by the expanding envelope becomes similar to that of the ejected envelope. For standard conditions, the free expansion phase can last ~ 60 yr [97]. In the second phase, the temperature of the gas is too high ($> 10^6$ K) to radiate, and thus the only energy losses are those relative to the adiabatic expansion of the gas (adiabatic phase). When the gas temperature drops to $\sim 10^6$ K ($\sim 4 \times 10^4$ yr, when the velocity of the remnant is ~ 85 km s^{-1}), the heavy elements present in the interstellar medium (such as C, N, and O) begin to recombine with electrons. The recombination emission lines and the collisional excitation of the ions cool the gas. Contrary to the previous phase, the propagation of the remnant is not sustained by thermal energy, but by the momentum of the gas (isothermal phase). During this phase, the kinetic energy of the remnant is transferred to the interstellar medium. This phase ends when the velocity of the remnant reaches the velocity of the interstellar medium (~ 9 km s^{-1}, $\sim 10^6$ yr).

To a first approximation, the efficiency with which the energy of the supernova is transferred to the interstellar medium is on the order of 3% [75].

Supernova remnants can combine with stellar winds produced by massive stars to form expanding bubbles, whose total kinetic energy can reach 10^{53} erg and are able to expel matter in the direction perpendicular to the galactic disk [734]. Massive stars lose matter at a rate of $\dot{M}_{SW} \sim 3 \times 10^{-6}\,M_\odot\,yr^{-1}$ on time scales on the order of $t_{SW} \sim 3 \times 10^6$ yr, with velocities $V_{SW} \sim 2000\,km^{-1}$, and thus provide kinetic energies ($E_{SW} \sim 4 \times 10^{50}$ erg) comparable to those of supernovae [75] to the interstellar medium during their lifetime. The number of massive stars and the number of supernovae are proportional to the star formation rate. Thus, the resulting energy transferred to the interstellar medium by the feedback of massive stars is proportional to the star formation activity of the galaxy.

Since the seminal work of [735], several scientists have tried to model the feedback process of massive stars on the scale of galaxies [608, 702, 736]. As for the feedback of AGNs, however, this exercise is made difficult by the need to simultaneously consider the three different phases of the interstellar medium (the cold neutral phase, the warm medium and the hot gas) [737], and the small- and large-scale geometrical effects relative to the fractal structure of the interstellar medium and to the shape of galaxies of different type and luminosity. Presently, there is a general consensus that the feedback of massive stars can quench the star formation activity in dwarf galaxies. However, external processes related to the environment, such as ram pressure stripping (exerted by the intergalactic medium in rich clusters [563] or by the halo of massive galaxies), [738] are very convincing alternatives.

19
The Role of the Environment on Galaxy Evolution

If the universe is homogeneous and isotropic on large scales, then on small scales galaxies are inhomogeneously distributed. In the local universe, galaxies are located in regions of density spanning from $\sim 0.2\rho_{gal}$ in local voids, to $\sim 5\rho_{gal}$ in superclusters and filaments, to $\sim 100\rho_{gal}$ in the core of rich clusters, and finally to $\sim 1000\rho_{gal}$ in compact groups. Here ρ_{gal} is the mean galaxy density of the local universe, whose value is ~ 0.006 objects Mpc^{-3} down to $M_i < -19.5$ [9]. In a cold dark matter dominated universe, these density differences increase with cosmic time. Galaxies are formed within cold dark matter halos after the collapse of the baryonic matter. Several objects can be formed within large dark matter halos, such as those typical of clusters or groups. In high-density regions, the gravitational interactions between galaxies or with the potential well of the cluster might have induced dynamical perturbations that were able to remove the gas located in the outer part of the disk and induce the inner gas to collapse into the nuclear region of the galaxy. Rich clusters of galaxies are also characterized by the presence of a hot ($T_{IGM} = 10^7$–10^8 K) and dense ($\rho_{IGM} = 10^{-3}$ cm^{-3}) intergalactic medium [739] with which galaxies moving at high velocity (~ 1000 km s^{-1}) can interact, losing their own interstellar gas. These interactions might have major effects on the evolution of galaxies located in high-density environments, as extensively discussed in our review paper [10] or in [740]. After a possible, short increase of the star formation activity induced by the compression of the gas and its infall in the nuclear region, the lack of gas removed during the interaction quenches the star formation activity, making cluster galaxies passive objects. Indeed, clusters and groups are dominated by early-type galaxies, particularly in the local universe [11, 13, 741]. To quantify the role of the environment on galaxy evolution, it becomes crucial to study how galaxy properties change as a function of the environment to which they belong.

19.1
Tracers of Different Environments

Environmental studies first require the identification of regions of different galaxy density. This can be done either through the identification of cluster of galaxies, of loose and compact groups, or through the determination of the mean volume density of galaxies around any given object. The first method has the advantage that any single high-density region can be recognized and studied in detail through a multifrequency analysis, while the second technique is more adapted to statistically trace the continuous variation of any physical parameter as a function of galaxy density. To increase the statistics in the study of the mean properties of galaxies within different density regions, objects belonging to the same bin of density are artificially combined. This second approach, where single structures are mixed within a bin of density, is well adapted for wide-field monochromatic surveys with large numbers of detected sources.

19.1.1
Detection of High-Density Regions

The identification of clusters and groups can be done using different techniques, as summarized in [742]. Originally identified through the visual inspection of photographic plates [743–745], they are now recognized thanks to their physical properties or of those of their members using multifrequency data. Clearly, the different techniques adopted in the selection of the cluster lead to the definition of different samples.

As previously mentioned, clusters of galaxies, and to a lesser extent compact groups, are characterized by a diffuse X-ray emission (thermal bremsstrahlung) due to the hot and dense gas trapped within the deep potential well of the cluster. X-ray blind surveys are thus often used to detect X-ray emitting regions potentially associated with clusters or groups [746–748]. This technique, however, is limited by the fact that not all optically selected clusters of galaxies are identifiable in X-ray [749, 750], and by the fact that the apparent X-ray surface brightness decrease rapidly with distance, making at high redshifts the X-ray selection only adapted for very massive systems.

Another method for detecting clusters is through gravitational lensing. The cluster potential well can be detected from the strong gravitational lensing or the cosmic shear induced by weak gravitational lensing [751, 752]. The gravitational lensing cross-section of clusters also decreases very rapidly with z, thus this technique can, at high redshift, detect only massive clusters.

The hot intergalactic medium present in the potential well of rich clusters scatters, through inverse Compton, the photons of the cosmic microwave background (CMB) and produces distortions in the microwave spectrum [753, 754]. These distortions can be recognized and used to identify clusters of galaxies. This effect, generally called the Sunyaev–Zeldovich effect, does not depend on the redshift of the cluster.

Generally characterized by the presence of a powerful radio source associated with the brightest elliptical galaxy, clusters have also been searched for around strong radio galaxies detected in blind centimeter radio surveys [755].

Clusters can be also identified thanks to the physical properties of their members. Dominated by early-types, cluster galaxies are, on average, redder than field objects and can thus be searched for as a concentration on the plane of the sky of red galaxies [756]. This technique favors the detection of relaxed and evolved systems while it is probably biased against clusters in formation, where the violent interactions between the different members increases the star formation activity making galaxies bluer than the quiescent systems generally inhabiting nearby high-density regions (the Buchler–Oelmer effect, [757]). All these previously mentioned methods, however, require follow up spectroscopic observations for confirming the presence of a cluster of galaxies.

When spectroscopic or photometric redshifts are available, high-density regions can be detected using geometrical algorithms based on the three-dimensional spatial distribution of galaxies. These algorithms use the hierarchical method [758, 759], the friend of a friend method [760], the three-dimensional adaptive matched filter method [761], the C4 method [762], and the Voronoi–Delaunay method (VDM, [763]). Redshifts are crucial for the identification of high-density regions. Whenever spectroscopic redshifts are not available, photometric redshifts, if not sufficiently accurate ($\sigma(z_{phot}) \sim 0.1$) to resolve cluster structures ($\sigma_{cluster} \sim 1000$ km s^{-1}), are useful to reduce the uncertainty in statistical background subtractions and to increase the density contrast in a given distance range [764]. The typical uncertainty in the recessional velocity of standard spectroscopic redshift surveys is often larger than the velocity dispersion within groups ($\sigma_{group} \sim 200$ km s^{-1}), limiting the accuracy in the identification of regions of intermediate density.

A simple way to estimate the local three-dimensional density is that of counting the number of galaxies in cylinders of given radius and half-length centered on each galaxy. The choice of the size of the cylinder is crucial for identifying the high-density regions under study. Those typically used for large scale surveys such as the SDSS by [765] (2 Mpc, 500 km s^{-1}) or [9] (8 h^{-1} Mpc, $\sim$ 1000 km s^{-1}), for instance, are inadequate for characterizing simultaneously the galaxy density in a typical supercluster region which includes rich clusters (with their associated filamentary structures, with sizes on the order of $\sim$ 1 Mpc and velocity dispersions of $\sim$ 1000 km s^{-1}), compact and small groups ($\sim$ 100 kpc, 250 km s^{-1}), and voids. Small cylinders underestimate the local density of galaxies belonging to the extremities of the "Fingers of God"[1] in rich clusters, while large cylinders overestimate the density of isolated galaxies because they include background and foreground objects not physically associated to the studied galaxy. A good compromise to overcome this problem is that of artificially compressing the "Fingers of God" in rich clusters ($\sigma_{cluster} \sim 1000$ km s^{-1}) to the mean velocity of the massive cluster to which they belong so that galaxies belonging to these structures are correctly considered as cluster members. Then, by adopting cylinders of 1 h^{-1} Mpc radius and 1000 km s^{-1} half-length, which are large enough to comprise at the same time the dispersion of small groups and the compressed "Fingers of God." This technique has been succesfully applied to identify high-density regions such as Coma and A1367, and galaxy pairs in the Great Wall on the local supercluster by [11], as depicted in Figure 19.1. The environment surrounding each galaxy can then be

---

1) Because of their high velocity dispersion ($\sim$ 1000 km s^{-1}), clusters of galaxies are identified in velocity vs. right ascension wedge diagrams as high-density regions pointing the observer (see Figure 19.1b), and are thus commonly called "Fingers of God."

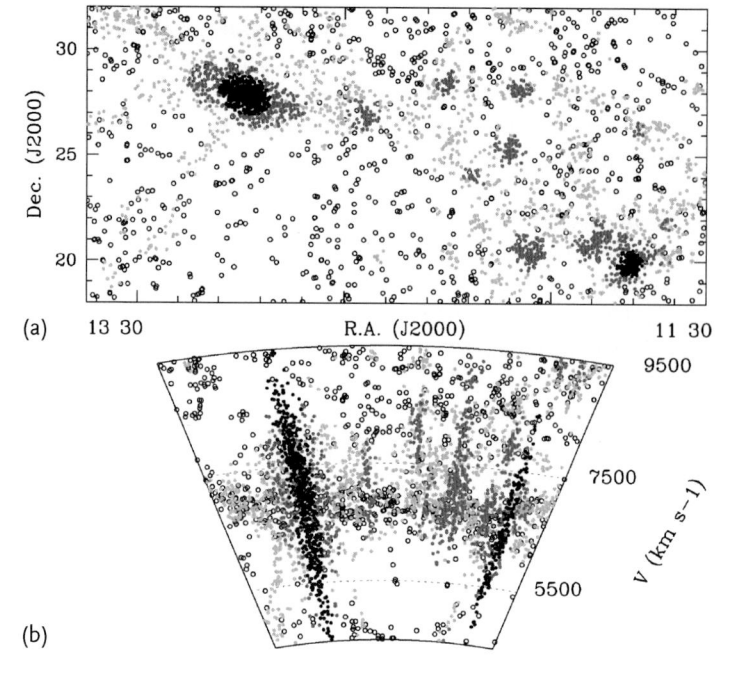

Figure 19.1 Celestial distribution of 4132 galaxies from SDSS in the Coma supercluster: (a) $11.5^h <$ RA $< 13.5^h$; $18° <$ Dec $< 32°$; $4000 < zc < 9500$ km s^{-1}, and (b) their wedge diagram, adapted from [11]. Galaxies are coded according to their over-density (defined in Eq. (19.1)) as follows: objects in the highest density regions ($\delta_{1\,Mpc,\,1000\,km/s} > 20$) are filled black dots; the medium density regions ($4 < \delta_{1\,Mpc,\,1000\,km/s} \leq 20$) are filled with dark gray dots (a); the medium density regions ($0 < \delta_{1\,Mpc,\,1000\,km/s} \leq 4$) are filled light gray dots (b); in the lowest density regions ($\delta_{1\,Mpc,\,1000\,km/s} \leq 0$) are black empty circles, where the mean density within the supercluster is $\langle \rho_{gal} \rangle = 0.05$ gal (h^{-1} Mpc^{-3}). These density intervals are well defined to resolve galaxies in the core (filled black dots) and in the outskirts (filled dark gray dots) of the 2 main clusters Coma and A1367, in groups (filled dark gray dots), in the supercluster filament connecting the two main clusters (filled light gray dots) and in voids (black empty circles). Reproduced with permission (c) ESO.

parametrized using the three-dimensional density contrast computed as [11]

$$\delta_{1\,Mpc,\,1000\,km/s} = \frac{\rho_{gal} - \langle \rho_{gal} \rangle}{\langle \rho_{gal} \rangle} \qquad (19.1)$$

where ρ_{gal} is the local galaxy density and $\langle \rho_{gal} \rangle$ is the mean galaxy density measured in the entire regions studied. Caution should be used to deal with the survey borders. A generalization of this technique adapted for high-redshift surveys can be found in [766].

19.1.2
Other Quantitative Tracers of High-Density Environments

Whenever high-density regions are identified, as is the case for nearby clusters, other more accurate tracers can be used to quantify the degree of influence that the environment might exert on any galaxy. The most simple of them, often used in the literature, is the angular distance from the cluster center, generally measured as the distance from the central cD galaxy or from the peak of the X-ray diffuse emission. Whenever the analysis is done by combining data for galaxies belonging to different clusters, the angular distance must be normalized to a characteristic size of the cluster. This normalization is generally done using the virial radius. Simple to measure, angular distances have the disadvantage (with respect to other indicators) that they are projected (not physical) distances, because they consider all galaxies at the same line of sight distance. For this reason, this technique smears the results and is only used for detecting very strong variations in the physical properties of the observed galaxies.

Hierarchical models of galaxy evolution in a Λ cold dark matter universe indicate that the most important parameter driving galaxy evolution is the mass of the dark matter halo inside which galaxies are formed and evolve. Whenever lensing is present, the mass of the high-density region can be directly measured by combining lensing effects with mass models [767]. If the high-density regions are virialized systems, then their mass is directly related to the velocity dispersion of galaxies within their potential well. The importance of the environment in which galaxies reside can be determined using the velocity dispersion of clusters or groups using redshift measurements. The perturbation induced on an individual galaxy can also be quantified by measuring the velocity of the galaxy with respect to the mean recessional velocity of the cluster itself. As for the angular distance, this technique smears the results since it only considers the line of sight velocity, completely neglecting the velocity of the galaxy on the plane of the sky.

The interactions of galaxies with the surrounding environment mainly have two different natures: gravitational nature, with nearby companions and with the whole potential of the high-density region under study (gravitational interactions, galaxy harassment), and dynamical or physical nature with the hot and dense intergalactic medium trapped in the potential well of the cluster (ram pressure, viscous stripping, thermal evaporation, etc.) [10]. Processes such as the ram pressure and viscous stripping depend on the relative velocity of the galaxy with respect to the cluster to the power of two, while only linearly on the density of the intergalactic medium, which generally decreases radially from the core of the cluster. Gravitational interactions depend on the density of galaxies within the cluster and on the shape of the cluster potential well, which in turn are tightly related to the distance from the cluster center. At the same time they depend on the duration of the gravitational perturbation, which is inversely proportional to the velocity dispersion of the cluster [10]. The significance and the distribution of the intracluster medium can be quantified by measuring the X-ray luminosity of the high-density region, where the hot and dense gas emits for thermal bremsstrahlung [739]. All these alternative tracers are thus very important to identify the physical process perturbing galaxies.

19.2
Measuring the Induced Perturbations

The perturbations induced by the environment on the physical, structural and kinematical properties of galaxies can be quantified either by tracing how these properties change as a function of any density tracer, or by comparing the statistical properties of galaxies selected according to similar criteria but belonging to different density regions. For instance, the first technique has been adopted to confirm (with a strong statistical basis) previous studies of local clusters [178, 768] using 2dF and SDSS data, showing that the star formation activity decreases with galaxy density [769, 770]. The most spectacular result based on this approach, however, is the discovery of the morphology segregation effect [13, 14], shown in Figure 19.2. The fraction of elliptical galaxies strongly increases, while that of spirals decreases, going from the periphery to the core of clusters with ratios strongly changing with redshift [741]. This is the clearest observational evidence that the environment plays a major role in shaping galaxy evolution.

The second approach has been used by [765] to show that the stellar mass distribution of galaxies shifts by almost a factor of two towards higher masses between low- and high-density regions and that the specific star formation rate and the fraction of galaxies hosting an AGN decrease with galaxy density. A recent review on the effects of the environment on galaxy evolution made using this technique can be found in [771]. An interesting example of how galaxy properties change as a function of galaxy density is shown in Figure 19.3, where the $g - i$ vs. M_i color–magnitude relations of an optically selected sample of objects in the Coma/A1367 supercluster, defined as indicated in the previous section, are plotted in four different bins of galaxy density. Figure 19.3 clearly shows that high-density regions (cluster cores) are dominated by galaxies belonging to the red sequence, thus composed of evolved stellar populations (red ellipticals and lenticulars, Figure 19.3a), while low-density regions (field and voids) are mainly formed by star-forming systems with blue colors and young stellar populations (spirals and irregulars, Figure 19.3c).

A complete and ideal way of quantifying the effects of the environment on galaxy evolution is that of defining different samples of galaxies belonging to different environments (clusters, groups, pairs), and compare their multifrequency statistical properties to those of field galaxies (reference sample) selected according to similar criteria. This method has been first proposed by [305] to study the HI properties of galaxies in the nearby universe and has been extensively adopted to show that cluster galaxies are generally devoid of atomic gas [772]. It has been later used for several multifrequency studies of cluster galaxies, as reviewed in [10]. The derived properties of the studied galaxies might depend on the criteria adopted for defining the reference samples. This can be done either using a complete sample of truly isolated objects selected using strict geometrical criteria, as initially done by [773] and lat-

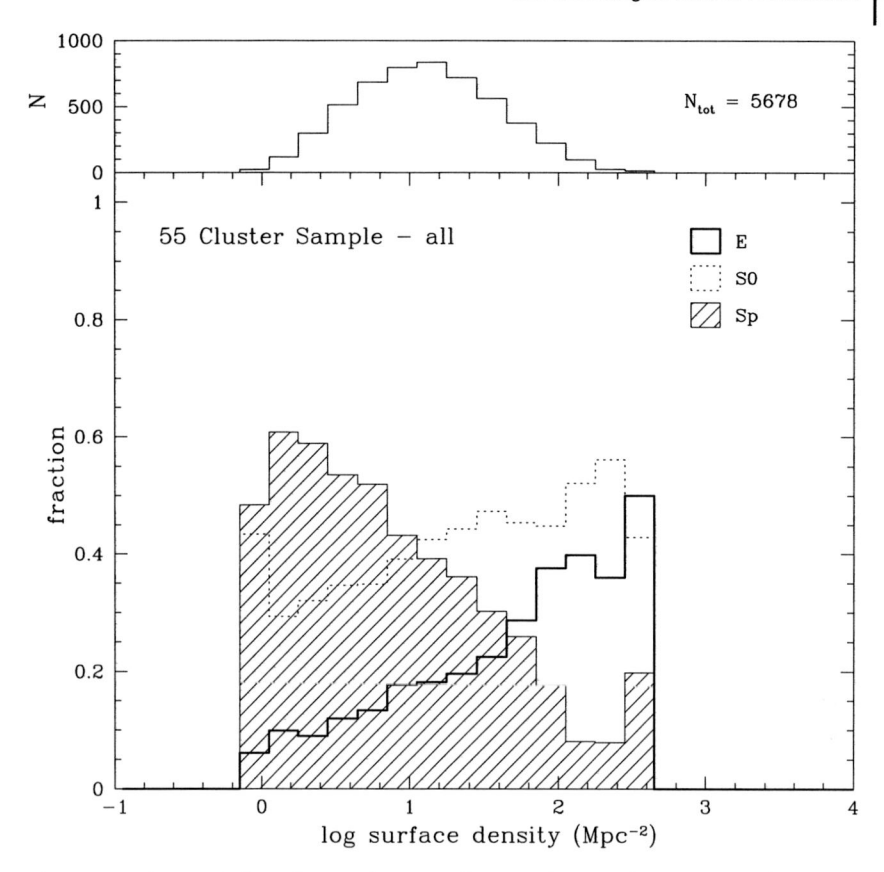

Figure 19.2 The morphology–density relation determined using 55 nearby galaxies by [13], reanalysed as described in [741]. Because these galaxies are cluster members, and are at the same distance, the X-axis gives the space density measured as number of galaxies per Mpc^2. Reproduced by permission of the AAS.

er adopted by the AMIGA project [774], or using a well-defined, complete, volume-limited sample of galaxies, as proposed by [775]. If the first method might select a sample that is not representative of the mean galaxy population of the slice of universe under study (the sample of isolated galaxies of [773] comprises only $\sim 3\%$ of the galaxies in the local universe), then the second one can include regions of different density whose statistical weight should be quantified before any analysis.

A useful parameter in statistical studies of the HI properties of galaxies belonging to different environments is the HI-deficiency parameter defined in Section 11.1.1. The atomic gas, which is located on a disk that is ~ 1.8 times more extended than the optical disk, is weakly linked to the potential well of the galaxy, and is thus easily removed in any gravitational or dynamical interaction with nearby companions or with the hot and dense intergalactic medium. Indeed cluster galaxies have, on average, an HI mass content ~ 10 times less significant than similar objects in the field (see Figure 19.4). For this reason, the HI-deficiency parameter

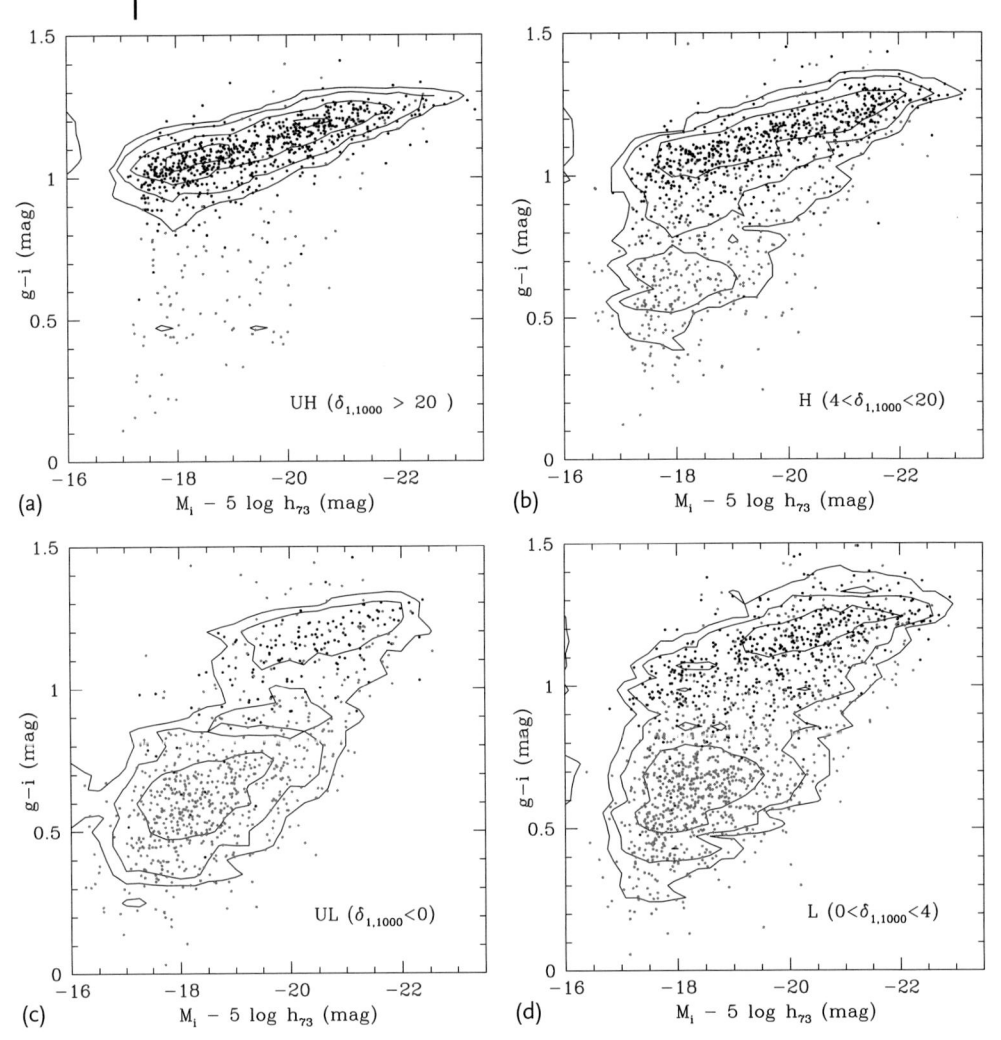

Figure 19.3 The $g-i$ vs. M_i color–magnitude relations for galaxies in the Coma/A1367 supercluster region for different bins of density as coded in Figure 19.1. (a) the highest density regions ($\delta_{1\,\text{Mpc},\,1000\,\text{km/s}} > 20$); (b) upper medium density regions ($4 < \delta_{1\,\text{Mpc},\,1000\,\text{km/s}} \leq 20$); (c) lower medium density regions ($0 < \delta_{1\,\text{Mpc},\,1000\,\text{km/s}} \leq 4$); (d) lowest density regions ($\delta_{1\,\text{Mpc},\,1000\,\text{km/s}} \leq 0$, where the mean density within the supercluster is $\langle \rho_{\text{gal}} \rangle = 0.05$ gal (h^{-1} Mpc^{-3}), from [11]. Filled dots are for early-type galaxies (ellipticals and lenticulars), gray dots are for late-type galaxies (spirals, irregulars). Reproduced with permission (c) ESO.

can be used as a direct and quantitative tracer of the perturbation that the hostile environment is inducing on different galaxies, as extensively done in [10]. The dispersion in the statistical scaling relations determined for defining this parameter in isolated galaxies, however, although relatively low (0.3 dex), limits its use in the study of single objects.

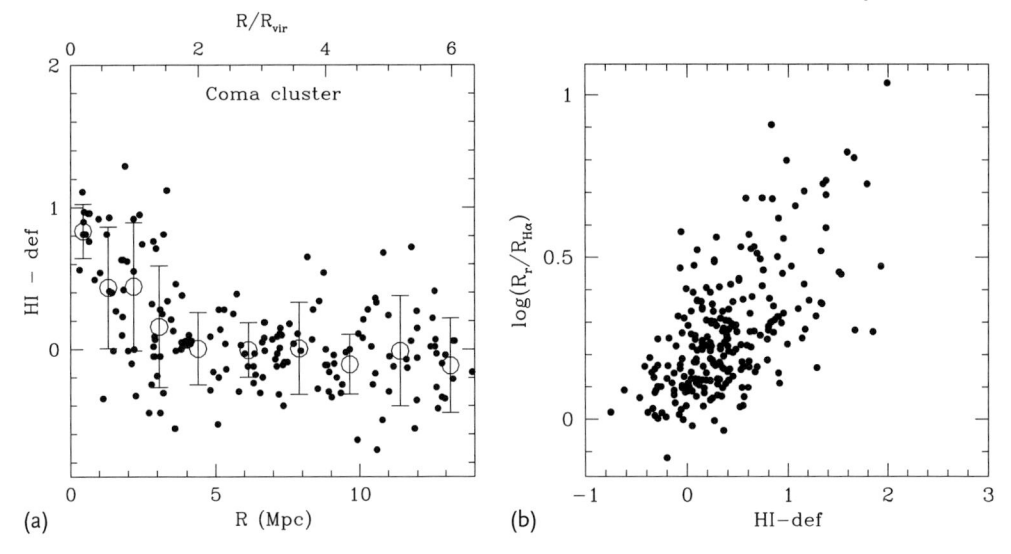

Figure 19.4 (a) Late-type galaxies located close to the center of the Coma cluster have, on average, an HI mass content ∼ ten times lower than similar objects in the field or at the periphery of the cluster, as indicated by the variation of the HI-deficiency parameter with the cluster-centric distance (filled dots indicate single galaxies, empty circles mean values within distance bins). For this reason, the HI-deficiency parameter is often used as a direct and quantitative tracer of the perturbation induced by the environment on galaxies. (b) As an example, the ratio of the optical r band to H_α radii (in logarithmic scale) is plotted vs. the HI-deficiency parameter. Galaxies with a radially truncated star formation activity, as traced by the H_α disks, are those most HI deficient (Adapted from [10]: this article originally appeared in the Publications of the Astronomical Society of the Pacific. Copyright 2004, Astronomical Society of the Pacific; reproduced with permission of the Editors.).

19.2.1
Other Tracers of Induced Perturbations

Other quantitative tracers are sometimes used to quantify the perturbation induced by the harsh environment on galaxy evolution. The most popular one is the asymmetry index A described in Section 14.3. This index, defined to quantify the degree of asymmetry in the light distribution of galaxies, is perfectly adapted for detecting any strong perturbation induced principally by strong gravitational interactions on the stellar distribution within galaxies. The availability of high resolution HI maps of galaxies in the nearby universe makes it a potential candidate for detecting and quantifying the importance of any kind of interaction, including those with the diffuse cluster intergalactic medium able to perturb only the atomic gas component of cluster galaxies. An alternative way to detect any ongoing perturbation is through the comparison of the simulated and observed HI or H_α velocity fields of galaxies, as successfully done on several nearby Virgo cluster galaxies by [776–779]. Other more qualitative tracers indicate an ongoing or past interaction. The presence of tidal tails, such as those observed in the Antennae galaxies or in NGC

4438 in the Virgo cluster, are clear examples of an ongoing tidal interaction between galaxies. Shells around elliptical galaxies, such as those observed by [780], are instead signs of a merging event that ended $\sim 10^8$ yr ago and lasted for more than 1 Gyr [781–783].

Appendix A
Photometric Redshifts and *K*-Corrections

A.1
The Photometric Redshifts

Photometric redshifts became, in recent years, the most widely used method for determining the distance of galaxies detected in deep cosmological surveys. Initially proposed by [784], they are based on the idea that, for a given photometric band, the observed range of the spectral energy distribution of galaxies shifts to shorter rest frame wavelengths as the redshift increases. Broad and narrow band imaging data are used, with very different techniques, to determine the observed low resolution spectra of the targets, which are required for an estimate of their redshift. The uncertainty on the photometric redshift determination is thus directly related to the accuracy and to the spectral resolution with which the spectra of the observed objects are measured. Photometric and spectroscopic redshifts are very complementary. Photometric redshifts (photo-z) can be, in principle, applied to any detected extragalactic source, provided that high quality photometric data covering a relatively large spectral domain are available. The success of photometric redshifts resides in the fact that they are directly applicable to large samples of very faint sources detected in deep cosmological surveys, whose measure of a spectroscopic redshift would be prohibitive. At the same time, they require spectroscopic redshifts on subsamples of objects to refine the different adopted techniques or to introduce priors used to minimize the number of erroneous redshift determinations, generally called "catastrophic failures." There exist in the literature a vast plethora of techniques and codes for measuring photometric redshifts. Describing all of them is beyond the scope of the present book; a few ideas on how they are generally determined are given here, and I refer the reader to more specialized articles for details.

A.1.1
UV-Optical-Near-Infrared Photo-z

The idea behind the determination of photometric redshift using UV-optical-near-infrared imaging data is that of comparing observed colors to those predicted for different kind of galaxies at different redshifts. This is clearly depicted in Fig-

A Panchromatic View of Galaxies, First Edition. Alessandro Boselli.
© 2012 WILEY-VCH Verlag GmbH & Co. KGaA. Published 2012 by WILEY-VCH Verlag GmbH & Co. KGaA.

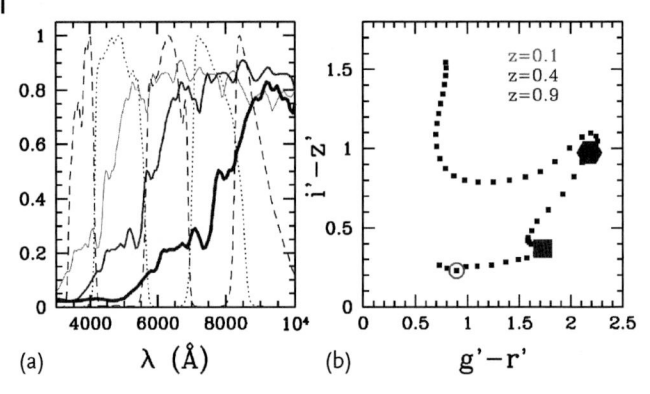

Figure A.1 The sampled range of the spectral energy distribution of a galaxy changes as a function of redshift (a), drastically modifying its observed color in the i'-z' vs. g'-r' color–color diagram (b). The SED of an observed early-type galaxy is represented by solid lines of increasing thickness, while the colors are represented by a large empty circle, a filled square, and a filled hexagon for the same galaxy at a redshift $z = 0.1$, 0.4, and 0.9, respectively. (Kindly made available by O. Ilbert.)

ure A.1, where the i'-z' vs. g'-r' color–color diagram for a template galaxy is determined as a function of its redshift. The predicted colors are measured by convolving the redshifted SED templates representative of the observed galaxies with the filter transmissivity. Templates can be empirical, determined as described in Chapter 9 using observed data, or synthetic, generated using population synthesis models assuming different star formation histories, IMF, metallicities and so on. Since the templates must be representative of the observed galaxies, they are corrected whenever necessary to account for the contribution of dust using the most adapted recipes described in Section 12.2. The effects induced by the opacity of the intergalactic medium, which might become important for very high redshift objects, should also be considered.

Data must be collected in several photometric bands to reduce any form of degeneracy as much as possible, for instance that induced by dust attenuation (highly obscured starbursts may have UV-optical-near-infrared SEDs and colors similar to those of early-type spirals). A good sampling of the UV to near-infrared spectral range is also required for constraining the position in wavelength of several well defined features in the stellar spectra of galaxies such as the Lyman break at 912 Å[1] or the $D_n(4000)$ discontinuity at ~ 4000 Å (see Section 3.3 and Chapter 10). Because of their sharp nature, these discontinuities are particularly useful for an accurate determination of photometric redshift. The non detection of those objects whose redshift brings the observed Lyman discontinuity outside the U band filter, for instance, permitted the detection of the first Lyman break galaxies (LBG) at $z \geq 3$ [785, 786].

---

1) The sharp decrease of the UV spectrum below the Lyman break is, however, smoothed in between 912 Å and the Ly-α line emission (1216 Å) because of the absorption due to the intergalactic medium.

If the methods for measuring photometric redshifts described in the literature and available on the web are varied and different, then they can be roughly divided in two main families. The first family is based on a SED fitting technique, which uses both observed and synthetic template SEDs modified to take into account the contribution of dust. The templates, which can vary in number of the different codes, are then fitted to the observed data using a simple χ^2. In this family of methods, the uncertainty on the determination of the photo-z and the number of catastrophic failures is minimized by optimizing the number and the representativity of the adopted SED templates (increasing their number might also increase the number of possible degeneracies). Among those codes based on a pure SED fitting technique, the most widely used are Hyperz [787] and GAZELLE [788]. In some of these photo-z codes, the uncertainty and the number of catastrophic failures can also be minimized by making some assumptions on the probability that a galaxy of a given magnitude has to have a redshift z. This technique is generally called Bayesian since it is based on a prior assumption. To have an idea on how a prior works, bright objects and elliptical galaxies are assumed unlikely to be at high redshift. The prior probability function can be determined, for instance, from the spectroscopic redshift distribution of galaxies of similar magnitude and color (BPZ, [789]; Kernelz), from a representative luminosity function (GOODZ [790]; LRT, [191]), or from the redshift distribution of objects predicted with semianalytical simulations (EAZY, [791]), a useful approach when spectroscopic surveys are not available. Other codes, such as Purger (template repair) [792] and ZEBRA [793], can also optimize their measurement by modifying the template SEDs to match the spectra of the observed galaxies, while others can adjust the zero point used to calibrate the photometric images to minimize any systematic difference between the template SEDs and the observed galaxies (Le Phare, [794]). Generally, all these options (prior functions, zero point matching, modification of the template SEDs) can be activated up on request in the online codes.

The second family of methods is completely empirical and based on artificial neural networks [795]. The idea behind this technique is that a galaxy with given observed parameters (different colors, magnitudes, etc.) has a given probability to have a given redshift, where the probability is determined from the observed distribution of similar objects with available redshifts. Using a logical network or decision tree algorithm, an ensemble of weak classifiers is combined into a single, powerful classifier. Among these photo-z estimators, we can remember ANNz [796] successfully applied to the SDSS [797, 798], Purger (nearest-neighbour fit) [799], Li (polynomial) [800], Wolf (empirical χ^2, which combines a SED fitting technique with neural nets) [801], ArborZ BDT [802] and Carliles (regression trees) [803]. A comparison of the performances of all these photo-z codes has been recently presented in [804].

A.1.2
Far Infrared-Radio Continuum Photo-z

Deep cosmological surveys in the infrared-submillimeter domain obtained with Spitzer, Herschel, and SCUBA revealed the presence of a dominant high redshift dust-rich galaxy population characterized by a strong infrared and a weak optical emission. These sources, which are generally called submillimeter galaxies (SMG), are ultra-luminous objects that dominate the star formation activity in the early universe [805–807]. Confusion remains severe at infrared and submillimeter wavelengths. Thus, the identification of their (faint) optical counterparts is very difficult. Because of the high dust attenuation, the measure of their distance through spectroscopic and photometric redshifts can be prohibitive. For this reason, alternative methods for measuring photo-z have been suggested in the literature. Among these, the most widely used is the far-infrared–radio photo-z method firstly proposed by [808, 809] and later adopted by [810–815]. This method is based on the far-infrared–radio correlation, one of the tightest correlations observed in almost all kind of extragalactic sources (see Section 13.8). The recent far-infrared–radio ratio vs. z calibration of [807] for submillimeter galaxies is given by the relation

$$\frac{S(850\,\mu m)}{S(1.4\,GHz)} = 11.1 + 35.2z \tag{A.1}$$

which is valid in the redshift range $1 \leq z \leq 4$, where $S(850\,\mu m)$ is the observed flux density at 850 μm and $S(1.4\,GHz)$ that at 20 cm. This calibration, which is significantly flatter than that proposed by [813] or determined using the SED of Arp 220 or of a normal, late-type galaxy as templates (see Figure A.2) can be used to measure photo-z with an uncertainty $\Delta z \sim 1$ [807]. This uncertainty is mainly dominated by the degeneracy between redshift and dust temperature [814].

Infrared data that samples both sides of the peak of the dust emission (50–200 μm) could be in principle used to measure the photometric redshift of any extragalactic source whenever radio continuum data are not available, as tentatively done by [816] using Herschel data. The strong degeneracy between the shape of the infrared SED and the dust temperature, however, make these estimates highly uncertain.

A.2
The K-Correction

Because of the expansion of the universe, imaging surveys in given photometric bands are sampling different regions of the emitted spectra of galaxies at different redshift: if λ_{em} and ν_{em} are the emitted wavelength and frequency, respectively, and λ_{obs} and ν_{obs} the observed ones, then

$$\lambda_{obs} = \lambda_{em}(1 + z) \tag{A.2}$$

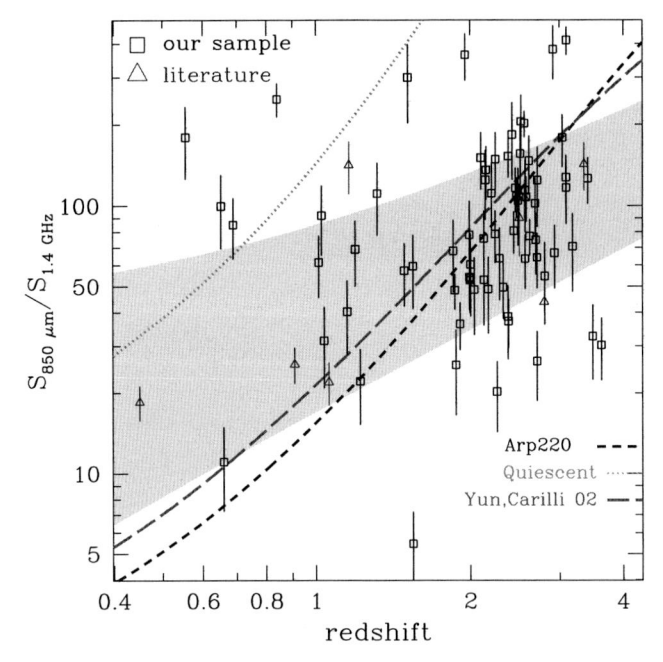

Figure A.2 The relationship between far-infrared to radio flux density ratio $S(850\,\mu m)/S(1.4\,GHz)$ and z for a sample of SMG galaxies with spectroscopic redshift, from [807]. The shaded region shows the $\pm 1\sigma$ envelope of the rms dispersion of the relation given in Eq. (A.1). The dotted, short-dashed, and long-dashed lines give the predicted variation in the flux density ratio for objects with a SED typical of a quiescent galaxy, of Arp 220 and the empirical calibration of [813], respectively. Reproduced by permission of the AAS.

and

$$\nu_{obs} = \frac{\nu_{em}}{(1+z)} \tag{A.3}$$

To be compared, imaging data of galaxies at different z are required to be corrected to the same rest frame wavelength. The corrections necessary to transform magnitudes observed in a filter R into magnitudes in a desired band Q for a galaxy at a redshift z are called K-corrections and are defined as [817–819]

$$m_R = M_Q + DM(z) + K_{QR}(z) - 5\log h \tag{A.4}$$

where

$$DM(z) = 25 + 5\log\left[\frac{D_L}{(h^{-1}\,Mpc)}\right] \tag{A.5}$$

is the bolometric distance modulus calculated from the luminosity distance D_L, and M_Q the absolute magnitude in the desired filter Q. Originally defined to calculate the correction of the colors observed in the same bands ($Q = R$), they are now commonly used in deep cosmological surveys to correct magnitudes observed in a

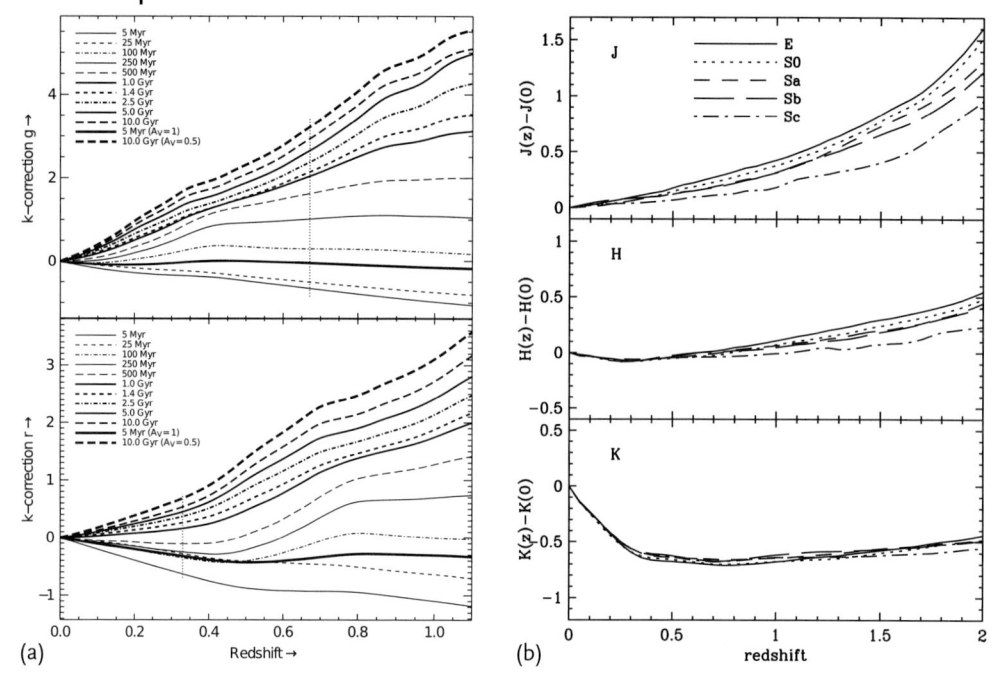

Figure A.3 (a) The K-correction in the g and r optical bands (from [825]: this article originally appeared in the Publications of the Astronomical Society of the Pacific. Copyright 2004, Astronomical Society of the Pacific; reproduced with permission of the Editors.), and (b) in the J, H and K near infrared bands (from [820]) for galaxies of different morphological type or mean age of the underlying stellar population. Galaxies with ages of 10 Gyr are representative of ellipticals, while those with ages $\leq$ 2 Gyr are representative of star-forming and starburst galaxies. Reproduced with the permission of John Wiley & Sons Ltd.

given band, R, into rest frame magnitudes in another band, Q. As clearly explained in the pedagogical paper of [818], where the complete analytical definition of K can be found, the K-correction depends on the transmissivity of the observed and desired filters R and Q, on the shape of the spectral energy distribution of the target galaxy, and on its redshift z.

As for the photometric redshifts, several codes for determining K-corrections in different photometric bands are available in the literature or on the web, as for instance the K-correct code of [819]. These codes have been written with approaches similar to those used for measuring photo-z, that is, adopting template SEDs representative of the observed sources. These templates are chosen according to the morphological type of the observed galaxy, where the morphology is determined using visual inspection [197, 820] or neural network classifications [821], by modeling the galaxy SED as a function of wavelength [819, 822] or, whenever spectra are available, directly using the spectral information, as done in early-type galaxies by [823]. Recently, various authors proposed K-correction approximations using simple analytical functions scaled on galaxy colors in SDSS or 2MASS filter bands [821, 824], or on spectral indices such as the $D_n(4000)$ [825]. As shown in Figure A.3, in each photometric band the K-correction generally increases with redshift and changes according to the morphological type or to the mean age of the stellar population of the galaxy.

Appendix B
Broad Band Photometry

Multiwavelength broad band observations provide fluxes in a given filter. With the exception of the UV-optical-near infrared bands, where they are transformed into magnitudes, fluxes are generally converted into flux densities (defined as fluxes per unit frequency or per unit wavelength) considering the transmissivity of the filter and making some assumptions on the spectral shape of the emitting sources. Flux densities per unit wavelength ($S(\lambda)$), measured in a given band, can be converted into flux densities per unit frequency ($S(\nu)$) by remembering that

$$S(\lambda)d\lambda - S(\nu)d\nu \tag{B.1}$$

and that

$$d\lambda = \frac{c}{\nu^2}d\nu \tag{B.2}$$

and

$$d\nu = \frac{c}{\lambda^2}d\lambda \tag{B.3}$$

where c is the speed of light. This implies, for example, that flux densities measured in Janskys can be converted into erg cm^{-2} s^{-1} Å^{-1} through the relation

$$S(\lambda)\,[\mathrm{erg\,cm^{-2}\,s^{-1}\,Å^{-1}}] = 2.997\,924\,58 \times 10^{-5}\,\frac{S(\nu)\,[\mathrm{Jy}]}{\lambda^2\,[\mathrm{Å}]} \tag{B.4}$$

B.1
Photometric Systems

UV, optical, and near-infrared magnitudes, m, are defined as

$$m = -2.5\log\frac{DN}{exptime} + Z_\mathrm{p} \tag{B.5}$$

where DN are the counts measured for a given source and the integration time, "$exptime$," while Z_p is the zero point necessary to transform counts into magnitudes in a given photometric system.

Appendix B Broad Band Photometry

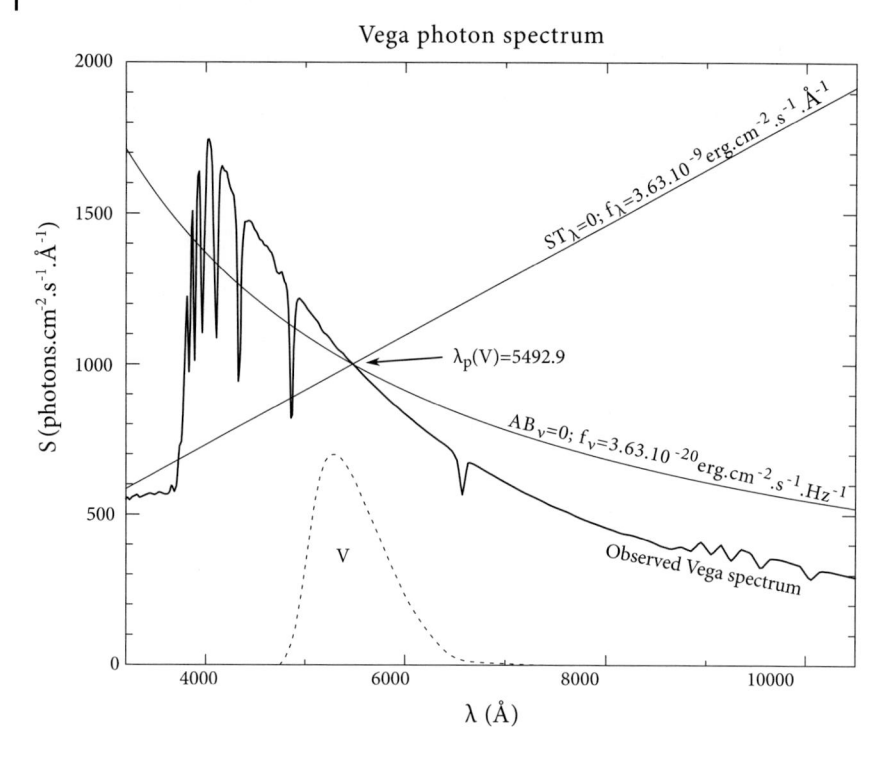

Figure B.1 The Vega photon spectrum is compared to that of the reference object for the AB and STMAG systems. The zero point of the AB and STMAG systems is chosen to have $m_{\text{Vega}}(V) = m_{\text{AB}}(V) = m_{\text{STMAG}}(V) = 0$ in the V band filter centered at $\lambda 5493$ Å (figure adapted from the HST data handbook). Reproduced with the permission of STScI.

In the Vega system, the zero point is chosen so that the magnitude of Vega (α Lyr) is equal to zero in all photometric bands[1]. Zero points are measured by means of standard stars with known magnitudes. In the recent years, two different photometric systems have been proposed, the AB system [826, 827] and the STMAG system (Space Telescope). In the AB system, now by far the most widely used, the reference object (equivalent to α Lyr in the Vega system) has a flat spectrum when the flux density $S(\nu)$ is expressed in erg cm^{-2} s^{-1} Hz^{-1}, with a zero point such as $m_{\text{AB}}(V) = m_{\text{Vega}}(V)$ (see Figure B.1), which implies that

$$m_{\text{AB}} = -2.5 \log S(\nu) - 48.60 \tag{B.6}$$

or equivalently,

$$S(\nu) \, [\text{erg cm}^{-2} \, \text{s}^{-1} \, \text{Hz}^{-1}] = 10^{-0.4(m_{\text{AB}} + 48.60)} \tag{B.7}$$

In mJy this is

$$S(\nu) \, [\text{mJy}] = 3.631 \times 10^6 \, 10^{-0.4 m_{\text{AB}}} \tag{B.8}$$

1) Recent calibrations enabled an accurate determination of the magnitude of Vega in the different bands, whose value might slightly change from 0.

Equation B.6–B.8 are valid for any photometric band. In the STMAG system, the zero point is chosen such that $m_{ST}(V) = m_{Vega}(V)$ for an object with a flat spectrum when the flux density $S(\lambda)$ is expressed in erg cm^{-2} s^{-1} Å^{-1} (see Figure B.1):

$$m_{AB} = -2.5 \log S(\lambda) - 21.10 \tag{B.9}$$

and

$$S(\lambda) \, [\text{erg cm}^{-2} \, \text{s}^{-1} \, \text{Å}^{-1}] = 10^{-0.4(m_{ST}+21.10)} = 3.631 \times 10^{-9} 10^{-0.4 m_{ST}} \tag{B.10}$$

Vega system magnitudes can be transformed into flux densities using the relation

$$f(\nu) \, [\text{erg cm}^{-2} \, \text{s}^{-1} \, \text{Hz}^{-1}] = k_{band}(\nu) \times 10^{-0.4 m_{Vega}(band)} \tag{B.11}$$

or equivalently,

$$f(\lambda) \, [\text{erg cm}^{-2} \, \text{s}^{-1} \, \text{Å}^{-1}] = k_{band}(\lambda) \times 10^{-0.4 m_{Vega}(band)} \tag{B.12}$$

Table B.1 The k_{band} values for different photometric bands.

Band	λ_{eff} μm	$k_{band}(\nu)$ 10^{-20} erg cm^{-2} s^{-1} Hz^{-1}	$k_{band}(\lambda)$ 10^{-11} erg cm^{-2} s^{-1} Å^{-1}	Ref
U Johnson–Cousins	0.366	1.790	417.5	[828]
B Johnson–Cousins	0.438	4.063	632.0	[828]
V Johnson–Cousins	0.545	3.636	363.1	[828]
R Johnson–Cousins	0.641	3.064	217.7	[828]
I Johnson–Cousins	0.798	2.416	112.6	[828]
J Johnson–Glass	1.22	1.589	31.47	[828]
H Johnson–Glass	1.63	1.021	11.38	[828]
K' Johnson–Glass	2.12	0.676	4.479	[828]
K Johnson–Glass	2.19	0.640	3.961	[828]
L Johnson–Glass	3.45	0.285	0.708	[828]
L* Johnson–Glass	3.80	0.238	0.489	[828]
J Johnson–Glass	1.22	1.631	33.14	[829]
H Johnson–Glass	1.65	1.050	11.51	[829]
K Johnson–Glass	2.18	0.655	4.139	[829]
L Johnson–Glass	3.55	0.276	0.659	[829]
L' Johnson–Glass	3.76	0.248	0.526	[829]
M Johnson–Glass	4.77	0.160	0.211	[829]
J 2MASS	1.23	1.594	31.29	[830]
H 2MASS	1.66	1.024	11.33	[830]
K_S 2MASS	2.16	0.667	4.283	[830]

Notes: different calibrations result form the slightly different transmissivity of the sets of filters at the different telescopes.

Table B.2 The c_{band} values for different photometric bands.

Band	c_{band}	Ref.
V	+0.044	[831]
B	+0.163	[831]
B_j	+0.139	[831]
R	−0.055	[831]
I	−0.309	[831]
g	+0.013	[831]
r	+0.226	[831]
i	+0.296	[831]
R_c	−0.117	[831]
I_c	−0.342	[831]
U	−0.770	[828]
B	+0.120	[828]
V	0.000	[828]
R	−0.186	[828]
I	−0.444	[828]
J	−0.899	[828]
H	−1.379	[828]
K	−1.886	[828]
K_P	−1.826	[828]
L	−2.765	[828]
L^*	−2.961	[828]

where m_{Vega}(band) is the Vega magnitude in a given photometric band and k_{band} is a wavelength-dependent variable whose value is given, for different photometric bands, in Table B.1.

Vega magnitudes can be converted into AB magnitudes using the following relation:

$$m_{AB}(\text{band}) = m_{Vega}(\text{band}) - c_{band} \qquad (B.13)$$

where c_{band} is a variable depending on the photometric band. Some values of c_{band} for different photometric bands are given in Table B.2.

Appendix C
Physical and Astronomical Constants and Unit Conversions

Table C.1 Physical constants.

Quantity	Symbol	Value	Units
Speed of light	c	$2.99792458 \times 10^{10}$	cm^{-1}
Gravitational constant	G	$6.67259(85) \times 10^{-8}$	$dyn\, cm^2\, g^{-2}$
Planck constant	h	$6.6260755(40) \times 10^{-27}$	$erg\, s^{-1}$
Boltzmann constant	k	$1.380658(12) \times 10^{-16}$	$erg\, K^{-1}$
Stefan–Boltzmann constant	σ	$5.67051(19) \times 10^{-5}$	$erg\, cm^{-2}\, K^{-4}\, s^{-1}$
Thomson cross-section	σ_T	$0.66524616 \times 10^{-24}$	cm^2
Electron charge	e	$4.8032068(15) \times 10^{-10}$	E.S.U.
Electron mass	m_e	$9.1093897(54) \times 10^{-28}$	g
Proton mass	m_p	$1.6726231(10) \times 10^{-24}$	g
Neutron mass	m_n	$1.6749286 \times 10^{-24}$	g
Atomic mass unit	m_u	$1.6605402 \times 10^{-24}$	g
Electron volt	eV	$1.6021733 \times 10^{-12}$	erg

Table C.2 Astronomical constants.

Quantity	Symbol	Value	Units
Astronomical unit	AU	1.496×10^{13}	cm
Parsec	pc	3.086×10^{18}	cm
Solar mass	$M_\odot$	1.989×10^{33}	g
Solar radius	$R_\odot$	6.955×10^{10}	cm
Solar luminosity	$L_\odot$	3.845×10^{33}	erg s^{-1}
Solar absolute bolometric magnitude	$M_{bol,\odot}$	4.72	mag
Solar absolute B magnitude	$M_{B,\odot}$	5.48	mag
Solar absolute V magnitude	$M_{V,\odot}$	4.83	mag
Solar absolute J magnitude	$M_{J,\odot}$	3.71	mag
Solar absolute H magnitude	$M_{H,\odot}$	3.37	mag
Solar absolute K magnitude	$M_{K,\odot}$	3.35	mag

Table C.3 Unit conversions.

Quantity	Symbol	Conversion
Ångström	Å	$1\,\text{Å} = 10^{-8}$ cm
Micron	μm	$1\,\mu\text{m} = 10^{-4}$ cm
Parsec	pc	$1\,\text{pc} = 3.086 \times 10^{18}$ cm
Light year	ly	$9.460\,530 \times 10^{17}$ cm
Kilo-electron volt	keV	$hc/E = 12.398\,54 \times 10^{-8}$ cm
Jansky	Jy	10^{-23} erg cm^{-2} s^{-1} Hz^{-1}

References

1. Hubble, E.P. (1926) Extragalactic nebulae. *ApJ*, **64**, 321–369, doi:10.1086/143018.
2. Hubble, E.P. (1936) *Realm of the Nebulae*, Yale University Press, New Haven.
3. de Vaucouleurs, G. (1959) Classification and Morphology of External Galaxies. *Handbuch der Physik*, **53**, 275.
4. van den Bergh, S. (1960) A preliminary luminosity clssification of late-type galaxies. *ApJ*, **131**, 215, doi:10.1086/146821.
5. Sandage, A. (1961) *The Hubble Atlas of Galaxies*.
6. van den Bergh, S. (1998) *Galaxy Morphology and Classification*.
7. Sandage, A. (2005) The classification of galaxies: early history and ongoing developments. *ARA&A*, **43**, 581–624, doi:10.1146/annurev.astro.43.112904.104839.
8. Buta, R.J., Corwin, H.G., and Odewahn, S.C. (2007) *The de Vaucouleurs Altlas of Galaxies*, Cambridge University Press.
9. Hogg, D.W., Blanton, M.R., Brinchmann, J., Eisenstein, D.J., Schlegel, D.J., Gunn, J.E., McKay, T.A., Rix, H., Bahcall, N.A., Brinkmann, J., and Meiksin, A. (2004) The Dependence on Environment of the Color-Magnitude Relation of Galaxies. *ApJ*, **601**, L29–L32, doi:10.1086/381749.
10. Boselli, A. and Gavazzi, G. (2006) Environmental effects on late-type galaxies in nearby clusters. *PASP*, **118**, 517–559, doi:10.1086/500691.
11. Gavazzi, G., Fumagalli, M., Cucciati, O., and Boselli, A. (2010) A snapshot on galaxy evolution occurring in the Great Wall: the role of Nurture at $z = 0$. *A&A*, **517**, A73, doi:10.1051/0004-6361/201014153.
12. Bamford, S.P., Nichol, R.C., Baldry, I.K., Land, K., Lintott, C.J., Schawinski, K., Slosar, A., Szalay, A.S., Thomas, D., Torki, M., Andreescu, D., Edmondson, E.M., Miller, C.J., Murray, P., Raddick, M.J., and Vandenberg, J. (2009) Galaxy zoo: the dependence of morphology and colour on environment. *MNRAS*, **393**, 1324–1352, doi:10.1111/j.1365-2966.2008.14252.x.
13. Dressler, A. (1980) Galaxy morphology in rich clusters – implications for the formation and evolution of galaxies. *ApJ*, **236**, 351–365, doi:10.1086/157753.
14. Whitmore, B.C., Gilmore, D.M., and Jones, C. (1993) What determines the morphological fractions in clusters of galaxies. *ApJ*, **407**, 489–509, doi:10.1086/172531.
15. Fukugita, M., Hogan, C.J., and Peebles, P.J.E. (1998) The cosmic baryon budget. *ApJ*, **503**, 518, doi:10.1086/306025.
16. Renzini, A. (2006) Stellar population diagnostics of elliptical galaxy formation. *ARA&A*, **44**, 141–192, doi:10.1146/annurev.astro.44.051905.092450.
17. Bennett, C.L., Bay, M., Halpern, M., Hinshaw, G., Jackson, C., Jarosik, N., Kogut, A., Limon, M., Meyer, S.S., Page, L., Spergel, D.N., Tucker, G.S., Wilkinson, D.T., Wollack, E., and Wright, E.L. (2003) The microwave anisotropy probe mission. *ApJ*, **583**, 1–23, doi:10.1086/345346.
18. Tauber, J.A., Mandolesi, N., Puget, J., Banos, T., Bersanelli, M., Bouchet,

F.R., Butler, R.C., Charra, J., Crone, G., Dodsworth, J. et al. (2010) Planck pre-launch status: the Planck mission. A&A, **520**, A1, doi:10.1051/0004-6361/200912983.

19. Spergel, D.N., Verde, L., Peiris, H.V., Komatsu, E., Nolta, M.R., Bennett, C.L., Halpern, M., Hinshaw, G., Jarosik, N., Kogut, A., Limon, M., Meyer, S.S., Page, L., Tucker, G.S., Weiland, J.L., Wollack, E., and Wright, E.L. (2003) First-year Wilkinson Microwave Anisotropy Probe (WMAP) observations: determination of cosmological parameters. *ApJS*, **148**, 175–194, doi:10.1086/377226.

20. Blanton, M.R., Hogg, D.W., Bahcall, N.A., Brinkmann, J., Britton, M., Connolly, A.J., Csabai, I., Fukugita, M., Loveday, J., Meiksin, A., Munn, J.A., Nichol, R.C., Okamura, S., Quinn, T., Schneider, D.P., Shimasaku, K., Strauss, M.A., Tegmark, M., Vogeley, M.S., and Weinberg, D.H. (2003) The galaxy luminosity function and luminosity density at redshift $z = 0.1$. *ApJ*, **592**, 819–838, doi:10.1086/375776.

21. Voges, W., Aschenbach, B., Boller, T., Bräuninger, H., Briel, U., Burkert, W., Dennerl, K., Englhauser, J., Gruber, R., Haberl, F., Hartner, G., Hasinger, G., Kürster, M., Pfeffermann, E., Pietsch, W., Predehl, P., Rosso, C., Schmitt, J.H.M.M., Trümper, J., and Zimmermann, H.U. (1999) The ROSAT all-sky survey bright source catalogue. *A&A*, **349**, 389–405.

22. Martin, D.C., Fanson, J., Schiminovich, D., Morrissey, P., Friedman, P.G., Barlow, T.A., Conrow, T., Grange, R., Jelinsky, P.N., Milliard, B., Siegmund, O.H.W., Bianchi, L., Byun, Y., Donas, J., Forster, K., Heckman, T.M., Lee, Y., Madore, B.F., Malina, R.F., Neff, S.G., Rich, R.M., Small, T., Surber, F., Szalay, A.S., Welsh, B., and Wyder, T.K. (2005) The galaxy evolution explorer: a space ultraviolet survey mission. *ApJ*, **619**, L1–L6, doi:10.1086/426387.

23. York, D.G., Adelman, J., Anderson, Jr., J.E., Anderson, S.F., Annis, J., Bahcall, N.A., Bakken, J.A., Barkhouser, R., Bastian, S., Berman, E., Boroski, W.N., Bracker, S., Briegel, C., Briggs, J.W., Brinkmann, J., Brunner, R., Burles, S., Carey, L., Carr, M.A., Castander, F.J., Chen, B., Colestock, P.L., Connolly, A.J., Crocker, J.H., Csabai, I., Czarapata, P.C., Davis, J.E., Doi, M., Dombeck, T., Eisenstein, D., Ellman, N., Elms, B.R., Evans, M.L., Fan, X., Federwitz, G.R., Fiscelli, L., Friedman, S., Frieman, J.A., Fukugita, M., Gillespie, B., Gunn, J.E., Gurbani, V.K., de Haas, E., Haldeman, M., Harris, F.H., Hayes, J., Heckman, T.M., Hennessy, G.S., Hindsley, R.B., Holm, S., Holmgren, D.J., Huang, C., Hull, C., Husby, D., Ichikawa, S., Ichikawa, T., Ivezić, Ž., Kent, S., Kim, R.S.J., Kinney, E., Klaene, M., Kleinman, A.N., Kleinman, S., Knapp, G.R., Korienek, J., Kron, R.G., Kunszt, P.Z., Lamb, D.Q., Lee, B., Leger, R.F., Limmongkol, S., Lindenmeyer, C., Long, D.C., Loomis, C., Loveday, J., Lucinio, R., Lupton, R.H., MacKinnon, B., Mannery, E.J., Mantsch, P.M., Margon, B., McGehee, P., McKay, T.A., Meiksin, A., Merelli, A., Monet, D.G., Munn, J.A., Narayanan, V.K., Nash, T., Neilsen, E., Neswold, R., Newberg, H.J., Nichol, R.C., Nicinski, T., Nonino, M., Okada, N., Okamura, S., Ostriker, J.P., Owen, R., Pauls, A.G., Peoples, J., Peterson, R.L., Petravick, D., Pier, J.R., Pope, A., Pordes, R., Prosapio, A., Rechenmacher, R., Quinn, T.R., Richards, G.T., Richmond, M.W., Rivetta, C.H., Rockosi, C.M., Ruthmansdorfer, K., Sandford, D., Schlegel, D.J., Schneider, D.P., Sekiguchi, M., Sergey, G., Shimasaku, K., Siegmund, W.A., Smee, S., Smith, J.A., Snedden, S., Stone, R., Stoughton, C., Strauss, M.A., Stubbs, C., SubbaRao, M., Szalay, A.S., Szapudi, I., Szokoly, G.P., Thakar, A.R., Tremonti, C., Tucker, D.L., Uomoto, A., Vanden Berk, D., Vogeley, M.S., Waddell, P., Wang, S., Watanabe, M., Weinberg, D.H., Yanny, B., and Yasuda, N. (2000) The Sloan digital sky survey: technical summary. *AJ*, **120**, 1579–1587, doi:10.1086/301513.

24. Skrutskie, M.F., Cutri, R.M., Stiening, R., Weinberg, M.D., Schneider, S., Carpenter, J.M., Beichman, C., Capps, R., Chester, T., Elias, J., Huchra, J., Liebert, J., Lonsdale, C., Monet, D.G., Price, S.,

Seitzer, P., Jarrett, T., Kirkpatrick, J.D., Gizis, J.E., Howard, E., Evans, T., Fowler, J., Fullmer, L., Hurt, R., Light, R., Kopan, E.L., Marsh, K.A., McCallon, H.L., Tam, R., Van Dyk, S., and Wheelock, S. (2006) The two micron all sky survey (2MASS). *AJ*, **131**, 1163–1183, doi:10.1086/498708.

25. Neugebauer, G., Habing, H.J., van Duinen, R., Aumann, H.H., Baud, B., Beichman, C.A., Beintema, D.A., Boggess, N., Clegg, P.E., de Jong, T., Emerson, J.P., Gautier, T.N., Gillett, F.C., Harris, S., Hauser, M.G., Houck, J.R., Jennings, R.E., Low, F.J., Marsden, P.L., Miley, G., Olnon, F.M., Pottasch, S.R., Raimond, E., Rowan-Robinson, M., Soifer, B.T., Walker, R.G., Wesselius, P.R., and Young, E. (1984) The Infrared Astronomical Satellite (IRAS) mission. *ApJ*, **278**, L1–L6, doi:10.1086/184209.

26. Murakami, H., Baba, H., Barthel, P., Clements, D.L., Cohen, M., Doi, Y., Enya, K., Figueredo, E., Fujishiro, N., Fujiwara, H., Fujiwara, M., Garcia-Lario, P., Goto, T., Hasegawa, S., Hibi, Y., Hirao, T., Hiromoto, N., Hong, S.S., Imai, K., Ishigaki, M., Ishiguro, M., Ishihara, D., Ita, Y., Jeong, W., Jeong, K.S., Kaneda, H., Kataza, H., Kawada, M., Kawai, T., Kawamura, A., Kessler, M.F., Kester, D., Kii, T., Kim, D.C., Kim, W., Kobayashi, H., Koo, B.C., Kwon, S.M., Lee, H.M., Lorente, R., Makiuti, S., Matsuhara, H., Matsumoto, T., Matsuo, H., Matsuura, S., Müller, T.G., Murakami, N., Nagata, H., Nakagawa, T., Naoi, T., Narita, M., Noda, M., Oh, S.H., Ohnishi, A., Ohyama, Y., Okada, Y., Okuda, H., Oliver, S., Onaka, T., Ootsubo, T., Oyabu, S., Pak, S., Park, Y., Pearson, C.P., Rowan-Robinson, M., Saito, T., Sakon, I., Salama, A., Sato, S., Savage, R.S., Serjeant, S., Shibai, H., Shirahata, M., Sohn, J., Suzuki, T., Takagi, T., Takahashi, H., Tanabé, T., Takeuchi, T.T., Takita, S., Thomson, M., Uemizu, K., Ueno, M., Usui, F., Verdugo, E., Wada, T., Wang, L., Watabe, T., Watarai, H., White, G.J., Yamamura, I., Yamauchi, C., and Yasuda, A. (2007) The infrared astronomical mission AKARI. *PASJ*, **59**, 369.

27. Becker, R.H., White, R.L., and Helfand, D.J. (1995) The FIRST survey: faint images of the radio sky at twenty centimeters. *ApJ*, **450**, 559, doi:10.1086/176166.

28. Condon, J.J., Cotton, W.D., Greisen, E.W., Yin, Q.F., Perley, R.A., Taylor, G.B., and Broderick, J.J. (1998) The NRAO VLA sky survey. *AJ*, **115**, 1693–1716, doi:10.1086/300337.

29. Barnes, D.G., Staveley-Smith, L., de Blok, W.J.G., Oosterloo, T., Stewart, I.M., Wright, A.E., Banks, G.D., Bhathal, R., Boyce, P.J., Calabretta, M.R., Disney, M.J., Drinkwater, M.J., Ekers, R.D., Freeman, K.C., Gibson, B.K., Green, A.J., Haynes, R.F., te Lintel Hekkert, P., Henning, P.A., Jerjen, H., Juraszek, S., Kesteven, M.J., Kilborn, V.A., Knezek, P.M., Koribalski, B., Kraan-Korteweg, R.C., Malin, D.F., Marquarding, M., Minchin, R.F., Mould, J.R., Price, R.M., Putman, M.E., Ryder, S.D., Sadler, E.M., Schröder, A., Stootman, F., Webster, R.L., Wilson, W.E., and Ye, T. (2001) The hi parkes all sky survey: southern observations, calibration and robust imaging. *MNRAS*, **322**, 486–498, doi:10.1046/j.1365-8711.2001.04102.x.

30. Lang, R.H., Boyce, P.J., Kilborn, V.A., Minchin, R.F., Disney, M.J., Jordan, C.A., Grossi, M., Garcia, D.A., Freeman, K.C., Phillipps, S., and Wright, A.E. (2003) First results from the HI Jodrell all sky survey: inclination-dependent selection effects in a 21-cm blind survey. *MNRAS*, **342**, 738–758, doi:10.1046/j.1365-8711.2003.06535.x.

31. Giovanelli, R., Haynes, M.P., Kent, B.R., Perillat, P., Saintonge, A., Brosch, N., Catinella, B., Hoffman, G.L., Stierwalt, S., Spekkens, K., Lerner, M.S., Masters, K.L., Momjian, E., Rosenberg, J.L., Springob, C.M., Boselli, A., Charmandaris, V., Darling, J.K., Davies, J., Garcia Lambas, D., Gavazzi, G., Giovanardi, C., Hardy, E., Hunt, L.K., Iovino, A., Karachentsev, I.D., Karachentseva, V.E., Koopmann, R.A., Marinoni, C., Minchin, R., Muller, E., Putman, M., Pantoja, C., Salzer, J.J., Scodeggio, M., Skillman, E., Solanes, J.M., Valotto, C., van Driel, W., and van Zee, L. (2005) The

arecibo legacy fast ALFA survey. I. Science goals, survey design, and strategy. *AJ*, **130**, 2598–2612, doi:10.1086/497431.

32. Colless, M., Dalton, G., Maddox, S., Sutherland, W., Norberg, P., Cole, S., Bland-Hawthorn, J., Bridges, T., Cannon, R., Collins, C., Couch, W., Cross, N., Deeley, K., De Propris, R., Driver, S.P., Efstathiou, G., Ellis, R.S., Frenk, C.S., Glazebrook, K., Jackson, C., Lahav, O., Lewis, I., Lumsden, S., Madgwick, D., Peacock, J.A., Peterson, B.A., Price, I., Seaborne, M., and Taylor, K. (2001) The 2dF galaxy redshift survey: spectra and redshifts. *MNRAS*, **328**, 1039–1063, doi:10.1046/j.1365-8711.2001.04902.x.

33. Giavalisco, M., Ferguson, H.C., Koekemoer, A.M., Dickinson, M., Alexander, D.M., Bauer, F.E., Bergeron, J., Biagetti, C., Brandt, W.N., Casertano, S., Cesarsky, C., Chatzichristou, E., Conselice, C., Cristiani, S., Da Costa, L., Dahlen, T., de Mello, D., Eisenhardt, P., Erben, T., Fall, S.M., Fassnacht, C., Fosbury, R., Fruchter, A., Gardner, J.P., Grogin, N., Hook, R.N., Hornschemeier, A.E., Idzi, R., Jogee, S., Kretchmer, C., Laidler, V., Lee, K.S., Livio, M., Lucas, R., Madau, P., Mobasher, B., Moustakas, L.A., Nonino, M., Padovani, P., Papovich, C., Park, Y., Ravindranath, S., Renzini, A., Richardson, M., Riess, A., Rosati, P., Schirmer, M., Schreier, E., Somerville, R.S., Spinrad, H., Stern, D., Stiavelli, M., Strolger, L., Urry, C.M., Vandame, B., Williams, R., and Wolf, C. (2004) The great observatories origins deep survey: initial results from optical and near-infrared imaging. *ApJ*, **600**, L93–L98, doi:10.1086/379232.

34. Scoville, N., Aussel, H., Brusa, M., Capak, P., Carollo, C.M., Elvis, M., Giavalisco, M., Guzzo, L., Hasinger, G., Impey, C., Kneib, J., LeFevre, O., Lilly, S.J., Mobasher, B., Renzini, A., Rich, R.M., Sanders, D.B., Schinnerer, E., Schminovich, D., Shopbell, P., Taniguchi, Y., and Tyson, N.D. (2007) The Cosmic Evolution Survey (COSMOS): overview. *ApJS*, **172**, 1–8, doi:10.1086/516585.

35. Kilgard, R.E., Cowan, J.J., Garcia, M.R., Kaaret, P., Krauss, M.I., McDowell, J.C., Prestwich, A.H., Primini, F.A., Stockdale, C.J., Trinchieri, G., Ward, M.J., and Zezas, A. (2005) A Chandra survey of nearby spiral galaxies. I. Point source catalogs. *ApJS*, **159**, 214–241, doi:10.1086/430443.

36. Boselli, A. and Gavazzi, G. (2002) H_α surface photometry of galaxies in the Virgo cluster. II. Observations with the OHP and Calar Alto 1.2 m telescopes. *A&A*, **386**, 124–133, doi:10.1051/0004-6361:20020215.

37. Gil de Paz, A., Boissier, S., Madore, B.F., Seibert, M., Joe, Y.H., Boselli, A., Wyder, T.K., Thilker, D., Bianchi, L., Rey, S., Rich, R.M., Barlow, T.A., Conrow, T., Forster, K., Friedman, P.G., Martin, D.C., Morrissey, P., Neff, S.G., Schiminovich, D., Small, T., Donas, J., Heckman, T.M., Lee, Y., Milliard, B., Szalay, A.S., and Yi, S. (2007) The GALEX ultraviolet atlas of nearby galaxies. *ApJS*, **173**, 185–255, doi:10.1086/516636.

38. Calzetti, D., Kennicutt, Jr., R.C., Bianchi, L., Thilker, D.A., Dale, D.A., Engelbracht, C.W., Leitherer, C., Meyer, M.J., Sosey, M.L., Mutchler, M., Regan, M.W., Thornley, M.D., Armus, L., Bendo, G.J., Boissier, S., Boselli, A., Draine, B.T., Gordon, K.D., Helou, G., Hollenbach, D.J., Kewley, L., Madore, B.F., Martin, D.C., Murphy, E.J., Rieke, G.H., Rieke, M.J., Roussel, H., Sheth, K., Smith, J.D., Walter, F., White, B.A., Yi, S., Scoville, N.Z., Polletta, M., and Lindler, D. (2005) Star formation in NGC 5194 (M51a): the panchromatic view from GALEX to Spitzer. *ApJ*, **633**, 871–893, doi:10.1086/466518.

39. Guelin, M., Zylka, R., Mezger, P.G., Haslam, C.G.T., and Kreysa, E. (1995) Cold dust emission from spiral arms of M 51. *A&A*, **298**, L29.

40. Braun, R., Oosterloo, T.A., Morganti, R., Klein, U., and Beck, R. (2007) The Westerbork SINGS survey. I. Overview and image atlas. *A&A*, **461**, 455–470, doi:10.1051/0004-6361:20066092.

41. Walter, F., Brinks, E., de Blok, W.J.G., Bigiel, F., Kennicutt, R.C., Thornley, M.D., and Leroy, A. (2008) THINGS: The H I nearby galaxy survey. *AJ*, **136**, 2563–2647, doi:10.1088/0004-6256/136/6/2563.

42. Muñoz-Mateos, J.C., Gil de Paz, A., Zamorano, J., Boissier, S., Dale, D.A., Pérez-González, P.G., Gallego, J., Madore, B.F., Bendo, G., Boselli, A., Buat, V., Calzetti, D., Moustakas, J., and Kennicutt, R.C. (2009) Radial distribution of stars, gas, and dust in SINGS galaxies. I. Surface photometry and morphology. ApJ, **703**, 1569–1596, doi:10.1088/0004-637X/703/2/1569.
43. Moustakas, J., Kennicutt, Jr., R.C., Tremonti, C.A., Dale, D.A., Smith, J., and Calzetti, D. (2010) Optical spectroscopy and nebular oxygen abundances of the Spitzer/SINGS galaxies. ApJS, **190**, 233–266, doi:10.1088/0067-0049/190/2/233.
44. Smith, J.D.T., Draine, B.T., Dale, D.A., Moustakas, J., Kennicutt, Jr., R.C., Helou, G., Armus, L., Roussel, H., Sheth, K., Bendo, G.J., Buckalew, B.A., Calzetti, D., Engelbracht, C.W., Gordon, K.D., Hollenbach, D.J., Li, A., Malhotra, S., Murphy, E.J., and Walter, F. (2007) The mid-infrared spectrum of star-forming galaxies: global properties of polycyclic aromatic hydrocarbon emission. ApJ, **656**, 770–791, doi:10.1086/510549.
45. Sanders, D.B., Soifer, B.T., Elias, J.H., Madore, B.F., Matthews, K., Neugebauer, G., and Scoville, N.Z. (1988) Ultraluminous infrared galaxies and the origin of quasars. ApJ, **325**, 74–91, doi:10.1086/165983.
46. Sanders, D.B. and Mirabel, I.F. (1996) Luminous infrared galaxies. ARA&A, **34**, 749, doi:10.1146/annurev.astro.34.1.749.
47. Giacconi, R., Branduardi, G., Briel, U., Epstein, A., Fabricant, D., Feigelson, E., Forman, W., Gorenstein, P., Grindlay, J., Gursky, H., Harnden, F.R., Henry, J.P., Jones, C., Kellogg, E., Koch, D., Murray, S., Schreier, E., Seward, F., Tananbaum, H., Topka, K., Van Speybroeck, L., Holt, S.S., Becker, R.H., Boldt, E.A., Serlemitsos, P.J., Clark, G., Canizares, C., Markert, T., Novick, R., Helfand, D., and Long, K. (1979) The Einstein /HEAO 2/ X-ray observatory. ApJ, **230**, 540–550, doi:10.1086/157110.
48. Truemper, J. (1982) The ROSAT mission. Adv. Space Res., **2**, 241–249, doi:10.1016/0273-1177(82)90070-9.
49. Tanaka, Y., Inoue, H., and Holt, S.S. (1994) The X-ray astronomy satellite ASCA. PASJ, **46**, L37–L41.
50. Weisskopf, M.C., Tananbaum, H.D., Van Speybroeck, L.P., and O'Dell, S.L. (2000) Chandra X-ray Observatory (CXO): overview, in Society of Photo-Optical Instrumentation Engineers (SPIE) Conference Series, vol. 4012 (eds J.E. Truemper and B. Aschenbach), pp. 2–16.
51. Jansen, F., Lumb, D., Altieri, B., Clavel, J., Ehle, M., Erd, C., Gabriel, C., Guainazzi, M., Gondoin, P., Much, R., Munoz, R., Santos, M., Schartel, N., Texier, D., and Vacanti, G. (2001) XMM-Newton observatory. I. The spacecraft and operations. A&A, **365**, L1–L6, doi:10.1051/0004-6361:20000036.
52. Fabbiano, G. (1989) X-rays from normal galaxies. ARA&A, **27**, 87–138, doi:10.1146/annurev.aa.27.090189.000511.
53. Mathews, W.G. and Brighenti, F. (2003) Hot gas in and around elliptical galaxies. ARA&A, **41**, 191–239, doi:10.1146/annurev.astro.41.090401.094542.
54. Fabbiano, G. (2006) Populations of X-ray sources in galaxies. ARA&A, **44**, 323–366, doi:10.1146/annurev.astro.44.051905.092519.
55. Fabbiano, G. (2008) X-ray observations of galaxies: the importance of deep high-resolution observations, in X-rays From Nearby Galaxies (eds S. Carpano, M. Ehle, and W. Pietsch), pp. 87–92.
56. Trinchieri, G. (2010) X-ray properties of normal galaxies in the local universe, in American Institute of Physics Conference Series, vol. 1248 (eds A. Comastri, L. Angelini, and M. Cappi), pp. 223–228, doi:10.1063/1.3475216.
57. Zezas, A., Alonso-Herrero, A., and Ward, M.J. (2001) Searching for X-ray luminous starburst galaxies. Ap&SS, **276**, 601–607, doi:10.1023/A:1017526118945.

58. Colbert, E.J.M. and Mushotzky, R.F. (1999) The nature of accreting black holes in nearby galaxy nuclei. *ApJ*, **519**, 89–107, doi:10.1086/307356.
59. Gladstone, J.C. (2010) Observations of ultraluminous X-ray sources, in *American Institute of Physics Conference Series*, vol. 1314 (eds V. Kologera and M. van der Sluys), pp. 353–360, doi:10.1063/1.3536400.
60. Hamilton, A.J.S., Chevalier, R.A., and Sarazin, C.L. (1983) X-ray line emission from supernova remnants. I – models for adiabatic remnants. *ApJS*, **51**, 115–147, doi:10.1086/190841.
61. Borkowski, K.J., Lyerly, W.J., and Reynolds, S.P. (2001) Supernova remnants in the Sedov expansion phase: thermal X-ray emission. *ApJ*, **548**, 820–835, doi:10.1086/319011.
62. McKee, C.F. (1974) X-ray emission from an inward-propagating shock in young supernova remnants. *ApJ*, **188**, 335–340, doi:10.1086/152721.
63. Reynolds, S.P. (1998) Models of synchrotron X-rays from shell supernova remnants. *ApJ*, **493**, 375, doi:10.1086/305103.
64. Pringle, J.E. and Rees, M.J. (1972) Accretion disk models for compact X-ray sources. *A&A*, **21**, 1.
65. Shakura, N.I. and Sunyaev, R.A. (1976) A theory of the instability of disk accretion on to black holes and the variability of binary X-ray sources, galactic nuclei and quasars. *MNRAS*, **175**, 613–632.
66. Shapiro, S.L. and Teukolsky, S.A. (1983) *Black Holes, White Dwarfs, and Neutron Stars: the Physics of Compact Objects*.
67. Prialnik, D. (2000) *An Introduction to the Theory of Stellar Structure and Evolution*.
68. Rees, M.J. (1984) Black hole models for active galactic nuclei. *ARA&A*, **22**, 471–506, doi:10.1146/annurev.aa.22.090184.002351.
69. Pringle, J.E. (1981) Accretion disks in astrophysics. *ARA&A*, **19**, 137–162, doi:10.1146/annurev.aa.19.090181.001033.
70. Begelman, M.C., Blandford, R.D., and Rees, M.J. (1984) Theory of extragalactic radio sources. *Rev. Mod. Phys.*, **56**, 255–351, doi:10.1103/RevModPhys.56.255.
71. Mewe, R., Lemen, J.R., and van den Oord, G.H.J. (1986) Calculated X-radiation from optically thin plasmas. VI – improved calculations for continuum emission and approximation formulae for nonrelativistic average Gaunt factors. *A&AS*, **65**, 511–536.
72. Raymond, J.C. and Smith, B.W. (1977) Soft X-ray spectrum of a hot plasma. *ApJS*, **35**, 419–439, doi:10.1086/190486.
73. Mewe, R., Gronenschild, E.H.B.M., and van den Oord, G.H.J. (1985) Calculated X-radiation from optically thin plasmas. V. *A&AS*, **62**, 197–254.
74. Lang, K.R. (1999) *Astrophysical Formulae*.
75. Lequeux, J. (2005) *The Interstellar Medium*.
76. Rybicki, G.B. and Lightman, A.P. (1979) *Radiative Processes in Astrophysics*.
77. Maraston, C. (2005) Evolutionary population synthesis: models, analysis of the ingredients and application to high-z galaxies. *MNRAS*, **362**, 799–825, doi:10.1111/j.1365-2966.2005.09270.x.
78. O'Connell, R.W. (1999) Far-ultraviolet radiation from elliptical galaxies. *ARA&A*, **37**, 603–648, doi:10.1146/annurev.astro.37.1.603.
79. Bressan, A., Fagotto, F., Bertelli, G., and Chiosi, C. (1993) Evolutionary sequences of stellar models with new radiative opacities. II – $Z = 0.02$. *A&AS*, **100**, 647–664.
80. Lejeune, T., Cuisinier, F., and Buser, R. (1997) Standard stellar library for evolutionary synthesis. I. Calibration of theoretical spectra. *A&AS*, **125**, 229–246, doi:10.1051/aas:1997373.
81. Charlot, S. and Bruzual, A.G. (1991) Stellar population synthesis revisited. *ApJ*, **367**, 126–140, doi:10.1086/169608.
82. Osterbrock, D.E. and Ferland, G.J. (2006) *Astrophysics of Gaseous Nebulae and Active Galactic Nuclei*.
83. Dopita, M.A. and Sutherland, R.S. (2003) *Astrophysics of the Diffuse Universe*.
84. Gavazzi, G., Zaccardo, A., Sanvito, G., Boselli, A., and Bonfanti, C. (2004) Spectrophotometry of galaxies in the Virgo cluster. II. The data. *A&A*, **417**, 499–514, doi:10.1051/0004-6361:20034105.

85 Vanden Berk, D.E., Richards, G.T., Bauer, A., Strauss, M.A., Schneider, D.P., Heckman, T.M., York, D.G., Hall, P.B., Fan, X., Knapp, G.R., Anderson, S.F., Annis, J., Bahcall, N.A., Bernardi, M., Briggs, J.W., Brinkmann, J., Brunner, R., Burles, S., Carey, L., Castander, F.J., Connolly, A.J., Crocker, J.H., Csabai, I., Doi, M., Finkbeiner, D., Friedman, S., Frieman, J.A., Fukugita, M., Gunn, J.E., Hennessy, G.S., Ivezić, Ž., Kent, S., Kunszt, P.Z., Lamb, D.Q., Leger, R.F., Long, D.C., Loveday, J., Lupton, R.H., Meiksin, A., Merelli, A., Munn, J.A., Newberg, H.J., Newcomb, M., Nichol, R.C., Owen, R., Pier, J.R., Pope, A., Rockosi, C.M., Schlegel, D.J., Siegmund, W.A., Smee, S., Snir, Y., Stoughton, C., Stubbs, C., SubbaRao, M., Szalay, A.S., Szokoly, G.P., Tremonti, C., Uomoto, A., Waddell, P., Yanny, B., and Zheng, W. (2001) Composite quasar spectra from the Sloan digital sky survey. *AJ*, **122**, 549–564, doi:10.1086/321167.

86 Hummer, D.G. and Storey, P.J. (1987) Recombination-line intensities for hydrogenic ions. I – case B calculations for HI and HeII. *MNRAS*, **224**, 801–820.

87 Pagel, B.E.J. (1997) *Nucleosynthesis and Chemical Evolution of Galaxies*.

88 Binney, J. and Merrifield, M. (1998) *Galactic Astronomy*.

89 Worthey, G. (1994) Comprehensive stellar population models and the disentanglement of age and metallicity effects. *ApJS*, **95**, 107–149, doi:10.1086/192096.

90 Worthey, G. and Ottaviani, D.L. (1997) H gamma and H delta absorption features in stars and stellar populations. *ApJS*, **111**, 377, doi:10.1086/313021.

91 Maraston, C. and Thomas, D. (2000) Strong Balmer lines in old stellar populations: no need for young ages in ellipticals. *ApJ*, **541**, 126–133, doi:10.1086/309433.

92 Rauch, M. (1998) The Lyman alpha forest in the spectra of QSOs. *ARA&A*, **36**, 267–316, doi:10.1146/annurev.astro.36.1.267.

93 Burstein, D., Faber, S.M., Gaskell, C.M., and Krumm, N. (1984) Old stellar populations. I – a spectroscopic comparison of galactic globular clusters, M31 globular clusters, and elliptical galaxies. *ApJ*, **287**, 586–609, doi:10.1086/162718.

94 Faber, S.M., Friel, E.D., Burstein, D., and Gaskell, C.M. (1985) Old stellar populations. II – an analysis of K-giant spectra. *ApJS*, **57**, 711–741, doi:10.1086/191024.

95 Worthey, G., Faber, S.M., Gonzalez, J.J., and Burstein, D. (1994) Old stellar populations. 5: absorption feature indices for the complete LICK/IDS sample of stars. *ApJS*, **94**, 687–722, doi:10.1086/192087.

96 Kauffmann, G., Heckman, T.M., White, S.D.M., Charlot, S., Tremonti, C., Brinchmann, J., Bruzual, G., Peng, E.W., Seibert, M., Bernardi, M., Blanton, M., Brinkmann, J., Castander, F., Csábai, I., Fukugita, M., Ivezic, Z., Munn, J.A., Nichol, R.C., Padmanabhan, N., Thakar, A.R., Weinberg, D.H., and York, D. (2003) Stellar masses and star formation histories for 10^5 galaxies from the Sloan digital sky survey. *MNRAS*, **341**, 33–53, doi:10.1046/j.1365-8711.2003.06291.x.

97 Spitzer, L. (1978) *Physical Processes in the Interstellar Medium*.

98 Draine, B.T. (2003) Interstellar dust grains. *ARA&A*, **41**, 241–289, doi:10.1146/annurev.astro.41.011802.094840.

99 Desert, F.X., Boulanger, F., and Puget, J.L. (1990) Interstellar dust models for extinction and emission. *A&A*, **237**, 215–236.

100 Dwek, E., Arendt, R.G., Fixsen, D.J., Sodroski, T.J., Odegard, N., Weiland, J.L., Reach, W.T., Hauser, M.G., Kelsall, T., Moseley, S.H., Silverberg, R.F., Shafer, R.A., Ballester, J., Bazell, D., and Isaacman, R. (1997) Detection and characterization of cold interstellar dust and polycyclic aromatic hydrocarbon emission, from COBE observations. *ApJ*, **475**, 565, doi:10.1086/303568.

101 Mathis, J.S., Rumpl, W., and Nordsieck, K.H. (1977) The size distribution of interstellar grains. *ApJ*, **217**, 425–433, doi:10.1086/155591.

102 Weingartner, J.C. and Draine, B.T. (2001) Dust grain-size distributions and extinction in the Milky Way, Large Magellanic Cloud, and Small Magellanic Cloud. *ApJ*, **548**, 296–309, doi:10.1086/318651.

103 Hildebrand, R.H. (1983) The determination of cloud masses and dust characteristics from submillimetre thermal emission. *QJRAS*, **24**, 267.
104 Draine, B.T. (2004) Interstellar dust, in *Origin and Evolution of the Elements*, pp. 317.
105 Joblin, C., Leger, A., and Martin, P. (1992) Contribution of polycyclic aromatic hydrocarbon molecules to the interstellar extinction curve. *ApJ*, **393**, L79–L82, doi:10.1086/186456.
106 Mathis, J.S. (1994) The origin of variations in the 2175 Å extinction bump. *ApJ*, **422**, 176–186, doi:10.1086/173715.
107 Boselli, A., Lequeux, J., and Gavazzi, G. (2004) Mid-IR emission of galaxies in the Virgo cluster and in the Coma supercluster. IV. The nature of the dust heating sources. *A&A*, **428**, 409–423, doi:10.1051/0004-6361:20041316.
108 Armus, L. (2009) Starbursts and AGN in luminous infrared galaxies, in *Astronomical Society of the Pacific Conference Series*, vol. 408 (eds W. Wang, Z. Yang, Z. Luo, and Z. Chen), pp. 105.
109 Desai, V., Armus, L., Spoon, H.W.W., Charmandaris, V., Bernard-Salas, J., Brandl, B.R., Farrah, D., Soifer, B.T., Teplitz, H.I., Ogle, P.M., Devost, D., Higdon, S.J.U., Marshall, J.A., and Houck, J.R. (2007) PAH emission from ultraluminous infrared galaxies. *ApJ*, **669**, 810–820, doi:10.1086/522104.
110 Wu, Y., Charmandaris, V., Huang, J., Spinoglio, L., and Tommasin, S. (2009) Spitzer/IRS 5–35 μm low-resolution spectroscopy of the 12 μm Seyfert sample. *ApJ*, **701**, 658–676, doi:10.1088/0004-637X/701/1/658.
111 Beirão, P., Brandl, B.R., Appleton, P.N., Groves, B., Armus, L., Förster Schreiber, N.M., Smith, J.D., Charmandaris, V., and Houck, J.R. (2008) Spatially resolved Spitzer IRS spectroscopy of the central region of M82. *ApJ*, **676**, 304–316, doi:10.1086/527343.
112 Hollenbach, D.J. and Tielens, A.G.G.M. (1999) Photodissociation regions in the interstellar medium of galaxies. *Rev. Mod. Phys.*, **71**, 173–230, doi:10.1103/RevModPhys.71.173.
113 Wolfire, M.G., Hollenbach, D., McKee, C.F., Tielens, A.G.G.M., and Bakes, E.L.O. (1995) The neutral atomic phases of the interstellar medium. *ApJ*, **443**, 152–168, doi:10.1086/175510.
114 Kaufman, M.J., Wolfire, M.G., Hollenbach, D.J., and Luhman, M.L. (1999) Far-infrared and submillimeter emission from galactic and extragalactic photodissociation regions. *ApJ*, **527**, 795–813, doi:10.1086/308102.
115 Johnstone, R.M., Hatch, N.A., Ferland, G.J., Fabian, A.C., Crawford, C.S., and Wilman, R.J. (2007) Discovery of atomic and molecular mid-infrared emission lines in off-nuclear regions of NGC 1275 and NGC 4696 with the Spitzer Space Telescope. *MNRAS*, **382**, 1246–1260, doi:10.1111/j.1365-2966.2007.12460.x.
116 Hummer, D.G. and Kunasz, P.B. (1980) Energy loss by resonance line photons in an absorbing medium. *ApJ*, **236**, 609–618, doi:10.1086/157779.
117 Charlot, S. and Fall, S.M. (1993) Lyman-alpha emission from galaxies. *ApJ*, **415**, 580, doi:10.1086/173187.
118 Hartmann, L.W., Huchra, J.P., Geller, M.J., O'Brien, P., and Wilson, R. (1988) Lyman-alpha emission in star-forming galaxies. *ApJ*, **326**, 101–109, doi:10.1086/166072.
119 Williams, P.K.G. and Bower, G.C. (2010) Evaluating the calorimeter model with broadband, continuous spectra of starburst galaxies observed with the Allen telescope array. *ApJ*, **710**, 1462–1479, doi:10.1088/0004-637X/710/2/1462.
120 Niklas, S., Klein, U., and Wielebinski, R. (1997) A radio continuum survey of Shapley-Ames galaxies at λ 2.8 cm. II. Separation of thermal and non-thermal radio emission. *A&A*, **322**, 19–28.
121 Wilson, T.L., Rohlfs, K., and Hüttemeister, S. (2009) *Tools of Radio Astronomy*, Springer-Verlag, doi:10.1007/978-3-540-85122-6.
122 Catinella, B., Haynes, M.P., Giovanelli, R., Gardner, J.P., and Connolly, A.J. (2008) A pilot survey of HI in field galaxies at redshift $z \sim 0.2$. *ApJ*, **685**, L13–L17, doi:10.1086/592328.

123 Wilson, C.D., Warren, B.E., Israel, F.P., Serjeant, S., Bendo, G., Brinks, E., Clements, D., Courteau, S., Irwin, J., Knapen, J.H., Leech, J., Matthews, H.E., Mühle, S., Mortier, A.M.J., Petitpas, G., Sinukoff, E., Spekkens, K., Tan, B.K., Tilanus, R.P.J., Usero, A., van der Werf, P., Wiegert, T., and Zhu, M. (2009) The James Clerk Maxwell Telescope nearby galaxies legacy survey. I. Star-forming molecular gas in Virgo cluster spiral galaxies. *ApJ*, **693**, 1736–1748, doi:10.1088/0004-637X/693/2/1736.

124 Gao, Y. and Solomon, P.M. (2004) HCN Survey of normal spiral, infrared-luminous, and ultraluminous galaxies. *ApJS*, **152**, 63–80, doi:10.1086/383003.

125 Gao, Y. and Solomon, P.M. (2004) The star formation rate and dense molecular gas in galaxies. *ApJ*, **606**, 271–290, doi:10.1086/382999.

126 Lovas, F.J. (1992) Recommended rest frequencies for observed interstellar molecular microwave transitions – 1991 revision. *J. Phys. Chem. Ref. Data*, **21**, 181–272.

127 Giovanelli, R. and Haynes, M.P. (1988) *Extragalactic Neutral Hydrogen*, pp. 522–562.

128 Braun, R. (1997) The temperature and opacity of atomic hydrogen in spiral galaxies. *ApJ*, **484**, 637, doi:10.1086/304346.

129 de Bruyn, A.G., O'Dea, C.P., and Baum, S.A. (1996) WSRT detection of HI absorption in the $z = 3.4$ damped Ly$_\alpha$ system in PKS 0201+113. *A&A*, **305**, 450.

130 Dickey, J.M., Strasser, S., Gaensler, B.M., Haverkorn, M., Kavars, D., McClure-Griffiths, N.M., Stil, J., and Taylor, A.R. (2009) The outer disk of the Milky Way seen in $\lambda 21$ cm absorption. *ApJ*, **693**, 1250–1260, doi:10.1088/0004-637X/693/2/1250.

131 Ciotti, L., D'Ercole, A., Pellegrini, S., and Renzini, A. (1991) Winds, outflows, and inflows in X-ray elliptical galaxies. *ApJ*, **376**, 380–403, doi:10.1086/170289.

132 Nagino, R. and Matsushita, K. (2009) Gravitational potential and X-ray luminosities of early-type galaxies observed with XMM-Newton and Chandra. *A&A*, **501**, 157–169, doi:10.1051/0004-6361/200810978.

133 Helou, G., Soifer, B.T., and Rowan-Robinson, M. (1985) Thermal infrared and nonthermal radio – remarkable correlation in disks of galaxies. *ApJ*, **298**, L7–L11, doi:10.1086/184556.

134 Dale, D.A., Helou, G., Contursi, A., Silbermann, N.A., and Kolhatkar, S. (2001) The infrared spectral energy distribution of normal star-forming galaxies. *ApJ*, **549**, 215–227, doi:10.1086/319077.

135 Dale, D.A. and Helou, G. (2002) The infrared spectral energy distribution of normal star-forming galaxies: calibration at far-infrared and submillimeter wavelengths. *ApJ*, **576**, 159–168, doi:10.1086/341632.

136 Draine, B.T. and Li, A. (2007) Infrared emission from interstellar dust. IV. The silicate-graphite-PAH model in the post-Spitzer era. *ApJ*, **657**, 810–837, doi:10.1086/511055.

137 Bothun, G.D., Lonsdale, C.J., and Rice, W. (1989) IRAS observations of optically selected galaxies. I – the properties of the UGC redshift sample. *ApJ*, **341**, 129–150, doi:10.1086/167478.

138 de Jong, T., Clegg, P.E., Rowan-Robinson, M., Soifer, B.T., Habing, H.J., Houck, J.R., Aumann, H.H., and Raimond, E. (1984) IRAS observations of Shapley-Ames galaxies. *ApJ*, **278**, L67–L70, doi:10.1086/184225.

139 Soifer, B.T., Neugebauer, G., and Houck, J.R. (1987) The IRAS view of the extragalactic sky. *ARA&A*, **25**, 187–230, doi:10.1146/annurev.aa.25.090187.001155.

140 Chary, R. and Elbaz, D. (2001) Interpreting the cosmic infrared background: constraints on the evolution of the dust-enshrouded star formation rate. *ApJ*, **556**, 562–581, doi:10.1086/321609.

141 Takeuchi, T.T., Buat, V., Iglesias-Páramo, J., Boselli, A., and Burgarella, D. (2005) Mid-infrared luminosity as an indicator of the total infrared luminosity of galaxies. *A&A*, **432**, 423–429, doi:10.1051/0004-6361:20042189.

142 Papovich, C. and Bell, E.F. (2002) On measuring the infrared luminosity of distant galaxies with the space in-

frared telescope facility. *ApJ*, **579**, L1–L4, doi:10.1086/344814.

143 Bavouzet, N., Dole, H., Le Floc'h, E., Caputi, K.I., Lagache, G., and Kochanek, C.S. (2008) Estimating the total infrared luminosity of galaxies up to $z \sim 2$ from mid- and far-infrared observations. *A&A*, **479**, 83–96, doi:10.1051/0004-6361:20077896.

144 Boquien, M., Bendo, G., Calzetti, D., Dale, D., Engelbracht, C., Kennicutt, R., Lee, J.C., van Zee, L., and Moustakas, J. (2010) Total infrared luminosity estimation of resolved and unresolved galaxies. *ApJ*, **713**, 626–636, doi:10.1088/0004-637X/713/1/626.

145 Elbaz, D., Hwang, H.S., Magnelli, B., Daddi, E., Aussel, H., Altieri, B., Amblard, A., Andreani, P., Arumugam, V., Auld, R., Babbedge, T., Berta, S., Blain, A., Bock, J., Bongiovanni, A., Boselli, A., Buat, V., Burgarella, D., Castro-Rodriguez, N., Cava, A., Cepa, J., Chanial, P., Chary, R., Cimatti, A., Clements, D.L., Conley, A., Conversi, L., Cooray, A., Dickinson, M., Dominguez, H., Dowell, C.D., Dunlop, J.S., Dwek, E., Eales, S., Farrah, D., Förster Schreiber, N., Fox, M., Franceschini, A., Gear, W., Genzel, R., Glenn, J., Griffin, M., Gruppioni, C., Halpern, M., Hatziminaoglou, E., Ibar, E., Isaak, K., Ivison, R.J., Lagache, G., Le Borgne, D., Le Floc'h, E., Levenson, L., Lu, N., Lutz, D., Madden, S., Maffei, B., Magdis, G., Mainetti, G., Maiolino, R., Marchetti, L., Mortier, A.M.J., Nguyen, H.T., Nordon, R., O'Halloran, B., Okumura, K., Oliver, S.J., Omont, A., Page, M.J., Panuzzo, P., Papageorgiou, A., Pearson, C.P., Perez Fournon, I., Pérez García, A.M., Poglitsch, A., Pohlen, M., Popesso, P., Pozzi, F., Rawlings, J.I., Rigopoulou, D., Riguccini, L., Rizzo, D., Rodighiero, G., Roseboom, I.G., Rowan-Robinson, M., Saintonge, A., Sanchez Portal, M., Santini, P., Sauvage, M., Schulz, B., Scott, D., Seymour, N., Shao, L., Shupe, D.L., Smith, A.J., Stevens, J.A., Sturm, E., Symeonidis, M., Tacconi, L., Trichas, M., Tugwell, K.E., Vaccari, M., Valtchanov, I., Vieira, J., Vigroux, L., Wang, L., Ward, R., Wright, G., Xu, C.K., and Zemcov, M. (2010) Herschel unveils a puzzling uniformity of distant dusty galaxies. *A&A*, **518**, L29, doi:10.1051/0004-6361/201014687.

146 Engelbracht, C.W., Gordon, K.D., Rieke, G.H., Werner, M.W., Dale, D.A., and Latter, W.B. (2005) Metallicity effects on mid-infrared colors and the 8 µm PAH emission in galaxies. *ApJ*, **628**, L29–L32, doi:10.1086/432613.

147 Engelbracht, C.W., Rieke, G.H., Gordon, K.D., Smith, J., Werner, M.W., Moustakas, J., Willmer, C.N.A., and Vanzi, L. (2008) Metallicity effects on dust properties in starbursting galaxies. *ApJ*, **678**, 804–827, doi:10.1086/529513.

148 Devereux, N.A. and Young, J.S. (1990) The gas/dust ratio in spiral galaxies. *ApJ*, **359**, 42–56, doi:10.1086/169031.

149 Draine, B.T. (1985) Tabulated optical properties of graphite and silicate grains. *ApJS*, **57**, 587–594, doi:10.1086/191016.

150 Spinoglio, L., Andreani, P., and Malkan, M.A. (2002) The far-infrared energy distributions of Seyfert and starburst galaxies in the local universe: infrared space observatory photometry of the 12 micron active galaxy sample. *ApJ*, **572**, 105–123, doi:10.1086/340302.

151 Lagache, G., Puget, J.L., and Dole, H. (2005) Dusty infrared galaxies: sources of the cosmic infrared background. *ARA&A*, **43**, 727–768, doi:10.1146/annurev.astro.43.072103.150606.

152 Boselli, A., Gavazzi, G., and Sanvito, G. (2003) UV to radio centimetric spectral energy distributions of optically-selected late-type galaxies in the Virgo cluster. *A&A*, **402**, 37–51, doi:10.1051/0004-6361:20030219.

153 Dale, D.A., Bendo, G.J., Engelbracht, C.W., Gordon, K.D., Regan, M.W., Armus, L., Cannon, J.M., Calzetti, D., Draine, B.T., Helou, G., Joseph, R.D., Kennicutt, R.C., Li, A., Murphy, E.J., Roussel, H., Walter, F., Hanson, H.M., Hollenbach, D.J., Jarrett, T.H., Kewley, L.J., Lamanna, C.A., Leitherer, C., Meyer, M.J., Rieke, G.H., Rieke, M.J., Sheth, K., Smith, J.D.T., and Thornley, M.D. (2005) Infrared spectral energy distributions of nearby galaxies. *ApJ*, **633**, 857–870, doi:10.1086/491642.

154 Li, A. and Draine, B.T. (2001) Infrared emission from interstellar dust. II. The diffuse interstellar medium. *ApJ*, **554**, 778–802, doi:10.1086/323147.

155 Vlahakis, C., Dunne, L., and Eales, S. (2005) The SCUBA local universe galaxy survey – III. Dust along the Hubble sequence. *MNRAS*, **364**, 1253–1285, doi:10.1111/j.1365-2966.2005.09666.x.

156 Draine, B.T. and Lee, H.M. (1984) Optical properties of interstellar graphite and silicate grains. *ApJ*, **285**, 89–108, doi:10.1086/162480.

157 Hughes, D.H., Robson, E.I., Dunlop, J.S., and Gear, W.K. (1993) Thermal dust emission from quasars – part one – submillimetre spectral indices of radio quiet quasars. *MNRAS*, **263**, 607.

158 James, A., Dunne, L., Eales, S., and Edmunds, M.G. (2002) SCUBA observations of galaxies with metallicity measurements: a new method for determining the relation between submillimetre luminosity and dust mass. *MNRAS*, **335**, 753–761, doi:10.1046/j.1365-8711.2002.05660.x.

159 da Cunha, E., Charlot, S., and Elbaz, D. (2008) A simple model to interpret the ultraviolet, optical and infrared emission from galaxies. *MNRAS*, **388**, 1595–1617, doi:10.1111/j.1365-2966.2008.13535.x.

160 Silva, L., Granato, G.L., Bressan, A., and Danese, L. (1998) Modeling the effects of dust on galactic spectral energy distributions from the ultraviolet to the millimeter band. *ApJ*, **509**, 103–117, doi:10.1086/306476.

161 Popescu, C.C., Misiriotis, A., Kylafis, N.D., Tuffs, R.J., and Fischera, J. (2000) Modelling the spectral energy distribution of galaxies. I. Radiation fields and grain heating in the edge-on spiral NGC 891. *A&A*, **362**, 138–150.

162 Draine, B.T., Dale, D.A., Bendo, G., Gordon, K.D., Smith, J.D.T., Armus, L., Engelbracht, C.W., Helou, G., Kennicutt, Jr., R.C., Li, A., Roussel, H., Walter, F., Calzetti, D., Moustakas, J., Murphy, E.J., Rieke, G.H., Bot, C., Hollenbach, D.J., Sheth, K., and Teplitz, H.I. (2007) Dust masses, PAH abundances, and starlight intensities in the SINGS galaxy sample. *ApJ*, **663**, 866–894, doi:10.1086/518306.

163 Muñoz-Mateos, J.C., Gil de Paz, A., Boissier, S., Zamorano, J., Dale, D.A., Pérez-González, P.G., Gallego, J., Madore, B.F., Bendo, G., Thornley, M.D., Draine, B.T., Boselli, A., Buat, V., Calzetti, D., Moustakas, J., and Kennicutt, R.C. (2009) Radial distribution of stars, gas, and dust in sings galaxies. II. Derived dust properties. *ApJ*, **701**, 1965–1991, doi:10.1088/0004-637X/701/2/1965.

164 Popescu, C.C., Tuffs, R.J., Völk, H.J., Pierini, D., and Madore, B.F. (2002) Cold dust in late-type Virgo cluster galaxies. *ApJ*, **567**, 221–236, doi:10.1086/338383.

165 Sadler, E.M., Slee, O.B., Reynolds, J.E., and Roy, A.L. (1995) Parsec-scale radio cores in spiral galaxies. *MNRAS*, **276**, 1373–1381.

166 Condon, J.J. (1992) Radio emission from normal galaxies. *ARA&A*, **30**, 575–611, doi:10.1146/annurev.aa.30.090192.003043.

167 Duric, N., Bourneuf, E., and Gregory, P.C. (1988) The separation of synchrotron and bremsstrahlung radio emission in spiral galaxies. *AJ*, **96**, 81–91, doi:10.1086/114791.

168 Lequeux, J., Maucherat-Joubert, M., Deharveng, J.M., and Kunth, D. (1981) Star formation and extinction in extragalactic HII regions. *A&A*, **103**, 305–318.

169 Caplan, J. and Deharveng, L. (1986) Extinction and reddening of HII regions in the Large Magellanic Cloud. *A&A*, **155**, 297–313.

170 Dumke, M., Braine, J., Krause, M., Zylka, R., Wielebinski, R., and Guelin, M. (1997) The interstellar medium in the edge-on galaxy NGC 5907. Cold dust and molecular line emission. *A&A*, **325**, 124–134.

171 Braine, J., Kruegel, E., Sievers, A., and Wielebinski, R. (1995) 1.3 mm continuum emission in the late-type spiral NGC 4631. Dust in NGC 4631. *A&A*, **295**, L55.

172 Braine, J. and Hughes, D.H. (1999) The 10 GHz–10 THz spectrum of a normal spiral galaxy. *A&A*, **344**, 779–786.

173 Condon, J.J., Cotton, W.D., and Broderick, J.J. (2002) Radio sources and star formation in the local universe. *AJ*, **124**, 675–689, doi:10.1086/341650.

174 Condon, J.J., Huang, Z., Yin, Q.F., and Thuan, T.X. (1991) Compact starbursts in ultraluminous infrared galaxies. *ApJ*, **378**, 65–76, doi:10.1086/170407.

175 Longair, M.S. (1981) *High Energy Astrophysics*.

176 Lequeux, J. (1971) The radio continuum of galaxies. II. The origin of the continuum emission in spiral galaxies. *A&A*, **15**, 42.

177 Kennicutt, R. (1983) The origin of the nonthermal radio emission in normal disk galaxies. *A&A*, **120**, 219–222.

178 Gavazzi, G., Boselli, A., and Kennicutt, R. (1991) Multifrequency windows on spiral galaxies. I – UBV and H-alpha aperture photometry. *AJ*, **101**, 1207–1230, doi:10.1086/115758.

179 Beck, R. (2005) Magnetic fields in galaxies, in *Cosmic Magnetic Fields*, Lecture Notes in Physics, vol. 664 (eds R. Wielebinski and R. Beck), Springer Verlag, Berlin, pp. 41, doi:10.1007/11369875_3.

180 Beck, R., Brandenburg, A., Moss, D., Shukurov, A., and Sokoloff, D. (1996) Galactic magnetism: recent developments and perspectives. *ARA&A*, **34**, 155–206, doi:10.1146/annurev.astro.34.1.155.

181 Boselli, A., Cortese, L., Deharveng, J.M., Gavazzi, G., Yi, K.S., Gil de Paz, A., Seibert, M., Boissier, S., Donas, J., Lee, Y., Madore, B.F., Martin, D.C., Rich, R.M., and Sohn, Y. (2005) UV properties of early-type galaxies in the Virgo cluster. *ApJ*, **629**, L29–L32, doi:10.1086/444534.

182 Edelson, R.A. and Malkan, M.A. (1986) Spectral energy distributions of active galactic nuclei between 0.1 and 100 microns. *ApJ*, **308**, 59–77, doi:10.1086/164479.

183 Schmitt, H.R., Kinney, A.L., Calzetti, D., and Storchi Bergmann, T. (1997) The spectral energy distribution of normal, starburst, and active galaxies. *AJ*, **114**, 592–612, doi:10.1086/118496.

184 Ho, L.C. (1999) The spectral energy distributions of low-luminosity active galactic nuclei. *ApJ*, **516**, 672–682, doi:10.1086/307137.

185 Dale, D.A., Gil de Paz, A., Gordon, K.D., Hanson, H.M., Armus, L., Bendo, G.J., Bianchi, L., Block, M., Boissier, S., Boselli, A., Buckalew, B.A., Buat, V., Burgarella, D., Calzetti, D., Cannon, J.M., Engelbracht, C.W., Helou, G., Hollenbach, D.J., Jarrett, T.H., Kennicutt, R.C., Leitherer, C., Li, A., Madore, B.F., Martin, D.C., Meyer, M.J., Murphy, E.J., Regan, M.W., Roussel, H., Smith, J.D.T., Sosey, M.L., Thilker, D.A., and Walter, F. (2007) An ultraviolet-to-radio broadband spectral atlas of nearby galaxies. *ApJ*, **655**, 863–884, doi:10.1086/510362.

186 Polletta, M., Tajer, M., Maraschi, L., Trinchieri, G., Lonsdale, C.J., Chiappetti, L., Andreon, S., Pierre, M., Le Fèvre, O., Zamorani, G., Maccagni, D., Garcet, O., Surdej, J., Franceschini, A., Alloin, D., Shupe, D.L., Surace, J.A., Fang, F., Rowan-Robinson, M., Smith, H.E., and Tresse, L. (2007) Spectral energy distributions of hard X-ray selected active galactic nuclei in the XMM-Newton medium deep survey. *ApJ*, **663**, 81–102, doi:10.1086/518113.

187 Faber, S.M. (1973) Variations in spectral-energy distributions and absorption-line strengths among elliptical galaxies. *ApJ*, **179**, 731–754, doi:10.1086/151912.

188 Krueger, H., Fritze-v. Alvensleben, U., and Loose, H. (1995) Optical and near infrared spectral energy distributions. Of blue compact galaxies from evolutionary synthesis. *A&A*, **303**, 41.

189 Gavazzi, G., Bonfanti, C., Sanvito, G., Boselli, A., and Scodeggio, M. (2002) Spectrophotometry of galaxies in the Virgo cluster. I. The star formation history. *ApJ*, **576**, 135–151, doi:10.1086/341730.

190 Boselli, A., Lequeux, J., Sauvage, M., Boulade, O., Boulanger, F., Cesarsky, D., Dupraz, C., Madden, S., Viallefond, F., and Vigroux, L. (1998) Mid-IR emission of galaxies in the Virgo cluster. II. Integrated properties. *A&A*, **335**, 53–68.

191 Assef, R.J., Kochanek, C.S., Brodwin, M., Brown, M.J.I., Caldwell, N., Cool, R.J., Eisenhardt, P., Eisenstein, D., Gonzalez, A.H., Jannuzi, B.T., Jones, C., McKenzie, E., Murray, S.S., and Stern, D. (2008) Low-Resolution spectral templates for galaxies from 0.2 to 10 μm. *ApJ*, **676**, 286–303, doi:10.1086/527533.

192 Assef, R.J., Kochanek, C.S., Brodwin, M., Cool, R., Forman, W., Gonzalez, A.H., Hickox, R.C., Jones, C., Le Floc'h, E., Moustakas, J., Murray, S.S., and Stern, D. (2010) Low-resolution spectral templates for active galactic nuclei and galaxies from 0.03 to 30 μm. *ApJ*, **713**, 970–985, doi:10.1088/0004-637X/713/2/970.

193 Boselli, A. (2008) Combining UV to radio continuum photometric and spectroscopic data of galaxies for scientific purposes, in *Astronomical Spectroscopy and Virtual Observatory* (eds M. Guainazzi and P. Osuna), pp. 147.

194 Boselli, A., Ciesla, L., Buat, V., Cortese, L., Auld, R., Baes, M., Bendo, G.J., Bianchi, S., Bock, J., Bomans, D.J., Bradford, M., Castro-Rodriguez, N., Chanial, P., Charlot, S., Clemens, M., Clements, D., Corbelli, E., Cooray, A., Cormier, D., Dariush, A., Davies, J., de Looze, I., di Serego Alighieri, S., Dwek, E., Eales, S., Elbaz, D., Fadda, D., Fritz, J., Galametz, M., Galliano, F., Garcia-Appadoo, D.A., Gavazzi, G., Gear, W., Giovanardi, C., Glenn, J., Gomez, H., Griffin, M., Grossi, M., Hony, S., Hughes, T.M., Hunt, L., Isaak, K., Jones, A., Levenson, L., Lu, N., Madden, S.C., O'Halloran, B., Okumura, K., Oliver, S., Page, M., Panuzzo, P., Papageorgiou, A., Parkin, T., Perez-Fournon, I., Pierini, D., Pohlen, M., Rangwala, N., Rigby, E., Roussel, H., Rykala, A., Sabatini, S., Sacchi, N., Sauvage, M., Schulz, B., Schirm, M., Smith, M.W.L., Spinoglio, L., Stevens, J., Sundar, S., Symeonidis, M., Trichas, M., Vaccari, M., Verstappen, J., Vigroux, L., Vlahakis, C., Wilson, C., Wozniak, H., Wright, G., Xilouris, E.M., Zeilinger, W., and Zibetti, S. (2010) FIR colours and SEDs of nearby galaxies observed with Herschel. *A&A*, **518**, L61, doi:10.1051/0004-6361/201014534.

195 Larson, R.B. and Tinsley, B.M. (1978) Star formation rates in normal and peculiar galaxies. *ApJ*, **219**, 46–59, doi:10.1086/155753.

196 Kennicutt, Jr., R.C., Tamblyn, P., and Congdon, C.E. (1994) Past and future star formation in disk galaxies. *ApJ*, **435**, 22–36, doi:10.1086/174790.

197 Fukugita, M., Shimasaku, K., and Ichikawa, T. (1995) Galaxy colors in various photometric band systems. *PASP*, **107**, 945, doi:10.1086/133643.

198 Gavazzi, G., Pierini, D., and Boselli, A. (1996) The phenomenology of disk galaxies. *A&A*, **312**, 397–408.

199 Richards, G.T., Fan, X., Schneider, D.P., Vanden Berk, D.E., Strauss, M.A., York, D.G., Anderson, Jr., J.E., Anderson, S.F., Annis, J., Bahcall, N.A., Bernardi, M., Briggs, J.W., Brinkmann, J., Brunner, R., Burles, S., Carey, L., Castander, F.J., Connolly, A.J., Crocker, J.H., Csabai, I., Doi, M., Finkbeiner, D., Friedman, S.D., Frieman, J.A., Fukugita, M., Gunn, J.E., Hindsley, R.B., Ivezić, Ž., Kent, S., Knapp, G.R., Lamb, D.Q., Leger, R.F., Long, D.C., Loveday, J., Lupton, R.H., McKay, T.A., Meiksin, A., Merrelli, A., Munn, J.A., Newberg, H.J., Newcomb, M., Nichol, R.C., Owen, R., Pier, J.R., Pope, A., Richmond, M.W., Rockosi, C.M., Schlegel, D.J., Siegmund, W.A., Smee, S., Snir, Y., Stoughton, C., Stubbs, C., SubbaRao, M., Szalay, A.S., Szokoly, G.P., Tremonti, C., Uomoto, A., Waddell, P., Yanny, B., and Zheng, W. (2001) Colors of 2625 quasars at $0 < z < 5$ measured in the Sloan digital sky survey photometric system. *AJ*, **121**, 2308–2330, doi:10.1086/320392.

200 Walcher, J., Groves, B., Budavári, T., and Dale, D. (2011) Fitting the integrated spectral energy distributions of galaxies. *Ap&SS*, **331**, 1–52, doi:10.1007/s10509-010-0458-z.

201 Bertelli, G., Bressan, A., Chiosi, C., Fagotto, F., and Nasi, E. (1994) Theoretical isochrones from models with new radiative opacities. *A&AS*, **106**, 275–302.

202 Girardi, L., Bressan, A., Bertelli, G., and Chiosi, C. (2000) Evolutionary tracks and isochrones for low- and intermediate-mass stars: from 0.15 to $7M_{sun}$, and from $Z = 0.0004$ to 0.03. *A&AS*, **141**, 371–383, doi:10.1051/aas:2000126.

203 Girardi, L., Bertelli, G., Bressan, A., Chiosi, C., Groenewegen, M.A.T., Marigo, P., Salasnich, B., and Weiss, A. (2002) Theoretical isochrones in several photometric systems. I. Johnson–Cousins–

Glass, HST/WFPC2, HST/NICMOS, Washington, and ESO imaging survey filter sets. *A&A*, **391**, 195–212, doi:10.1051/0004-6361:20020612.

204 Maeder, A. and Meynet, G. (1988) Tables of evolutionary star models from 0.85 to 120 solar masses with overshooting and mass loss. *A&AS*, **76**, 411–425.

205 Schaller, G., Schaerer, D., Meynet, G., and Maeder, A. (1992) New grids of stellar models from 0.8 to 120 solar masses at $Z = 0.020$ and $Z = 0.001$. *A&AS*, **96**, 269–331.

206 Meynet, G., Maeder, A., Schaller, G., Schaerer, D., and Charbonnel, C. (1994) Grids of massive stars with high mass loss rates. V. From 12 to $120 M_{sun}$ at $Z = 0.001, 0.004, 0.008, 0.020$ and 0.040. *A&AS*, **103**, 97–105.

207 Pickles, A.J. (1998) A stellar spectral flux library: 1150–25 000 Å. *PASP*, **110**, 863–878, doi:10.1086/316197.

208 Prugniel, P. and Soubiran, C. (2001) A database of high and medium-resolution stellar spectra. *A&A*, **369**, 1048–1057, doi:10.1051/0004-6361:20010163.

209 Le Borgne, J., Bruzual, G., Pelló, R., Lançon, A., Rocca-Volmerange, B., Sanahuja, B., Schaerer, D., Soubiran, C., and Vílchez-Gómez, R. (2003) STELIB: A library of stellar spectra at $R \sim 2000$. *A&A*, **402**, 433–442, doi:10.1051/0004-6361:20030243.

210 Sánchez-Blázquez, P., Peletier, R.F., Jiménez-Vicente, J., Cardiel, N., Cenarro, A.J., Falcón-Barroso, J., Gorgas, J., Selam, S., and Vazdekis, A. (2006) Medium-resolution Isaac Newton telescope library of empirical spectra. *MNRAS*, **371**, 703–718, doi:10.1111/j.1365-2966.2006.10699.x.

211 Lejeune, T., Cuisinier, F., and Buser, R. (1998) A standard stellar library for evolutionary synthesis. II. The M dwarf extension. *A&AS*, **130**, 65–75, doi:10.1051/aas:1998405.

212 Palacios, A., Gebran, M., Josselin, E., Martins, F., Plez, B., Belmas, M., and Lèbre, A. (2010) POLLUX: a database of synthetic stellar spectra. *A&A*, **516**, A13, doi:10.1051/0004-6361/200913932.

213 Bruzual, G. and Charlot, S. (2003) Stellar population synthesis at the resolution of 2003. *MNRAS*, **344**, 1000–1028, doi:10.1046/j.1365-8711.2003.06897.x.

214 Maraston, C. (1998) Evolutionary synthesis of stellar populations: a modular tool. *MNRAS*, **300**, 872–892, doi:10.1046/j.1365-8711.1998.01947.x.

215 Thomas, D., Maraston, C., and Bender, R. (2003) Stellar population models of Lick indices with variable element abundance ratios. *MNRAS*, **339**, 897–911, doi:10.1046/j.1365-8711.2003.06248.x.

216 Fioc, M. and Rocca-Volmerange, B. (1997) PEGASE: a UV to NIR spectral evolution model of galaxies. Application to the calibration of bright galaxy counts. *A&A*, **326**, 950–962.

217 Le Borgne, D., Rocca-Volmerange, B., Prugniel, P., Lançon, A., Fioc, M., and Soubiran, C. (2004) Evolutionary synthesis of galaxies at high spectral resolution with the code PEGASE-HR. Metallicity and age tracers. *A&A*, **425**, 881–897, doi:10.1051/0004-6361:200400044.

218 Leitherer, C., Schaerer, D., Goldader, J.D., González Delgado, R.M., Robert, C., Kune, D.F., de Mello, D.F., Devost, D., and Heckman, T.M. (1999) Starburst99: synthesis models for galaxies with active star formation. *ApJS*, **123**, 3–40, doi:10.1086/313233.

219 Vázquez, G.A. and Leitherer, C. (2005) Optimization of Starburst99 for intermediate-age and old stellar populations. *ApJ*, **621**, 695–717, doi:10.1086/427866.

220 Sandage, A. (1986) Star formation rates, galaxy morphology, and the Hubble sequence. *A&A*, **161**, 89–101.

221 Boissier, S. and Prantzos, N. (2000) Chemo-spectrophotometric evolution of spiral galaxies – II. Main properties of present-day disk galaxies. *MNRAS*, **312**, 398–416, doi:10.1046/j.1365-8711.2000.03133.x.

222 Hatziminaoglou, E., Omont, A., Stevens, J.A., Amblard, A., Arumugam, V., Auld, R., Aussel, H., Babbedge, T., Blain, A., Bock, J., Boselli, A., Buat, V., Burgarella, D., Castro-Rodríguez, N., Cava, A., Chanial, P., Clements, D.L., Conley, A., Conversi, L., Cooray, A., Dowell, C.D., Dwek, E., Dye, S., Eales, S., Elbaz, D., Farrah, D., Fox, M., Franceschini, A.,

Gear, W., Glenn, J., González Solares, E.A., Griffin, M., Halpern, M., Ibar, E., Isaak, K., Ivison, R.J., Lagache, G., Levenson, L., Lu, N., Madden, S., Maffei, B., Mainetti, G., Marchetti, L., Mortier, A.M.J., Nguyen, H.T., O'Halloran, B., Oliver, S.J., Page, M.J., Panuzzo, P., Papageorgiou, A., Pearson, C.P., Pérez-Fournon, I., Pohlen, M., Rawlings, J.I., Rigopoulou, D., Rizzo, D., Roseboom, I.G., Rowan-Robinson, M., Sanchez Portal, M., Schulz, B., Scott, D., Seymour, N., Shupe, D.L., Smith, A.J., Symeonidis, M., Trichas, M., Tugwell, K.E., Vaccari, M., Valtchanov, I., Vigroux, L., Wang, L., Ward, R., Wright, G., Xu, C.K., and Zemcov, M. (2010) HerMES: far infrared properties of known AGN in the HerMES fields. *A&A*, **518**, L33, doi:10.1051/0004-6361/201014679.

223 Rex, M., Rawle, T.D., Egami, E., Pérez-González, P.G., Zemcov, M., Aretxaga, I., Chung, S.M., Fadda, D., Gonzalez, A.H., Hughes, D.H., Horellou, C., Johansson, D., Kneib, J., Richard, J., Altieri, B., Fiedler, A.K., Pereira, M.J., Rieke, G.H., Smail, I., Valtchanov, I., Blain, A.W., Bock, J.J., Boone, F., Bridge, C.R., Clement, B., Combes, F., Dowell, C.D., Dessauges-Zavadsky, M., Ilbert, O., Ivison, R.J., Jauzac, M., Lutz, D., Omont, A., Pelló, R., Rodighiero, G., Schaerer, D., Smith, G.P., Walth, G.L., van der Werf, P., Werner, M.W., Austermann, J.E., Ezawa, H., Kawabe, R., Kohno, K., Perera, T.A., Scott, K.S., Wilson, G.W., and Yun, M.S. (2010) The far-infrared/submillimeter properties of galaxies located behind the bullet cluster. *A&A*, **518**, L13, doi:10.1051/0004-6361/201014693.

224 Rodighiero, G., Cimatti, A., Gruppioni, C., Popesso, P., Andreani, P., Altieri, B., Aussel, H., Berta, S., Bongiovanni, A., Brisbin, D., Cava, A., Cepa, J., Daddi, E., Dominguez-Sanchez, H., Elbaz, D., Fontana, A., Förster Schreiber, N., Franceschini, A., Genzel, R., Grazian, A., Lutz, D., Magdis, G., Magliocchetti, M., Magnelli, B., Maiolino, R., Mancini, C., Nordon, R., Perez Garcia, A.M., Poglitsch, A., Santini, P., Sanchez-Portal, M., Pozzi, F., Riguccini, L., Saintonge, A., Shao, L., Sturm, E., Tacconi, L., Valtchanov, I., Wetzstein, M., and Wieprecht, E. (2010) The first Herschel view of the mass-SFR link in high-z galaxies. *A&A*, **518**, L25, doi:10.1051/0004-6361/201014624.

225 Lacy, M., Storrie-Lombardi, L.J., Sajina, A., Appleton, P.N., Armus, L., Chapman, S.C., Choi, P.I., Fadda, D., Fang, F., Frayer, D.T., Heinrichsen, I., Helou, G., Im, M., Marleau, F.R., Masci, F., Shupe, D.L., Soifer, B.T., Surace, J., Teplitz, H.I., Wilson, G., and Yan, L. (2004) Obscured and unobscured active galactic nuclei in the Spitzer space telescope first look survey. *ApJS*, **154**, 166–169, doi:10.1086/422816.

226 Stern, D., Eisenhardt, P., Gorjian, V., Kochanek, C.S., Caldwell, N., Eisenstein, D., Brodwin, M., Brown, M.J.I., Cool, R., Dey, A., Green, P., Jannuzi, B.T., Murray, S.S., Pahre, M.A., and Willner, S.P. (2005) Mid-Infrared selection of active galaxies. *ApJ*, **631**, 163–168, doi:10.1086/432523.

227 Brand, K., Dey, A., Weedman, D., Desai, V., Le Floc'h, E., Jannuzi, B.T., Soifer, B.T., Brown, M.J.I., Eisenhardt, P., Gorjian, V., Papovich, C., Smith, H.A., Willner, S.P., and Cool, R.J. (2006) The active galactic nuclei contribution to the mid-infrared emission of luminous infrared galaxies. *ApJ*, **644**, 143–147, doi:10.1086/503416.

228 Schulz, B., Pearson, C.P., Clements, D.L., Altieri, B., Amblard, A., Arumugam, V., Auld, R., Aussel, H., Babbedge, T., Blain, A., Bock, J., Boselli, A., Buat, V., Burgarella, D., Castro-Rodríguez, N., Cava, A., Chanial, P., Conley, A., Conversi, L., Cooray, A., Dowell, C.D., Dwek, E., Eales, S., Elbaz, D., Fox, M., Franceschini, A., Gear, W., Giovannoli, E., Glenn, J., Griffin, M., Halpern, M., Hatziminaoglou, E., Ibar, E., Isaak, K., Ivison, R.J., Lagache, G., Levenson, L., Lu, N., Madden, S., Maffei, B., Mainetti, G., Marchetti, L., Marsden, G., Mortier, A.M.J., Nguyen, H.T., O'Halloran, B., Oliver, S.J., Omont, A., Page, M.J., Panuzzo, P., Papageorgiou, A., Pérez-Fournon, I., Pohlen, M., Rangwala, N., Rawlings, J.I., Ray-

mond, G., Rigopoulou, D., Rizzo, D., Roseboom, I.G., Rowan-Robinson, M., Sánchez Portal, M., Scott, D., Seymour, N., Shupe, D.L., Smith, A.J., Stevens, J.A., Symeonidis, M., Trichas, M., Tugwell, K.E., Vaccari, M., Valiante, E., Valtchanov, I., Vigroux, L., Wang, L., Ward, R., Wright, G., Xu, C.K., and Zemcov, M. (2010) HerMES: The submillimeter spectral energy distributions of Herschel/SPIRE-detected galaxies. A&A, **518**, L32, doi:10.1051/0004-6361/201014673.

229 Rowan-Robinson, M. and Crawford, J. (1989) Models for infrared emission from IRAS galaxies. MNRAS, **238**, 523–558.

230 Gioia, I.M., Gregorini, L., and Klein, U. (1982) High frequency radio continuum observations of bright spiral galaxies. A&A, **116**, 164–174.

231 Klein, U., Weiland, H., and Brinks, E. (1991) A radio-optical study of blue compact dwarf galaxies. I. Radio continuum observations. A&A, **246**, 323–340.

232 Urry, C.M. and Padovani, P. (1995) Unified schemes for radio-loud active galactic nuclei. PASP, **107**, 803, doi:10.1086/133630.

233 Kewley, L.J., Heisler, C.A., Dopita, M.A., and Lumsden, S. (2001) Optical classification of southern warm infrared galaxies. ApJS, **132**, 37–71, doi:10.1086/318944.

234 Heckman, T.M. (1980) An optical and radio survey of the nuclei of bright galaxies – activity in normal galactic nuclei. A&A, **87**, 152–164.

235 Ho, L.C., Filippenko, A.V., and Sargent, W.L.W. (1997) A search for "dwarf" Seyfert nuclei. V. Demographics of nuclear activity in nearby galaxies. ApJ, **487**, 568, doi:10.1086/304638.

236 Poggianti, B.M., Smail, I., Dressler, A., Couch, W.J., Barger, A.J., Butcher, H., Ellis, R.S., and Oemler, Jr., A. (1999) The star formation histories of galaxies in distant clusters. ApJ, **518**, 576–593, doi:10.1086/307322.

237 Véron-Cetty, M. and Véron, P. (2006) A catalogue of quasars and active nuclei: 12th edition. A&A, **455**, 773–777, doi:10.1051/0004-6361:20065177.

238 Khachikian, E.Y. and Weedman, D.W. (1974) An atlas of Seyfert galaxies. ApJ, **192**, 581–589, doi:10.1086/153093.

239 Osterbrock, D.E. (1977) Spectrophotometry of Seyfert 1 galaxies. ApJ, **215**, 733–745, doi:10.1086/155407.

240 Winkler, H. (1992) Variability studies of Seyfert galaxies. II – spectroscopy. MNRAS, **257**, 677–688.

241 Hao, L., Strauss, M.A., Fan, X., Tremonti, C.A., Schlegel, D.J., Heckman, T.M., Kauffmann, G., Blanton, M.R., Gunn, J.E., Hall, P.B., Ivezić, Ž., Knapp, G.R., Krolik, J.H., Lupton, R.H., Richards, G.T., Schneider, D.P., Strateva, I.V., Zakamska, N.L., Brinkmann, J., and Szokoly, G.P. (2005) Active galactic nuclei in the Sloan digital sky survey. II. Emission-line luminosity function. AJ, **129**, 1795–1808, doi:10.1086/428486.

242 Dressler, A. and Gunn, J.E. (1983) Spectroscopy of galaxies in distant clusters. II – the population of the 3C 295 cluster. ApJ, **270**, 7–19, doi:10.1086/161093.

243 Couch, W.J. and Sharples, R.M. (1987) A spectroscopic study of three rich galaxy clusters at $Z = 0.31$. MNRAS, **229**, 423–456.

244 Poggianti, B.M., Aragón-Salamanca, A., Zaritsky, D., De Lucia, G., Milvang-Jensen, B., Desai, V., Jablonka, P., Halliday, C., Rudnick, G., Varela, J., Bamford, S., Best, P., Clowe, D., Noll, S., Saglia, R., Pelló, R., Simard, L., von der Linden, A., and White, S. (2009) The environments of starburst and post-starburst galaxies at $z = 0.4$–0.8. ApJ, **693**, 112–131, doi:10.1088/0004-637X/693/1/112.

245 Dressler, A., Smail, I., Poggianti, B.M., Butcher, H., Couch, W.J., Ellis, R.S., and Oemler, Jr., A. (1999) A spectroscopic catalog of 10 distant rich clusters of galaxies. ApJS, **122**, 51–80, doi:10.1086/313213.

246 Goto, T., Nichol, R.C., Okamura, S., Sekiguchi, M., Miller, C.J., Bernardi, M., Hopkins, A., Tremonti, C., Connolly, A., Castander, F.J., Brinkmann, J., Fukugita, M., Harvanek, M., Ivezic, Z., Kleinman, S.J., Krzesinski, J., Long, D., Loveday, J., Neilsen, E.H., Newman, P.R., Nitta, A., Snedden, S.A., and Subbarao, M. (2003) H_δ-strong galaxies in the Sloan

digital sky survey: I. The catalog. *PASJ*, **55**, 771–787.

247 Poggianti, B.M. (2004) Modeling stellar populations in cluster galaxies, in *Clusters of Galaxies: Probes of Cosmological Structure and Galaxy Evolution* (eds J. S. Mulchaey, A. Dressler, and A. Oemler), pp. 245.

248 Quintero, A.D., Hogg, D.W., Blanton, M.R., Schlegel, D.J., Eisenstein, D.J., Gunn, J.E., Brinkmann, J., Fukugita, M., Glazebrook, K., and Goto, T. (2004) Selection and photometric properties of K+A galaxies. *ApJ*, **602**, 190–199, doi:10.1086/380601.

249 Goto, T. (2005) 266 E+A galaxies selected from the Sloan digital sky survey data release 2: the origin of E+A galaxies. *MNRAS*, **357**, 937–944, doi:10.1111/j.1365-2966.2005.08701.x.

250 Baldwin, J.A., Phillips, M.M., and Terlevich, R. (1981) Classification parameters for the emission-line spectra of extragalactic objects. *PASP*, **93**, 5–19, doi:10.1086/130766.

251 Veilleux, S. and Osterbrock, D.E. (1987) Spectral classification of emission-line galaxies. *ApJS*, **63**, 295–310, doi:10.1086/191166.

252 Ho, L.C., Filippenko, A.V., and Sargent, W.L.W. (1997) A search for "dwarf" Seyfert nuclei. III. Spectroscopic parameters and properties of the host galaxies. *ApJS*, **112**, 315, doi:10.1086/313041.

253 Kauffmann, G., Heckman, T.M., Tremonti, C., Brinchmann, J., Charlot, S., White, S.D.M., Ridgway, S.E., Brinkmann, J., Fukugita, M., Hall, P.B., Ivezić, Ž., Richards, G.T., and Schneider, D.P. (2003) The host galaxies of active galactic nuclei. *MNRAS*, **346**, 1055–1077, doi:10.1111/j.1365-2966.2003.07154.x.

254 Kewley, L.J., Groves, B., Kauffmann, G., and Heckman, T. (2006) The host galaxies and classification of active galactic nuclei. *MNRAS*, **372**, 961–976, doi:10.1111/j.1365-2966.2006.10859.x.

255 Schawinski, K., Thomas, D., Sarzi, M., Maraston, C., Kaviraj, S., Joo, S., Yi, S.K., and Silk, J. (2007) Observational evidence for AGN feedback in early-type galaxies. *MNRAS*, **382**, 1415–1431, doi:10.1111/j.1365-2966.2007.12487.x.

256 Dessauges-Zavadsky, M., Pindao, M., Maeder, A., and Kunth, D. (2000) Spectral classification of emission-line galaxies. *A&A*, **355**, 89–98.

257 Stasińska, G., Cid Fernandes, R., Mateus, A., Sodré, L., and Asari, N.V. (2006) Semi-empirical analysis of Sloan digital sky survey galaxies – III. How to distinguish AGN hosts. *MNRAS*, **371**, 972–982, doi:10.1111/j.1365-2966.2006.10732.x.

258 Cid Fernandes, R., Stasińska, G., Schlickmann, M.S., Mateus, A., Vale Asari, N., Schoenell, W., and Sodré, L. (2010) Alternative diagnostic diagrams and the "forgotten" population of weak line galaxies in the SDSS. *MNRAS*, **403**, 1036–1053, doi:10.1111/j.1365-2966.2009.16185.x.

259 Dale, D.A., Smith, J.D.T., Schlawin, E.A., Armus, L., Buckalew, B.A., Cohen, S.A., Helou, G., Jarrett, T.H., Johnson, L.C., Moustakas, J., Murphy, E.J., Roussel, H., Sheth, K., Staudaher, S., Bot, C., Calzetti, D., Engelbracht, C.W., Gordon, K.D., Hollenbach, D.J., Kennicutt, R.C., and Malhotra, S. (2009) The Spitzer infrared nearby galaxies survey: a high-resolution spectroscopy anthology. *ApJ*, **693**, 1821–1834, doi:10.1088/0004-637X/693/2/1821.

260 Charmandaris, V. (2008) Mid-infrared spectroscopic diagnostics of galactic nuclei, in *Infrared Diagnostics of Galaxy Evolution*, Astronomical Society of the Pacific Conference Series, vol. 381 (eds R.-R. Chary, H.I. Teplitz, and K. Sheth), pp. 3.

261 Hao, L., Wu, Y., Charmandaris, V., Spoon, H.W.W., Bernard-Salas, J., Devost, D., Lebouteiller, V., and Houck, J.R. (2009) Probing the excitation of extreme starbursts: high-resolution mid-infrared spectroscopy of blue compact dwarfs. *ApJ*, **704**, 1159–1173, doi:10.1088/0004-637X/704/2/1159.

262 Lutz, D., Spoon, H.W.W., Rigopoulou, D., Moorwood, A.F.M., and Genzel, R. (1998) The nature and evolution of ultraluminous infrared galaxies: a mid-

infrared spectroscopic survey. *ApJ*, **505**, L103–L107, doi:10.1086/311614.

263 Peeters, E., Spoon, H.W.W., and Tielens, A.G.G.M. (2004) Polycyclic aromatic hydrocarbons as a tracer of star formation. *ApJ*, **613**, 986–1003, doi:10.1086/423237.

264 Dale, D.A., Smith, J.D.T., Armus, L., Buckalew, B.A., Helou, G., Kennicutt, Jr., R.C., Moustakas, J., Roussel, H., Sheth, K., Bendo, G.J., Calzetti, D., Draine, B.T., Engelbracht, C.W., Gordon, K.D., Hollenbach, D.J., Jarrett, T.H., Kewley, L.J., Leitherer, C., Li, A., Malhotra, S., Murphy, E.J., and Walter, F. (2006) Mid-infrared spectral diagnostics of nuclear and extranuclear regions in nearby galaxies. *ApJ*, **646**, 161–173, doi:10.1086/504835.

265 Sturm, E., Rupke, D., Contursi, A., Kim, D., Lutz, D., Netzer, H., Veilleux, S., Genzel, R., Lehnert, M., Tacconi, L.J., Maoz, D., Mazzarella, J., Lord, S., Sanders, D., and Sternberg, A. (2006) Mid-infrared diagnostics of LINERS. *ApJ*, **653**, L13–L16, doi:10.1086/510381.

266 Weedman, D., Polletta, M., Lonsdale, C.J., Wilkes, B.J., Siana, B., Houck, J.R., Surace, J., Shupe, D., Farrah, D., and Smith, H.E. (2006) Active galactic nucleus and starburst classification from Spitzer mid-infrared spectra for high-redshift SWIRE sources. *ApJ*, **653**, 101–111, doi:10.1086/508647.

267 O'Halloran, B., Satyapal, S., and Dudik, R.P. (2006) The polycyclic aromatic hydrocarbon emission deficit in low-metallicity galaxies – a Spitzer view. *ApJ*, **641**, 795–800, doi:10.1086/500529.

268 Brandl, B.R., Bernard-Salas, J., Spoon, H.W.W., Devost, D., Sloan, G.C., Guilles, S., Wu, Y., Houck, J.R., Weedman, D.W., Armus, L., Appleton, P.N., Soifer, B.T., Charmandaris, V., Hao, L., Higdon, J.A., Marshall, S.J., and Herter, T.L. (2006) The mid-infrared properties of starburst galaxies from Spitzer-IRS spectroscopy. *ApJ*, **653**, 1129–1144, doi:10.1086/508849.

269 Spoon, H.W.W., Marshall, J.A., Houck, J.R., Elitzur, M., Hao, L., Armus, L., Brandl, B.R., and Charmandaris, V. (2007) Mid-infrared galaxy classification based on silicate obscuration and PAH equivalent width. *ApJ*, **654**, L49–L52, doi:10.1086/511268.

270 Armus, L., Charmandaris, V., Bernard-Salas, J., Spoon, H.W.W., Marshall, J.A., Higdon, S.J.U., Desai, V., Teplitz, H.I., Hao, L., Devost, D., Brandl, B.R., Wu, Y., Sloan, G.C., Soifer, B.T., Houck, J.R., and Herter, T.L. (2007) Observations of ultraluminous infrared galaxies with the infrared spectrograph on the Spitzer space telescope. II. The IRAS bright galaxy sample. *ApJ*, **656**, 148–167, doi:10.1086/510107.

271 Farrah, D., Bernard-Salas, J., Spoon, H.W.W., Soifer, B.T., Armus, L., Brandl, B., Charmandaris, V., Desai, V., Higdon, S., Devost, D., and Houck, J. (2007) High-resolution mid-infrared spectroscopy of ultraluminous infrared galaxies. *ApJ*, **667**, 149–169, doi:10.1086/520834.

272 Sajina, A., Yan, L., Armus, L., Choi, P., Fadda, D., Helou, G., and Spoon, H. (2007) Spitzer mid-infrared spectroscopy of infrared luminous galaxies at $z \sim 2$. II. Diagnostics. *ApJ*, **664**, 713–737, doi:10.1086/519446.

273 Kennicutt, Jr., R.C., Bresolin, F., and Garnett, D.R. (2003) The composition gradient in M101 revisited. II. Electron temperatures and implications for the nebular abundance scale. *ApJ*, **591**, 801–820, doi:10.1086/375398.

274 Stasińska, G. (2005) Biases in abundance derivations for metal-rich nebulae. *A&A*, **434**, 507–520, doi:10.1051/0004-6361:20042216.

275 Garnett, D.R., Kennicutt, Jr., R.C., and Bresolin, F. (2004) The first measured electron temperatures for metal-rich HII regions in M51. *ApJ*, **607**, L21–L24, doi:10.1086/421489.

276 Pettini, M. and Pagel, B.E.J. (2004) [OIII]/[NII] as an abundance indicator at high redshift. *MNRAS*, **348**, L59–L63, doi:10.1111/j.1365-2966.2004.07591.x.

277 Pilyugin, L.S. and Thuan, T.X. (2005) Oxygen abundance determination in HII regions: the strong line intensities-abundance calibration revisited. *ApJ*, **631**, 231–243, doi:10.1086/432408.

278 McGaugh, S.S. (1991) HII region abundances – model oxygen line ratios. *ApJ*, **380**, 140–150, doi:10.1086/170569.

279 Zaritsky, D., Kennicutt, Jr., R.C., and Huchra, J.P. (1994) HII regions and the abundance properties of spiral galaxies. *ApJ*, **420**, 87–109, doi:10.1086/173544.

280 Kewley, L.J. and Dopita, M.A. (2002) Using strong lines to estimate abundances in extragalactic HII regions and starburst galaxies. *ApJS*, **142**, 35–52, doi:10.1086/341326.

281 Tremonti, C.A., Heckman, T.M., Kauffmann, G., Brinchmann, J., Charlot, S., White, S.D.M., Seibert, M., Peng, E.W., Schlegel, D.J., Uomoto, A., Fukugita, M., and Brinkmann, J. (2004) The origin of the mass–metallicity relation: insights from 53 000 star-forming galaxies in the Sloan digital sky survey. *ApJ*, **613**, 898–913, doi:10.1086/423264.

282 Kobulnicky, H.A. and Kewley, L.J. (2004) Metallicities of $0.3 < z < 1.0$ galaxies in the GOODS-North Field. *ApJ*, **617**, 240–261, doi:10.1086/425299.

283 Denicoló, G., Terlevich, R., and Terlevich, E. (2002) New light on the search for low-metallicity galaxies – I. The N2 calibrator. *MNRAS*, **330**, 69–74, doi:10.1046/j.1365-8711.2002.05041.x.

284 Kewley, L.J. and Ellison, S.L. (2008) Metallicity calibrations and the mass–metallicity relation for star-forming galaxies. *ApJ*, **681**, 1183–1204, doi:10.1086/587500.

285 Pilyugin, L.S. (2001) Oxygen abundances in dwarf irregular galaxies and the metallicity–luminosity relationship. *A&A*, **374**, 412–420, doi:10.1051/0004-6361:20010732.

286 Asplund, M., Grevesse, N., Sauval, A.J., and Scott, P. (2009) The chemical composition of the Sun. *ARA&A*, **47**, 481–522, doi:10.1146/annurev.astro.46.060407.145222.

287 Vilchez, J.M. and Esteban, C. (1996) The chemical composition of HII regions in the outer Galaxy. *MNRAS*, **280**, 720–734.

288 Garnett, D.R., Edmunds, M.G., Henry, R.B.C., Pagel, B.E.J., and Skillman, E.D. (2004) NO abundances in metal-rich environments: infrared space observatory spectroscopy of ionized gas in M51. *AJ*, **128**, 2772–2782, doi:10.1086/425883.

289 Leitherer, C., Tremonti, C.A., Heckman, T.M., and Calzetti, D. (2011) An ultraviolet spectroscopic atlas of local starbursts and star-forming galaxies: the legacy of FOS and GHRS. *AJ*, **141**, 37, doi:10.1088/0004-6256/141/2/37.

290 Trager, S.C., Worthey, G., Faber, S.M., Burstein, D., and Gonzalez, J.J. (1998) Old stellar populations. VI. Absorption-line spectra of galaxy nuclei and globular clusters. *ApJS*, **116**, 1, doi:10.1086/313099.

291 Vazdekis, A. (1999) Evolutionary stellar population synthesis at 2 Å spectral resolution. *ApJ*, **513**, 224–241, doi:10.1086/306843.

292 Vazdekis, A., Sánchez-Blázquez, P., Falcón-Barroso, J., Cenarro, A.J., Beasley, M.A., Cardiel, N., Gorgas, J., and Peletier, R.F. (2010) Evolutionary stellar population synthesis with MILES – I. The base models and a new line index system. *MNRAS*, **404**, 1639–1671, doi:10.1111/j.1365-2966.2010.16407.x.

293 Thomas, D., Maraston, C., and Korn, A. (2004) Higher-order Balmer line indices in α/Fe-enhanced stellar population models. *MNRAS*, **351**, L19–L23, doi:10.1111/j.1365-2966.2004.07944.x.

294 Thomas, D., Maraston, C., Bender, R., and Mendes de Oliveira, C. (2005) The epochs of early-type galaxy formation as a function of environment. *ApJ*, **621**, 673–694, doi:10.1086/426932.

295 Rose, J.A. (1994) The integrated spectra of M32 and of 47 Tuc: a comparative study at high spectral resolution. *AJ*, **107**, 206–229, doi:10.1086/116845.

296 Bower, R.G., Ellis, R.S., Rose, J.A., and Sharples, R.M. (1990) The stellar populations of early-type galaxies as a function of their environment. *AJ*, **99**, 530–539, doi:10.1086/115347.

297 Rose, J.A., Bower, R.G., Caldwell, N., Ellis, R.S., Sharples, R.M., and Teague, P. (1994) Stellar population in early-type galaxies: further evidence for environmental influences. *AJ*, **108**, 2054–2068, doi:10.1086/117218.

298 Cenarro, A.J., Cardiel, N., Gorgas, J., Peletier, R.F., Vazdekis, A., and Pra-

da, F. (2001) Empirical calibration of the near-infrared CaII triplet – I. The stellar library and index definition. *MNRAS*, **326**, 959–980, doi:10.1046/j.1365-8711.2001.04688.x.

299 Gavazzi, G., O'Neil, K., Boselli, A., and van Driel, W. (2006) HI observations of galaxies. II. The Coma supercluster. *A&A*, **449**, 929–935, doi:10.1051/0004-6361:20053844.

300 Springob, C.M., Haynes, M.P., Giovanelli, R., and Kent, B.R. (2005) A digital archive of HI 21 centimeter line spectra of optically targeted galaxies. *ApJS*, **160**, 149–162, doi:10.1086/431550.

301 Heidmann, J., Heidmann, N., and de Vaucouleurs, G. (1972) Inclination and absorption effects on the apparent diameters, optical luminosities and neutral hydrogen radiation of galaxies-I, optical and 21-cm line data. *MmRAS*, **75**, 85.

302 Springob, C.M., Haynes, M.P., and Giovanelli, R. (2005) Morphology, environment, and the HI mass function. *ApJ*, **621**, 215–226, doi:10.1086/427432.

303 Zwaan, M.A., Staveley-Smith, L., Koribalski, B.S., Henning, P.A., Kilborn, V.A., Ryder, S.D., Barnes, D.G., Bhathal, R., Boyce, P.J., de Blok, W.J.G., Disney, M.J., Drinkwater, M.J., Ekers, R.D., Freeman, K.C., Gibson, B.K., Green, A.J., Haynes, R.F., Jerjen, H., Juraszek, S., Kesteven, M.J., Knezek, P.M., Kraan-Korteweg, R.C., Mader, S., Marquarding, M., Meyer, M., Minchin, R.F., Mould, J.R., O'Brien, J., Oosterloo, T., Price, R.M., Putman, M.E., Ryan-Weber, E., Sadler, E.M., Schröder, A., Stewart, I.M., Stootman, F., Warren, B., Waugh, M., Webster, R.L., and Wright, A.E. (2003) The 1000 brightest HIPASS galaxies: The HI mass function and Ω_{HI}. *AJ*, **125**, 2842–2858, doi:10.1086/374944.

304 Grossi, M., di Serego Alighieri, S., Giovanardi, C., Gavazzi, G., Giovanelli, R., Haynes, M.P., Kent, B.R., Pellegrini, S., Stierwalt, S., and Trinchieri, G. (2009) The HI content of early-type galaxies from the ALFALFA survey. II. The case of low density environments. *A&A*, **498**, 407–417, doi:10.1051/0004-6361/200810823.

305 Haynes, M.P. and Giovanelli, R. (1984) Neutral hydrogen in isolated galaxies. IV – results for the Arecibo sample. *AJ*, **89**, 758–800, doi:10.1086/113573.

306 Haynes, M.P. and Giovanelli, R. (1984) Neutral hydrogen in isolated galaxies. IV – results for the Arecibo sample. *AJ*, **89**, 758–800, doi:10.1086/113573.

307 Boselli, A. and Gavazzi, G. (2009) The HI properties of galaxies in the Coma I cloud revisited. *A&A*, **508**, 201–207, doi:10.1051/0004-6361/200912658.

308 Solanes, J.M., Giovanelli, R., and Haynes, M.P. (1996) The HI content of spirals. I. Field galaxy HI mass functions and HI mass–optical size regressions. *ApJ*, **461**, 609, doi:10.1086/177089.

309 Charlton, J., Churchill, C., and Murdin, P. (2000) *Quasistellar Objects: Intervening Absorption Lines*, doi:10.1888/0333750888/2366.

310 Lanzetta, K. and Murdin, P. (2000) *Lyman Alpha Absorption: the Damped Systems*, doi:10.1888/0333750888/2141.

311 Wolfe, A.M., Gawiser, E., and Prochaska, J.X. (2005) Damped Ly$_\alpha$ systems. *ARA&A*, **43**, 861–918, doi:10.1146/annurev.astro.42.053102.133950.

312 Mebold, U., Duesterberg, C., Dickey, J.M., Staveley-Smith, L., and Kalberla, P. (1997) HI clouds in the large magellanic cloud, cooler than in the galaxy. *ApJ*, **490**, L65, doi:10.1086/311000.

313 Young, J.S. and Scoville, N.Z. (1991) Molecular gas in galaxies. *ARA&A*, **29**, 581–625, doi:10.1146/annurev.aa.29.090191.003053.

314 Wilson, C.D., Walker, C.E., and Thornley, M.D. (1997) The density and temperature of molecular clouds in M33. *ApJ*, **483**, 210, doi:10.1086/304216.

315 Boselli, A., Lequeux, J., and Gavazzi, G. (2002) Molecular gas in normal late-type galaxies. *A&A*, **384**, 33–47, doi:10.1051/0004-6361:20011747.

316 Wild, W., Harris, A.I., Eckart, A., Genzel, R., Graf, U.U., Jackson, J.M., Russell, A.P.G., and Stutzki, J. (1992) A multiline study of the molecular interstellar medium in M 82's starburst nucleus. *A&A*, **265**, 447–464.

317 Hunter, S.D., Bertsch, D.L., Catelli, J.R., Dame, T.M., Digel, S.W., Dingus, B.L., Esposito, J.A., Fichtel, C.E., Hartman, R.C., Kanbach, G., Kniffen, D.A., Lin, Y.C., Mayer-Hasselwander, H.A., Michelson, P.F., von Montigny, C., Mukherjee, R., Nolan, P.L., Schneid, E., Sreekumar, P., Thaddeus, P., and Thompson, D.J. (1997) EGRET observations of the diffuse gamma-ray emission from the galactic plane. *ApJ*, **481**, 205, doi:10.1086/304012.

318 Tacconi, L.J., Genzel, R., Smail, I., Neri, R., Chapman, S.C., Ivison, R.J., Blain, A., Cox, P., Omont, A., Bertoldi, F., Greve, T., Förster Schreiber, N.M., Genel, S., Lutz, D., Swinbank, A.M., Shapley, A.E., Erb, D.K., Cimatti, A., Daddi, E., and Baker, A.J. (2008) Submillimeter galaxies at $z \sim 2$: evidence for major mergers and constraints on lifetimes, IMF, and CO-H_2 conversion factor. *ApJ*, **680**, 246–262, doi:10.1086/587168.

319 Shetty, R., Glover, S.C., Dullemond, C.P., and Klessen, R.S. (2011) Modelling CO emission – I. CO as a column density tracer and the X factor in molecular clouds. *MNRAS*, **412**, 1686, 11, doi:10.1111/j.1365-2966.2010.18005.x.

320 Rubio, M., Lequeux, J., Boulanger, F., Booth, R.S., Garay, G., de Graauw, T., Israel, F.P., Johansson, L.E.B., Kutner, M.L., and Nyman, L.A. (1993) Results of the ESO/SEST key programme – Co/ in the Magellanic Clouds – Part Two – Co/ in the southwest region of the Small Magellanic Cloud. *A&A*, **271**, 1.

321 Lequeux, J., Le Bourlot, J., Pineau des Forets, G., Roueff, E., Boulanger, F., and Rubio, M. (1994) Results of the ESO-SEST key program: CO in the Magellanic Clouds. 4: physical properties of molecular clouds in the Small Magellanic Cloud. *A&A*, **292**, 371–380.

322 Allen, R.J., Le Bourlot, J., Lequeux, J., Pineau des Forets, G., and Roueff, E. (1995) Physica properties of molecular clouds in the inner disk of M31. *ApJ*, **444**, 157–164, doi:10.1086/175590.

323 Boselli, A., Gavazzi, G., Lequeux, J., Buat, V., Casoli, F., Dickey, J., and Donas, J. (1997) The molecular gas content of spiral galaxies in the Coma/A1367 supercluster. *A&A*, **327**, 522–538.

324 Wilson, C.D. (1995) The metallicity dependence of the CO-to-H_2 conversion factor from observations of local group galaxies. *ApJ*, **448**, L97, doi:10.1086/309615.

325 Arimoto, N., Sofue, Y., and Tsujimoto, T. (1996) CO-to-H_2 conversion factor in galaxies. *PASJ*, **48**, 275–284.

326 Israel, F.P. (2005) Molecular gas in compact galaxies. *A&A*, **438**, 855–866, doi:10.1051/0004-6361:20042237.

327 Bolatto, A.D., Leroy, A.K., Rosolowsky, E., Walter, F., and Blitz, L. (2008) The resolved properties of extragalactic giant molecular clouds. *ApJ*, **686**, 948–965, doi:10.1086/591513.

328 Kutner, M.L. and Ulich, B.L. (1981) Recommendations for calibration of millimeter-wavelength spectral line data. *ApJ*, **250**, 341–348, doi:10.1086/159380.

329 Gordon, M.A., Baars, J.W.M., and Cocke, W.J. (1992) Observations of radio lines from unresolved sources – telescope coupling, Doppler effects, and cosmological corrections. *A&A*, **264**, 337–344.

330 Dannerbauer, H., Daddi, E., Riechers, D.A., Walter, F., Carilli, C.L., Dickinson, M., Elbaz, D., and Morrison, G.E. (2009) Low Milky-Way-like molecular gas excitation of massive disk galaxies at $z \sim 1.5$. *ApJ*, **698**, L178–L182, doi:10.1088/0004-637X/698/2/L178.

331 Solomon, P.M. and Vanden Bout, P.A. (2005) Molecular gas at high redshift. *ARA&A*, **43**, 677–725, doi:10.1146/annurev.astro.43.051804.102221.

332 Papadopoulos, P.P., van der Werf, P., Isaak, K., and Xilouris, E.M. (2010) CO spectral line energy distributions of infrared-luminous galaxies and active galactic nuclei. *ApJ*, **715**, 775–792, doi:10.1088/0004-637X/715/2/775.

333 Papadopoulos, P.P., Isaak, K., and van der Werf, P. (2010) CO $J = 6-5$ in Arp 220: strong effects of dust on high-J CO lines. *ApJ*, **711**, 757–763, doi:10.1088/0004-637X/711/2/757.

334 Solomon, P.M., Downes, D., Radford, S.J.E., and Barrett, J.W. (1997) The

molecular interstellar medium in ultraluminous infrared galaxies. *ApJ*, **478**, 144, doi:10.1086/303765.

335 Guelin, M., Zylka, R., Mezger, P.G., Haslam, C.G.T., Kreysa, E., Lemke, R., and Sievers, A.W. (1993) 1.3 MM emission in the disk of NGC 891: evidence of cold dust. *A&A*, **279**, L37–L40.

336 Sodroski, T.J., Bennett, C., Boggess, N., Dwek, E., Franz, B.A., Hauser, M.G., Kelsall, T., Moseley, S.H., Odegard, N., Silverberg, R.F., and Weiland, J.L. (1994) Large-scale characteristics of interstellar dust from COBE DIRBE observations. *ApJ*, **428**, 638–646, doi:10.1086/174274.

337 Koornneef, J. (1982) The gas to dust ratio and the near-infrared extinction law in the Large Magellanic Cloud. *A&A*, **107**, 247–251.

338 Bouchet, P., Lequeux, J., Maurice, E., Prevot, L., and Prevot-Burnichon, M.L. (1985) The visible and infrared extinction law and the gas-to-dust ratio in the Small Magellanic Cloud. *A&A*, **149**, 330–336.

339 Issa, M.R., MacLaren, I., and Wolfendale, A.W. (1990) Dust to gas ratio and metallicity variations in nearby galaxies. *A&A*, **236**, 237–241.

340 Valentijn, E.A. and van der Werf, P.P. (1999) First extragalactic direct detection of large-scale molecular hydrogen in the disk of NGC 891. *ApJ*, **522**, L29–L33, doi:10.1086/312208.

341 Roussel, H., Helou, G., Hollenbach, D.J., Draine, B.T., Smith, J.D., Armus, L., Schinnerer, E., Walter, F., Engelbracht, C.W., Thornley, M.D., Kennicutt, R.C., Calzetti, D., Dale, D.A., Murphy, E.J., and Bot, C. (2007) Warm molecular hydrogen in the Spitzer SINGS galaxy sample. *ApJ*, **669**, 959–981, doi:10.1086/521667.

342 Brunner, G., Sheth, K., Armus, L., Wolfire, M., Vogel, S., Schinnerer, E., Helou, G., Dufour, R., Smith, J., and Dale, D.A. (2008) Warm molecular gas in M51: mapping the excitation temperature and mass of H_2 with the Spitzer infrared spectrograph. *ApJ*, **675**, 316–329, doi:10.1086/524348.

343 Rigopoulou, D., Kunze, D., Lutz, D., Genzel, R., and Moorwood, A.F.M. (2002) An ISO-SWS survey of molecular hydrogen in starburst and Seyfert galaxies. *A&A*, **389**, 374–386, doi:10.1051/0004-6361:20020607.

344 Dale, D.A., Sheth, K., Helou, G., Regan, M.W., and Hüttemeister, S. (2005) Warm and cold molecular gas in galaxies. *AJ*, **129**, 2197–2202, doi:10.1086/429134.

345 Snow, T.P. and McCall, B.J. (2006) Diffuse atomic and molecular clouds. *ARA&A*, **44**, 367–414, doi:10.1146/annurev.astro.43.072103.150624.

346 Tumlinson, J., Shull, J.M., Rachford, B.L., Browning, M.K., Snow, T.P., Fullerton, A.W., Jenkins, E.B., Savage, B.D., Crowther, P.A., Moos, H.W., Sembach, K.R., Sonneborn, G., and York, D.G. (2002) A far ultraviolet spectroscopic explorer survey of interstellar molecular hydrogen in the small and Large Magellanic Clouds. *ApJ*, **566**, 857–879, doi:10.1086/338112.

347 Hoopes, C.G., Sembach, K.R., Heckman, T.M., Meurer, G.R., Aloisi, A., Calzetti, D., Leitherer, C., and Martin, C.L. (2004) Far-ultraviolet observations of molecular hydrogen in the diffuse interstellar medium of starburst galaxies. *ApJ*, **612**, 825–836, doi:10.1086/422830.

348 Whittet, D.C.B. (ed.) (2003) *Dust in the Galactic Environment*.

349 Disney, M., Davies, J., and Phillipps, S. (1989) Are galaxy disks optically thick. *MNRAS*, **239**, 939–976.

350 Burstein, D. and Heiles, C. (1982) Reddenings derived from HI and galaxy counts – accuracy and maps. *AJ*, **87**, 1165–1189, doi:10.1086/113199.

351 Schlegel, D.J., Finkbeiner, D.P., and Davis, M. (1998) Maps of dust infrared emission for use in estimation of reddening and cosmic microwave background radiation foregrounds. *ApJ*, **500**, 525, doi:10.1086/305772.

352 Burstein, D. and Heiles, C. (1978) HI, galaxy counts, and reddening – variation in the gas-to-dust ratio, the extinction at high galactic latitudes, and a new method for determining galactic reddening. *ApJ*, **225**, 40–55, doi:10.1086/156466.

353 Fukugita, M., Yasuda, N., Brinkmann, J., Gunn, J.E., Ivezić, Ž., Knapp, G.R., Lupton, R., and Schneider, D.P. (2004) Spatial variations of galaxy number counts in the Sloan digital sky survey. I. Extinction, large-scale structure, and photometric homogeneity. *AJ*, **127**, 3155–3160, doi:10.1086/420800.

354 Cardelli, J.A., Clayton, G.C., and Mathis, J.S. (1989) The relationship between infrared, optical, and ultraviolet extinction. *ApJ*, **345**, 245–256, doi:10.1086/167900.

355 Fitzpatrick, E.L. and Massa, D. (1990) An analysis of the shapes of ultraviolet extinction curves. III – an atlas of ultraviolet extinction curves. *ApJS*, **72**, 163–189, doi:10.1086/191413.

356 O'Donnell, J.E. (1994) R_v-dependent optical and near-ultraviolet extinction. *ApJ*, **422**, 158–163, doi:10.1086/173713.

357 Fitzpatrick, E.L. (1999) Correcting for the effects of interstellar extinction. *PASP*, **111**, 63–75, doi:10.1086/316293.

358 Fitzpatrick, E.L. and Massa, D. (2007) An analysis of the shapes of interstellar extinction curves. V. The IR-through-UV curve morphology. *ApJ*, **663**, 320–341, doi:10.1086/518158.

359 Fitzpatrick, E.L. (1986) An average interstellar extinction curve for the Large Magellanic Cloud. *AJ*, **92**, 1068–1073, doi:10.1086/114237.

360 Gordon, K.D., Clayton, G.C., Misselt, K.A., Landolt, A.U., and Wolff, M.J. (2003) A quantitative comparison of the Small Magellanic Cloud, Large Magellanic Cloud, and Milky Way ultraviolet to near-infrared extinction curves. *ApJ*, **594**, 279–293, doi:10.1086/376774.

361 Pei, Y.C. (1992) Interstellar dust from the Milky Way to the Magellanic Clouds. *ApJ*, **395**, 130–139, doi:10.1086/171637.

362 Buat, V. and Xu, C. (1996) Star formation and dust extinction in disk galaxies. Comparison between the UV non-ionizing and the FIR emissions. *A&A*, **306**, 61.

363 Gordon, K.D., Clayton, G.C., Witt, A.N., and Misselt, K.A. (2000) The flux ratio method for determining the dust attenuation of starburst galaxies. *ApJ*, **533**, 236–244, doi:10.1086/308668.

364 Witt, A.N. and Gordon, K.D. (2000) Multiple scattering in clumpy media. II. Galactic environments. *ApJ*, **528**, 799–816, doi:10.1086/308197.

365 Charlot, S. and Fall, S.M. (2000) A simple model for the absorption of starlight by dust in galaxies. *ApJ*, **539**, 718–731, doi:10.1086/309250.

366 Misiriotis, A., Popescu, C.C., Tuffs, R., and Kylafis, N.D. (2001) Modeling the spectral energy distribution of galaxies. II. Disk opacity and star formation in 5 edge-on spirals. *A&A*, **372**, 775–783, doi:10.1051/0004-6361:20010568.

367 Panuzzo, P., Bressan, A., Granato, G.L., Silva, L., and Danese, L. (2003) Dust and nebular emission. I. Models for normal galaxies. *A&A*, **409**, 99–114, doi:10.1051/0004-6361:20031094.

368 Tuffs, R.J., Popescu, C.C., Völk, H.J., Kylafis, N.D., and Dopita, M.A. (2004) Modelling the spectral energy distribution of galaxies. III. Attenuation of stellar light in spiral galaxies. *A&A*, **419**, 821–835, doi:10.1051/0004-6361:20035689.

369 Pierini, D., Gordon, K.D., Witt, A.N., and Madsen, G.J. (2004) Dust attenuation in late-type galaxies. I. Effects on bulge and disk components. *ApJ*, **617**, 1022–1046, doi:10.1086/425651.

370 Panuzzo, P., Granato, G.L., Buat, V., Inoue, A.K., Silva, L., Iglesias-Páramo, J., and Bressan, A. (2007) Ultraviolet dust attenuation in spiral galaxies: the role of age-dependent extinction and the initial mass function. *MNRAS*, **375**, 640–648, doi:10.1111/j.1365-2966.2006.11337.x.

371 Buat, V., Iglesias-Páramo, J., Seibert, M., Burgarella, D., Charlot, S., Martin, D.C., Xu, C.K., Heckman, T.M., Boissier, S., Boselli, A., Barlow, T., Bianchi, L., Byun, Y., Donas, J., Forster, K., Friedman, P.G., Jelinski, P., Lee, Y., Madore, B.F., Malina, R., Milliard, B., Morrissey, P., Neff, S., Rich, M., Schiminovitch, D., Siegmund, O., Small, T., Szalay, A.S., Welsh, B., and Wyder, T.K. (2005) Dust attenuation in the nearby universe: a comparison between galaxies selected in the ultraviolet and in the far-infrared. *ApJ*, **619**, L51–L54, doi:10.1086/423241.

372 Calzetti, D., Armus, L., Bohlin, R.C., Kinney, A.L., Koornneef, J., and Storchi-Bergmann, T. (2000) The dust content and opacity of actively star-forming galaxies. *ApJ*, **533**, 682–695, doi:10.1086/308692.

373 Kong, X., Charlot, S., Brinchmann, J., and Fall, S.M. (2004) Star formation history and dust content of galaxies drawn from ultraviolet surveys. *MNRAS*, **349**, 769–778, doi:10.1111/j.1365-2966.2004.07556.x.

374 Iglesias-Páramo, J., Buat, V., Takeuchi, T.T., Xu, K., Boissier, S., Boselli, A., Burgarella, D., Madore, B.F., Gil de Paz, A., Bianchi, L., Barlow, T.A., Byun, Y., Donas, J., Forster, K., Friedman, P.G., Heckman, T.M., Jelinski, P.N., Lee, Y., Malina, R.F., Martin, D.C., Milliard, B., Morrissey, P.F., Neff, S.G., Rich, R.M., Schiminovich, D., Seibert, M., Siegmund, O.H.W., Small, T., Szalay, A.S., Welsh, B.Y., and Wyder, T.K. (2006) Star formation in the nearby universe: the ultraviolet and infrared points of view. *ApJS*, **164**, 38–51, doi:10.1086/502628.

375 Cortese, L., Boselli, A., Franzetti, P., Decarli, R., Gavazzi, G., Boissier, S., and Buat, V. (2008) Ultraviolet dust attenuation in star-forming galaxies – II. Calibrating the A(UV) versus L_{TIR}/L_{UV} relation. *MNRAS*, **386**, 1157–1168, doi:10.1111/j.1365-2966.2008.13118.x.

376 Meurer, G.R., Heckman, T.M., and Calzetti, D. (1999) Dust absorption and the ultraviolet luminosity density at $Z \sim 3$ as calibrated by local starburst galaxies. *ApJ*, **521**, 64–80, doi:10.1086/307523.

377 Calzetti, D. (1997) Reddening and star formation in starburst galaxies. *AJ*, **113**, 162–184, doi:10.1086/118242.

378 Calzetti, D. (2001) The dust opacity of star-forming galaxies. *PASP*, **113**, 1449–1485, doi:10.1086/324269.

379 Fischera, J., Dopita, M.A., and Sutherland, R.S. (2003) Starburst galaxies: why the Calzetti dust extinction law. *ApJ*, **599**, L21–L24, doi:10.1086/381190.

380 Fischera, J. and Dopita, M. (2005) Attenuation caused by a distant isothermal turbulent screen. *ApJ*, **619**, 340–356, doi:10.1086/426185.

381 Overzier, R.A., Heckman, T.M., Wang, J., Armus, L., Buat, V., Howell, J., Meurer, G., Seibert, M., Siana, B., Basu-Zych, A., Charlot, S., Gonçalves, T.S., Martin, D.C., Neill, J.D., Rich, R.M., Salim, S., and Schiminovich, D. (2011) Dust attenuation in UV-selected starbursts at high redshift and their local counterparts: implications for the cosmic star formation rate density. *ApJ*, **726**, L7, doi:10.1088/2041-8205/726/1/L7.

382 Salpeter, E.E. (1955) The luminosity function and stellar evolution. *ApJ*, **121**, 161, doi:10.1086/145971.

383 Scalo, J.M. (1986) The stellar initial mass function. *Fund. Cosmic Phys.*, **11**, 1–278.

384 Scalo, J. (1998) The IMF revisited: a case for variations, in *The Stellar Initial Mass Function (38th Herstmonceux Conference)*, Astronomical Society of the Pacific Conference Series, vol. 142 (eds G. Gilmore and D. Howell), pp. 201.

385 Kroupa, P. (2001) On the variation of the initial mass function. *MNRAS*, **322**, 231–246, doi:10.1046/j.1365-8711.2001.04022.x.

386 Chabrier, G. (2003) Galactic stellar and substellar initial mass function. *PASP*, **115**, 763–795, doi:10.1086/376392.

387 Boselli, A., Boissier, S., Cortese, L., Buat, V., Hughes, T.M., and Gavazzi, G. (2009) High-mass star formation in normal late-type galaxies: observational constraints to the initial mass function. *ApJ*, **706**, 1527–1544, doi:10.1088/0004-637X/706/2/1527.

388 Pflamm-Altenburg, J., Weidner, C., and Kroupa, P. (2007) Converting H_α luminosities into star formation rates. *ApJ*, **671**, 1550–1558, doi:10.1086/523033.

389 Pflamm-Altenburg, J., Weidner, C., and Kroupa, P. (2009) Diverging UV and H_α fluxes of star-forming galaxies predicted by the IGIMF theory. *MNRAS*, **395**, 394–400, doi:10.1111/j.1365-2966.2009.14522.x.

390 Kennicutt, Jr., R.C. (1998) Star formation in galaxies along the Hubble sequence. *ARA&A*, **36**, 189–232, doi:10.1146/annurev.astro.36.1.189.

391 Bruzual A., G. and Charlot, S. (1993) Spectral evolution of stellar populations

using isochrone synthesis. *ApJ*, **405**, 538–553, doi:10.1086/172385.

392 Boselli, A., Gavazzi, G., Donas, J., and Scodeggio, M. (2001) 1.65 Micron (H band) surface photometry of galaxies. VI. The history of star formation in normal late-type galaxies. *AJ*, **121**, 753–767, doi:10.1086/318734.

393 Brinchmann, J., Charlot, S., White, S.D.M., Tremonti, C., Kauffmann, G., Heckman, T., and Brinkmann, J. (2004) The physical properties of star-forming galaxies in the low-redshift universe. *MNRAS*, **351**, 1151–1179, doi:10.1111/j.1365-2966.2004.07881.x.

394 Larson, R. (1992) Galaxy formation and evolution, in *Star Formation in Stellar Systems* (eds G. Tenorio-Tagle, M. Prieto, and F. Sanchez), pp. 125.

395 Wang, B. and Silk, J. (1994) Gravitational instability and disk star formation. *ApJ*, **427**, 759–769, doi:10.1086/174182.

396 Young, J.S., Allen, L., Kenney, J.D.P., Lesser, A., and Rownd, B. (1996) The global rate and efficiency of star formation in spiral galaxies as a function of morphology and environment. *AJ*, **112**, 1903, doi:10.1086/118152.

397 Roberts, M.S. (1963) The content of galaxies: stars and gas. *ARA&A*, **1**, 149, doi:10.1146/annurev.aa.01.090163.001053.

398 Kennicutt, R.C., Hao, C., Calzetti, D., Moustakas, J., Dale, D.A., Bendo, G., Engelbracht, C.W., Johnson, B.D., and Lee, J.C. (2009) Dust-corrected star formation rates of galaxies. I. Combinations of H_α and infrared tracers. *ApJ*, **703**, 1672–1695, doi:10.1088/0004-637X/703/2/1672.

399 Decarli, R., Gavazzi, G., Arosio, I., Cortese, L., Boselli, A., Bonfanti, C., and Colpi, M. (2007) The census of nuclear activity of late-type galaxies in the Virgo cluster. *MNRAS*, **381**, 136–150, doi:10.1111/j.1365-2966.2007.12208.x.

400 Boselli, A., Gavazzi, G., Lequeux, J., and Pierini, D. (2002) [CII] at 158 µm as a star formation tracer in late-type galaxies. *A&A*, **385**, 454–463, doi:10.1051/0004-6361:20020156.

401 Kewley, L.J., Jansen, R.A., and Geller, M.J. (2005) Aperture effects on star formation rate, metallicity, and reddening. *PASP*, **117**, 227–244, doi:10.1086/428303.

402 Inoue, A.K., Hirashita, H., and Kamaya, H. (2000) Conversion law of infrared luminosity to star-formation rate for galaxies. *PASJ*, **52**, 539–543.

403 Bicker, J. and Fritze-v. Alvensleben, U. (2005) Metallicity dependent calibrations of flux based SFR tracers. *A&A*, **443**, L19–L23, doi:10.1051/0004-6361:200500194.

404 Leitherer, C. (2008) Revision of star-formation measures, in *IAU Symposium*, vol. 255 (eds L.K. Hunt, S. Madden, and R. Schneider), pp. 305–309, doi:10.1017/S1743921308024988.

405 Charlot, S. and Longhetti, M. (2001) Nebular emission from star-forming galaxies. *MNRAS*, **323**, 887–903, doi:10.1046/j.1365-8711.2001.04260.x.

406 Bell, E.F. (2003) Estimating star formation rates from infrared and radio luminosities: the origin of the radio-infrared correlation. *ApJ*, **586**, 794–813, doi:10.1086/367829.

407 Sauvage, M. and Thuan, T.X. (1992) On the use of far-infrared luminosity as a star formation indicator in galaxies. *ApJ*, **396**, L69–L73, doi:10.1086/186519.

408 Alonso-Herrero, A., Rieke, G.H., Rieke, M.J., Colina, L., Pérez-González, P.G., and Ryder, S.D. (2006) Near-infrared and star-forming properties of local luminous infrared galaxies. *ApJ*, **650**, 835–849, doi:10.1086/506958.

409 Calzetti, D., Kennicutt, R.C., Engelbracht, C.W., Leitherer, C., Draine, B.T., Kewley, L., Moustakas, J., Sosey, M., Dale, D.A., Gordon, K.D., Helou, G.X., Hollenbach, D.J., Armus, L., Bendo, G., Bot, C., Buckalew, B., Jarrett, T., Li, A., Meyer, M., Murphy, E.J., Prescott, M., Regan, M.W., Rieke, G.H., Roussel, H., Sheth, K., Smith, J.D.T., Thornley, M.D., and Walter, F. (2007) The calibration of mid-infrared star formation rate indicators. *ApJ*, **666**, 870–895, doi:10.1086/520082.

410 Verbunt, F. and van den Heuvel, E.P.J. (1995) Formation and evolution of neutron stars and black holes in binaries, in *X-Ray Binaries* (eds W.H.G. Lewin, J. van

Paradijs, and E. P. J. van den Heuvel), pp. 457–494.

411 Griffiths, R.E. and Padovani, P. (1990) Star-forming galaxies and the X-ray background. *ApJ*, **360**, 483–489, doi:10.1086/169139.

412 Grimm, H., Gilfanov, M., and Sunyaev, R. (2003) High-mass X-ray binaries as a star formation rate indicator in distant galaxies. *MNRAS*, **339**, 793–809, doi:10.1046/j.1365-8711.2003.06224.x.

413 Ranalli, P., Comastri, A., and Setti, G. (2003) The 2–10 keV luminosity as a star formation rate indicator. *A&A*, **399**, 39–50, doi:10.1051/0004-6361:20021600.

414 Gallagher, J.S., Hunter, D.A., and Bushouse, H. (1989) Star-formation rates and forbidden OII emission in blue galaxies. *AJ*, **97**, 700–707, doi:10.1086/115015.

415 Kennicutt, Jr., R.C. (1992) The integrated spectra of nearby galaxies – general properties and emission-line spectra. *ApJ*, **388**, 310–327, doi:10.1086/171154.

416 Kewley, L.J., Geller, M.J., and Jansen, R.A. (2004) [OII] as a star formation rate indicator. *AJ*, **127**, 2002–2030, doi:10.1086/382723.

417 Moustakas, J., Kennicutt, Jr., R.C., and Tremonti, C.A. (2006) Optical star formation rate indicators. *ApJ*, **642**, 775–796, doi:10.1086/500964.

418 Stacey, G.J., Geis, N., Genzel, R., Lugten, J.B., Poglitsch, A., Sternberg, A., and Townes, C.H. (1991) The 158 micron CII line – a measure of global star formation activity in galaxies. *ApJ*, **373**, 423–444, doi:10.1086/170062.

419 Luhman, M.L., Satyapal, S., Fischer, J., Wolfire, M.G., Sturm, E., Dudley, C.C., Lutz, D., and Genzel, R. (2003) The [CII] 158 micron line deficit in ultraluminous infrared galaxies revisited. *ApJ*, **594**, 758–775, doi:10.1086/376965.

420 Shaver, P.A. (1978) Extragalactic radio recombination lines. *A&A*, **68**, 97–105.

421 Anantharamaiah, K.R., Viallefond, F., Mohan, N.R., Goss, W.M., and Zhao, J.H. (2000) Starburst in the ultraluminous galaxy Arp 220: constraints from observations of radio recombination lines and continuum. *ApJ*, **537**, 613–630, doi:10.1086/309063.

422 Poggianti, B.M. and Barbaro, G. (1997) Indicators of star formation: 4000 Å break and Balmer lines. *A&A*, **325**, 1025–1030.

423 Shioya, Y., Bekki, K., Couch, W.J., and De Propris, R. (2002) Spectrophotometric evolution of spiral galaxies with truncated star formation: an evolutionary link between spirals and S0s in distant clusters. *ApJ*, **565**, 223–237, doi:10.1086/324433.

424 Faber, S.M., Tremaine, S., Ajhar, E.A., Byun, Y., Dressler, A., Gebhardt, K., Grillmair, C., Kormendy, J., Lauer, T.R., and Richstone, D. (1997) The centers of early-type galaxies with HST. IV. Central parameter relations. *AJ*, **114**, 1771, doi:10.1086/118606.

425 Simard, L. (1998) GIM2D: an IRAF package for the quantitative morphology analysis of distant galaxies, in *Astronomical Data Analysis Software and Systems VII*, Astronomical Society of the Pacific Conference Series, vol. 145 (eds R. Albrecht, R.N. Hook, and H. A. Bushouse), pp. 108.

426 Peng, C.Y., Ho, L.C., Impey, C.D., and Rix, H. (2002) Detailed structural decomposition of galaxy images. *AJ*, **124**, 266–293, doi:10.1086/340952.

427 de Souza, R.E., Gadotti, D.A., and dos Anjos, S. (2004) BUDDA: a new two-dimensional bulge/disk decomposition code for detailed structural analysis of galaxies. *ApJS*, **153**, 411–427, doi:10.1086/421554.

428 de Vaucouleurs, G. (1948) Recherches sur les nebuleuses extragalactiques. *Annales d'Astrophysique*, **11**, 247.

429 Sersic, J.L. (1968) *Atlas de galaxias australes*.

430 Graham, A.W. and Driver, S.P. (2005) A concise reference to (projected) Sersic $R^{1/n}$ quantities, including concentration, profile slopes, Petrosian indices, and Kron magnitudes. *PASA*, **22**, 118–127, doi:10.1071/AS05001.

431 Graham, A.W. and Guzmán, R. (2003) HST photometry of dwarf elliptical galaxies in Coma, and an explanation for the alleged structural dichotomy between dwarf and bright

elliptical galaxies. *AJ*, **125**, 2936–2950, doi:10.1086/374992.

432 Gavazzi, G., Franzetti, P., Scodeggio, M., Boselli, A., and Pierini, D. (2000) 1.65 μm (H-band) surface photometry of galaxies. V. Profile decomposition of 1157 galaxies. *A&A*, **361**, 863–876.

433 Côté, P., Ferrarese, L., Jordán, A., Blakeslee, J.P., Chen, C., Infante, L., Merritt, D., Mei, S., Peng, E.W., Tonry, J.L., West, A.A., and West, M.J. (2007) The ACS Fornax cluster survey. II. The central brightness profiles of early-type galaxies: a characteristic radius on nuclear scales and the transition from central luminosity deficit to excess. *ApJ*, **671**, 1456–1465, doi:10.1086/522822.

434 Graham, A.W., Erwin, P., Trujillo, I., and Asensio Ramos, A. (2003) A new empirical model for the structural analysis of early-type galaxies, and a critical review of the nuker model. *AJ*, **125**, 2951–2963, doi:10.1086/375320.

435 Côté, P., Piatek, S., Ferrarese, L., Jordán, A., Merritt, D., Peng, E.W., Haşegan, M., Blakeslee, J.P., Mei, S., West, M.J., Milosavljević, M., and Tonry, J.L. (2006) The ACS Virgo cluster survey. VIII. The nuclei of early-type galaxies. *ApJS*, **165**, 57–94, doi:10.1086/504042.

436 Trujillo, I., Erwin, P., Asensio Ramos, A., and Graham, A.W. (2004) Evidence for a new elliptical-galaxy paradigm: Sérsic and core galaxies. *AJ*, **127**, 1917–1942, doi:10.1086/382712.

437 Ferrarese, L., Côté, P., Jordán, A., Peng, E.W., Blakeslee, J.P., Piatek, S., Mei, S., Merritt, D., Milosavljević, M., Tonry, J.L., and West, M.J. (2006) The ACS Virgo cluster survey. VI. Isophotal analysis and the structure of early-type galaxies. *ApJS*, **164**, 334–434, doi:10.1086/501350.

438 Lauer, T.R., Ajhar, E.A., Byun, Y., Dressler, A., Faber, S.M., Grillmair, C., Kormendy, J., Richstone, D., and Tremaine, S. (1995) The centers of early-type galaxies with HST. I. An observational survey. *AJ*, **110**, 2622, doi:10.1086/117719.

439 Rest, A., van den Bosch, F.C., Jaffe, W., Tran, H., Tsvetanov, Z., Ford, H.C., Davies, J., and Schafer, J. (2001) WFPC2 Images of the central regions of early-type galaxies. I. The data. *AJ*, **121**, 2431–2482, doi:10.1086/320370.

440 Lauer, T.R., Faber, S.M., Gebhardt, K., Richstone, D., Tremaine, S., Ajhar, E.A., Aller, M.C., Bender, R., Dressler, A., Filippenko, A.V., Green, R., Grillmair, C.J., Ho, L.C., Kormendy, J., Magorrian, J., Pinkney, J., and Siopis, C. (2005) The centers of early-type galaxies with Hubble Space Telescope. V. New WFPC2 photometry. *AJ*, **129**, 2138–2185, doi:10.1086/429565.

441 van der Kruit, P.C. and Searle, L. (1981) Surface photometry of edge-on spiral galaxies. I – a model for the three-dimensional distribution of light in galactic disks. *A&A*, **95**, 105–115.

442 Pohlen, M., Dettmar, R., Lütticke, R., and Schwarzkopf, U. (2000) Three-dimensional modelling of edge-on disk galaxies. *A&AS*, **144**, 405–428, doi:10.1051/aas:2000218.

443 Seth, A.C., Dalcanton, J.J., and de Jong, R.S. (2005) A study of edge-on galaxies with the Hubble Space Telescope advanced camera for surveys. II. Vertical distribution of the resolved stellar population. *AJ*, **130**, 1574–1592, doi:10.1086/444620.

444 Conselice, C.J. (2003) The relationship between stellar light distributions of galaxies and their formation histories. *ApJS*, **147**, 1–28, doi:10.1086/375001.

445 Lotz, J.M., Primack, J., and Madau, P. (2004) A new nonparametric approach to galaxy morphological classification. *AJ*, **128**, 163–182, doi:10.1086/421849.

446 de Vaucouleurs, G. (1977) Qualitative and quantitative classifications of galaxies, in *Evolution of Galaxies and Stellar Populations* (eds B.M. Tinsley and R. B. Larson), pp. 43.

447 Abraham, R.G., Valdes, F., Yee, H.K.C., and van den Bergh, S. (1994) The morphologies of distant galaxies. 1: an automated classification system. *ApJ*, **432**, 75–90, doi:10.1086/174550.

448 Kent, S.M. (1985) CCD surface photometry of field Galaxies. II – bulge/disk decompositions. *ApJS*, **59**, 115–159, doi:10.1086/191066.

449 Bershady, M.A., Jangren, A., and Conselice, C.J. (2000) Structural and photo-

metric classification of galaxies. I. Calibration based on a nearby galaxy sample. *AJ*, **119**, 2645–2663, doi:10.1086/301386.

450 Abraham, R.G., van den Bergh, S., Glazebrook, K., Ellis, R.S., Santiago, B.X., Surma, P., and Griffiths, R.E. (1996) The morphologies of distant galaxies. II. Classifications from the Hubble Space Telescope medium deep survey. *ApJS*, **107**, 1, doi:10.1086/192352.

451 Conselice, C.J., Bershady, M.A., and Jangren, A. (2000) The asymmetry of galaxies: physical morphology for nearby and high-redshift galaxies. *ApJ*, **529**, 886–910, doi:10.1086/308300.

452 Abraham, R.G., van den Bergh, S., and Nair, P. (2003) A new approach to galaxy morphology. I. Analysis of the Sloan digital sky survey early data release. *ApJ*, **588**, 218–229, doi:10.1086/373919.

453 Lisker, T. (2008) Is the Gini coefficient a stable measure of galaxy structure. *ApJS*, **179**, 319–325, doi:10.1086/591795.

454 Navarro, J.F., Frenk, C.S., and White, S.D.M. (1996) The Structure of cold dark matter halos. *ApJ*, **462**, 563, doi:10.1086/177173.

455 Hayashi, E., Navarro, J.F., and Springel, V. (2007) The shape of the gravitational potential in cold dark matter haloes. *MNRAS*, **377**, 50–62, doi:10.1111/j.1365-2966.2007.11599.x.

456 de Blok, W.J.G., McGaugh, S.S., Bosma, A., and Rubin, V.C. (2001) Mass density profiles of low surface brightness galaxies. *ApJ*, **552**, L23–L26, doi:10.1086/320262.

457 Blais-Ouellette, S., Amram, P., and Carignan, C. (2001) Accurate determination of the mass distribution in spiral galaxies. II. Testing the shape of dark halos. *AJ*, **121**, 1952–1964, doi:10.1086/319944.

458 de Blok, W.J.G. and Bosma, A. (2002) High-resolution rotation curves of low surface brightness galaxies. *A&A*, **385**, 816–846, doi:10.1051/0004-6361:20020080.

459 Swaters, R.A., Madore, B.F., van den Bosch, F.C., and Balcells, M. (2003) The central mass distribution in dwarf and low surface brightness galaxies. *ApJ*, **583**, 732–751, doi:10.1086/345426.

460 Graham, A.W., Merritt, D., Moore, B., Diemand, J., and Terzić, B. (2006) Empirical models for dark matter halos. II. Inner profile slopes, dynamical profiles, and ρ/σ^3. *AJ*, **132**, 2701–2710, doi:10.1086/508990.

461 Spano, M., Marcelin, M., Amram, P., Carignan, C., Epinat, B., and Hernandez, O. (2008) GHASP: an H_α kinematic survey of spiral and irregular galaxies – V. Dark matter distribution in 36 nearby spiral galaxies. *MNRAS*, **383**, 297–316, doi:10.1111/j.1365-2966.2007.12545.x.

462 Bell, E.F. and de Jong, R.S. (2001) Stellar mass-to-light ratios and the Tully–Fisher relation. *ApJ*, **550**, 212–229, doi:10.1086/319728.

463 Bell, E.F., McIntosh, D.H., Katz, N., and Weinberg, M.D. (2003) The optical and near-infrared properties of galaxies. I. Luminosity and stellar mass functions. *ApJS*, **149**, 289–312, doi:10.1086/378847.

464 Kodama, T. and Arimoto, N. (1997) Origin of the colour–magnitude relation of elliptical galaxies. *A&A*, **320**, 41–53.

465 Schulz, J., Fritze-v. Alvensleben, U., Möller, C.S., and Fricke, K.J. (2002) Spectral and photometric evolution of simple stellar populations at various metallicities. *A&A*, **392**, 1–11, doi:10.1051/0004-6361:20020657.

466 Vazdekis, A., Casuso, E., Peletier, R.F., and Beckman, J.E. (1996) A new chemo-evolutionary population synthesis model for early-type galaxies. I. Theoretical basis. *ApJS*, **106**, 307, doi:10.1086/192340.

467 Longair, M.S. (2008) *Galaxy Formation*.

468 Binney, J. and Tremaine, S. (2008) *Galactic Dynamics*, 2nd edn, Princeton University Press.

469 Burbidge, E.M. and Burbidge, G.R. (1975) *The Masses of Galaxies*, The University of Chicago Press, pp. 81.

470 Brandt, J.C. and Belton, M.J.S. (1962) On the distribution of mass in galaxies. III. Surface densities. *ApJ*, **136**, 352, doi:10.1086/147387.

471 Freeman, K.C. (1970) On the disks of spiral and so galaxies. *ApJ*, **160**, 811, doi:10.1086/150474.

472 Casertano, S. (1983) Rotation curve of the edge-on spiral galaxy NGC 5907: disk and halo masses. *MNRAS*, **203**, 735–747.

473 Bosma, A. (1981) 21-cm line studies of spiral galaxies. II. The distribution and kinematics of neutral hydrogen in spiral galaxies of various morphological types. *AJ*, **86**, 1825–1846, doi:10.1086/113063.

474 Bosma, A. (1995) Dark matter in external galaxies-HI observations, in *Dark Matter*, American Institute of Physics Conference Series, vol. 336 (eds S.S. Holt and C.L. Bennett), pp. 111–120, doi:10.1063/1.48317.

475 van Albada, T.S., Bahcall, J.N., Begeman, K., and Sancisi, R. (1985) Distribution of dark matter in the spiral galaxy NGC 3198. *ApJ*, **295**, 305–313, doi:10.1086/163375.

476 Dutton, A.A., Courteau, S., de Jong, R., and Carignan, C. (2005) Mass modeling of disk galaxies: degeneracies, constraints, and adiabatic contraction. *ApJ*, **619**, 218–242, doi:10.1086/426375.

477 Ostriker, J.P. and Peebles, P.J.E. (1973) A numerical study of the stability of flattened galaxies: or, can cold galaxies survive. *ApJ*, **186**, 467–480, doi:10.1086/152513.

478 Zhao, H. (1996) Analytical models for galactic nuclei. *MNRAS*, **278**, 488–496.

479 Navarro, J.F., Frenk, C.S., and White, S.D.M. (1997) A universal density profile from hierarchical clustering. *ApJ*, **490**, 493, doi:10.1086/304888.

480 Jimenez, R., Verde, L., and Oh, S.P. (2003) Dark halo properties from rotation curves. *MNRAS*, **339**, 243–259, doi:10.1046/j.1365-8711.2003.06165.x.

481 Spekkens, K., Giovanelli, R., and Haynes, M.P. (2005) The cusp/core problem in galactic halos: long-slit spectra for a large dwarf galaxy sample. *AJ*, **129**, 2119–2137, doi:10.1086/429592.

482 Kent, S.M. (1986) Dark matter in spiral galaxies. I – galaxies with optical rotation curves. *AJ*, **91**, 1301–1327, doi:10.1086/114106.

483 Kuzio de Naray, R., McGaugh, S.S., de Blok, W.J.G., and Bosma, A. (2006) High-resolution optical velocity fields of 11 low surface brightness galaxies. *ApJS*, **165**, 461–479, doi:10.1086/505345.

484 Persic, M. and Salucci, P. (1991) The universal galaxy rotation curve. *ApJ*, **368**, 60–65, doi:10.1086/169670.

485 Persic, M., Salucci, P., and Stel, F. (1996) The universal rotation curve of spiral galaxies – I. The dark matter connection. *MNRAS*, **281**, 27–47.

486 Catinella, B., Giovanelli, R., and Haynes, M.P. (2006) Template rotation curves for disk galaxies. *ApJ*, **640**, 751–761, doi:10.1086/500171.

487 Emsellem, E., Cappellari, M., Krajnović, D., van de Ven, G., Bacon, R., Bureau, M., Davies, R.L., de Zeeuw, P.T., Falcón-Barroso, J., Kuntschner, H., McDermid, R., Peletier, R.F., and Sarzi, M. (2007) The SAURON project – IX. A kinematic classification for early-type galaxies. *MNRAS*, **379**, 401–417, doi:10.1111/j.1365-2966.2007.11752.x.

488 Toloba, E., Boselli, A., Gorgas, J., Peletier, R.F., Cenarro, A.J., Gadotti, D.A., Gil de Paz, A., Pedraz, S., and Yıldız, U. (2009) Kinematic properties as probes of the evolution of dwarf galaxies in the Virgo cluster. *ApJ*, **707**, L17–L21, doi:10.1088/0004-637X/707/1/L17.

489 Côté, P., McLaughlin, D.E., Cohen, J.G., and Blakeslee, J.P. (2003) Dynamics of the globular cluster system associated with M49 (NGC 4472): cluster orbital properties and the distribution of dark matter. *ApJ*, **591**, 850–877, doi:10.1086/375488.

490 Mathews, W.G. (1978) The enormous mass of the elliptical galaxy M87 – a model for the extended X-ray source. *ApJ*, **219**, 413–423, doi:10.1086/155794.

491 Humphrey, P.J., Buote, D.A., Gastaldello, F., Zappacosta, L., Bullock, J.S., Brighenti, F., and Mathews, W.G. (2006) A Chandra view of dark matter in early-type galaxies. *ApJ*, **646**, 899–918, doi:10.1086/505019.

492 Gastaldello, F., Buote, D.A., Humphrey, P.J., Zappacosta, L., Bullock, J.S., Brighenti, F., and Mathews, W.G. (2007) Probing the dark matter and gas fraction in relaxed galaxy groups with X-ray observations from Chandra

and XMM-Newton. *ApJ*, **669**, 158–183, doi:10.1086/521519.

493 Kormendy, J. and Richstone, D. (1995) Inward bound – the search for supermassive black holes in galactic nuclei. *ARA&A*, **33**, 581, doi:10.1146/annurev.aa.33.090195.003053.

494 Ferrarese, L. and Ford, H. (2005) Supermassive black holes in galactic nuclei: past, present and future research. *Space Sci. Rev.*, **116**, 523–624, doi:10.1007/s11214-005-3947-6.

495 Miyoshi, M., Moran, J., Herrnstein, J., Greenhill, L., Nakai, N., Diamond, P., and Inoue, M. (1995) Evidence for a black hole from high rotation velocities in a sub-parsec region of NGC4258. *Nature*, **373**, 127–129, doi:10.1038/373127a0.

496 Macchetto, F., Marconi, A., Axon, D.J., Capetti, A., Sparks, W., and Crane, P. (1997) The Supermassive black hole of M87 and the kinematics of its associated gaseous disk. *ApJ*, **489**, 579, doi:10.1086/304823.

497 van der Marel, R.P. (1994) Velocity profiles of galaxies with claimed black-holes – part three – observations and models for M87. *MNRAS*, **270**, 271.

498 Metcalfe, N., Shanks, T., Fong, R., and Jones, L.R. (1991) Galaxy number counts. II – CCD observations to $B = 25$ mag. *MNRAS*, **249**, 498–522.

499 Metcalfe, N., Shanks, T., Campos, A., McCracken, H.J., and Fong, R. (2001) Galaxy number counts – V. Ultradeep counts: the Herschel and Hubble Deep Fields. *MNRAS*, **323**, 795–830, doi:10.1046/j.1365-8711.2001.04168.x.

500 Ellis, R.S. (1997) Faint blue galaxies. *ARA&A*, **35**, 389–443, doi:10.1146/annurev.astro.35.1.389.

501 Shanks, T., Stevenson, P.R.F., Fong, R., and MacGillivray, H.T. (1984) Galaxy number counts and cosmology. *MNRAS*, **206**, 767–800.

502 Fukugita, M., Yamashita, K., Takahara, F., and Yoshii, Y. (1990) Test for the cosmological constant with the number count of faint galaxies. *ApJ*, **361**, L1–L4, doi:10.1086/185813.

503 Longair, M.S. (1966) On the interpretation of radio source counts. *MNRAS*, **133**, 421.

504 Pozzetti, L., Madau, P., Zamorani, G., Ferguson, H.C., and Bruzual, A.G. (1998) High-redshift galaxies in the Hubble Deep Field – II. Colours and number counts. *MNRAS*, **298**, 1133–1144, doi:10.1046/j.1365-8711.1998.01724.x.

505 Bauer, F.E., Alexander, D.M., Brandt, W.N., Schneider, D.P., Treister, E., Hornschemeier, A.E., and Garmire, G.P. (2004) The fall of active galactic nuclei and the rise of star-forming galaxies: a close look at the Chandra deep field X-ray number counts. *AJ*, **128**, 2048–2065, doi:10.1086/424859.

506 Brunner, H., Cappelluti, N., Hasinger, G., Barcons, X., Fabian, A.C., Mainieri, V., and Szokoly, G. (2008) XMM-Newton observations of the Lockman hole: X-ray source catalogue and number counts. *A&A*, **479**, 283–300, doi:10.1051/0004-6361:20077687.

507 Oliver, S.J., Wang, L., Smith, A.J., Altieri, B., Amblard, A., Arumugam, V., Auld, R., Aussel, H., Babbedge, T., Blain, A., Bock, J., Boselli, A., Buat, V., Burgarella, D., Castro-Rodríguez, N., Cava, A., Chanial, P., Clements, D.L., Conley, A., Conversi, L., Cooray, A., Dowell, C.D., Dwek, E., Eales, S., Elbaz, D., Fox, M., Franceschini, A., Gear, W., Glenn, J., Griffin, M., Halpern, M., Hatziminaoglou, E., Ibar, E., Isaak, K., Ivison, R.J., Lagache, G., Levenson, L., Lu, N., Madden, S., Maffei, B., Mainetti, G., Marchetti, L., Mitchell-Wynne, K., Mortier, A.M.J., Nguyen, H.T., O'Halloran, B., Omont, A., Page, M.J., Panuzzo, P., Papageorgiou, A., Pearson, C.P., Pérez-Fournon, I., Pohlen, M., Rawlings, J.I., Raymond, G., Rigopoulou, D., Rizzo, D., Roseboom, I.G., Rowan-Robinson, M., Sánchez Portal, M., Savage, R., Schulz, B., Scott, D., Seymour, N., Shupe, D.L., Stevens, J.A., Symeonidis, M., Trichas, M., Tugwell, K.E., Vaccari, M., Valiante, E., Valtchanov, I., Vieira, J.D., Vigroux, L., Ward, R., Wright, G., Xu, C.K., and Zemcov, M. (2010) HerMES: SPIRE galaxy number counts at

250, 350, and 500 μm. A&A, **518**, L21, doi:10.1051/0004-6361/201014697.

508 Xu, C.K., Donas, J., Arnouts, S., Wyder, T.K., Seibert, M., Iglesias-Páramo, J., Blaizot, J., Small, T., Milliard, B., Schiminovich, D., Martin, D.C., Barlow, T.A., Bianchi, L., Byun, Y., Forster, K., Friedman, P.G., Heckman, T.M., Jelinsky, P.N., Lee, Y., Madore, B.F., Malina, R.F., Morrissey, P., Neff, S.G., Rich, R.M., Siegmund, O.H.W., Szalay, A.S., and Welsh, B.Y. (2005) Number counts of GALEX sources in far-ultraviolet (1530 Å) and near-ultraviolet (2310 Å) bands. ApJ, **619**, L11–L14, doi:10.1086/425252.

509 McCracken, H.J., Metcalfe, N., Shanks, T., Campos, A., Gardner, J.P., and Fong, R. (2000) Galaxy number counts – IV. Surveying the Herschel deep field in the near-infrared. MNRAS, **311**, 707–718, doi:10.1046/j.1365-8711.2000.03096.x.

510 Le Floc'h, E., Papovich, C., Dole, H., Bell, E.F., Lagache, G., Rieke, G.H., Egami, E., Pérez-González, P.G., Alonso-Herrero, A., Rieke, M.J., Blaylock, M., Engelbracht, C.W., Gordon, K.D., Hines, D.C., Misselt, K.A., Morrison, J.E., and Mould, J. (2005) Infrared luminosity functions from the Chandra deep field-south: the Spitzer view on the history of dusty star formation at $0 < z < 1$. ApJ, **632**, 169–190, doi:10.1086/432789.

511 Windhorst, R.A., Miley, G.K., Owen, F.N., Kron, R.G., and Koo, D.C. (1985) Sub-millijansky 1.4 GHz source counts and multicolor studies of weak radio galaxy populations. ApJ, **289**, 494–513, doi:10.1086/162911.

512 Smolčić, V., Schinnerer, E., Scodeggio, M., Franzetti, P., Aussel, H., Bondi, M., Brusa, M., Carilli, C.L., Capak, P., Charlot, S., Ciliegi, P., Ilbert, O., Ivezić, Ž., Jahnke, K., McCracken, H.J., Obrić, M., Salvato, M., Sanders, D.B., Scoville, N., Trump, J.R., Tremonti, C., Tasca, L., Walcher, C.J., and Zamorani, G. (2008) A new method to separate star-forming from AGN galaxies at intermediate redshift: the submillijansky radio population in the VLA-COSMOS survey. ApJS, **177**, 14–38, doi:10.1086/588028.

513 Taylor, A.C., Grainge, K., Jones, M.E., Pooley, G.G., Saunders, R.D.E., and Waldram, E.M. (2001) The radio source counts at 15 GHz and their implications for cm-wave CMB imaging. MNRAS, **327**, L1–L4, doi:10.1046/j.1365-8711.2001.04877.x.

514 Windhorst, R.A., Fomalont, E.B., Partridge, R.B., and Lowenthal, J.D. (1993) Microjansky source counts and spectral indices at 8.44 GHz. ApJ, **405**, 498–517, doi:10.1086/172382.

515 Henkel, B. and Partridge, R.B. (2005) Completing the counts of radio sources at 8.5 GHz. ApJ, **635**, 950–958, doi:10.1086/497588.

516 Prandoni, I., Parma, P., Wieringa, M.H., de Ruiter, H.R., Gregorini, L., Mignano, A., Vettolani, G., and Ekers, R.D. (2006) The ATESP 5 GHz radio survey. I. Source counts and spectral index properties of the faint radio population. A&A, **457**, 517–529, doi:10.1051/0004-6361:20054273.

517 Huynh, M.T., Jackson, C.A., Norris, R.P., and Prandoni, I. (2005) Radio observations of the Hubble Deep Field-south region. II. The 1.4 GHz catalog and source counts. AJ, **130**, 1373–1388, doi:10.1086/432873.

518 de Zotti, G., Massardi, M., Negrello, M., and Wall, J. (2010) Radio and millimeter continuum surveys and their astrophysical implications. A&A Rev., **18**, 1–65, doi:10.1007/s00159-009-0026-0.

519 Bondi, M., Ciliegi, P., Schinnerer, E., Smolčić, V., Jahnke, K., Carilli, C., and Zamorani, G. (2008) The VLA-COSMOS survey. III. Further catalog analysis and the radio source counts. ApJ, **681**, 1129–1135, doi:10.1086/589324.

520 Bundy, K., Ellis, R.S., and Conselice, C.J. (2005) The mass assembly histories of galaxies of various morphologies in the GOODS fields. ApJ, **625**, 621–632, doi:10.1086/429549.

521 Bell, E.F., Zheng, X.Z., Papovich, C., Borch, A., Wolf, C., and Meisenheimer, K. (2007) Star formation and the growth of stellar mass. ApJ, **663**, 834–843, doi:10.1086/518594.

522 Arnouts, S., Walcher, C.J., Le Fèvre, O., Zamorani, G., Ilbert, O., Le Brun, V., Pozzetti, L., Bardelli, S., Tresse, L., Zucca, E., Charlot, S., Lamareille, F., McCracken, H.J., Bolzonella, M., Iovino, A., Lonsdale, C., Polletta, M., Surace, J., Bottini, D., Garilli, B., Maccagni, D., Picat, J.P., Scaramella, R., Scodeggio, M., Vettolani, G., Zanichelli, A., Adami, C., Cappi, A., Ciliegi, P., Contini, T., de la Torre, S., Foucaud, S., Franzetti, P., Gavignaud, I., Guzzo, L., Marano, B., Marinoni, C., Mazure, A., Meneux, B., Merighi, R., Paltani, S., Pellò, R., Pollo, A., Radovich, M., Temporin, S., and Vergani, D. (2007) The SWIRE-VVDS-CFHTLS surveys: stellar mass assembly over the last 10 Gyr. Evidence for a major build up of the red sequence between $z = 2$ and $z = 1$. A&A, **476**, 137–150, doi:10.1051/0004-6361:20077632.

523 Pérez-González, P.G., Rieke, G.H., Villar, V., Barro, G., Blaylock, M., Egami, E., Gallego, J., Gil de Paz, A., Pascual, S., Zamorano, J., and Donley, J.L. (2008) The stellar mass assembly of galaxies from $z = 0$ to $z = 4$: analysis of a sample selected in the rest-frame near-infrared with Spitzer. ApJ, **675**, 234–261, doi:10.1086/523690.

524 Ilbert, O., Salvato, M., Le Floc'h, E., Aussel, H., Capak, P., McCracken, H.J., Mobasher, B., Kartaltepe, J., Scoville, N., Sanders, D.B., Arnouts, S., Bundy, K., Cassata, P., Kneib, J., Koekemoer, A., Le Fèvre, O., Lilly, S., Surace, J., Taniguchi, Y., Tasca, L., Thompson, D., Tresse, L., Zamojski, M., Zamorani, G., and Zucca, E. (2010) Galaxy stellar mass assembly between $0.2 < z < 2$ from the S-COSMOS survey. ApJ, **709**, 644–663, doi:10.1088/0004-637X/709/2/644.

525 Kauffmann, G., White, S.D.M., and Guiderdoni, B. (1993) The formation and evolution of galaxies within merging dark matter haloes. MNRAS, **264**, 201.

526 Somerville, R.S. and Primack, J.R. (1999) Semi-analytic modelling of galaxy formation: the local universe. MNRAS, **310**, 1087–1110, doi:10.1046/j.1365-8711.1999.03032.x.

527 Baugh, C.M., Cole, S., and Frenk, C.S. (1996) Evolution of the Hubble sequence in hierarchical models for galaxy formation. MNRAS, **283**, 1361–1378.

528 Schmidt, M. (1968) Space distribution and luminosity functions of quasi-stellar radio sources. ApJ, **151**, 393, doi:10.1086/149446.

529 Takeuchi, T.T., Yoshikawa, K., and Ishii, T.T. (2000) Tests of statistical methods for estimating galaxy luminosity function and applications to the Hubble Deep Field. ApJS, **129**, 1–31, doi:10.1086/313409.

530 Felten, J.E. (1976) On Schmidt's Vm estimator and other estimators of luminosity functions. ApJ, **207**, 700–709, doi:10.1086/154538.

531 Lynden-Bell, D. (1971) A method of allowing for known observational selection in small samples applied to 3CR quasars. MNRAS, **155**, 95.

532 Zucca, E., Zamorani, G., Vettolani, G., Cappi, A., Merighi, R., Mignoli, M., Stirpe, G.M., MacGillivray, H., Collins, C., Balkowski, C., Cayatte, V., Maurogordato, S., Proust, D., Chincarini, G., Guzzo, L., Maccagni, D., Scaramella, R., Blanchard, A., and Ramella, M. (1997) The ESO Slice Project (ESP) galaxy redshift survey. II. The luminosity function and mean galaxy density. A&A, **326**, 477–488.

533 Efstathiou, G., Ellis, R.S., and Peterson, B.A. (1988) Analysis of a complete galaxy redshift survey. II – the field-galaxy luminosity function. MNRAS, **232**, 431–461.

534 Sandage, A., Tammann, G.A., and Yahil, A. (1979) The velocity field of bright nearby galaxies. I – the variation of mean absolute magnitude with redshift for galaxies in a magnitude-limited sample. ApJ, **232**, 352–364, doi:10.1086/157295.

535 Ilbert, O., Tresse, L., Zucca, E., Bardelli, S., Arnouts, S., Zamorani, G., Pozzetti, L., Bottini, D., Garilli, B., Le Brun, V., Le Fèvre, O., Maccagni, D., Picat, J., Scaramella, R., Scodeggio, M., Vettolani, G., Zanichelli, A., Adami, C., Arnaboldi, M., Bolzonella, M., Cappi, A., Charlot, S., Contini, T., Foucaud, S., Franzetti,

P., Gavignaud, I., Guzzo, L., Iovino, A., McCracken, H.J., Marano, B., Marinoni, C., Mathez, G., Mazure, A., Meneux, B., Merighi, R., Paltani, S., Pello, R., Pollo, A., Radovich, M., Bondi, M., Bongiorno, A., Busarello, G., Ciliegi, P., Lamareille, F., Mellier, Y., Merluzzi, P., Ripepi, V., and Rizzo, D. (2005) The VIMOS-VLT deep survey. Evolution of the galaxy luminosity function up to $z = 2$ in first epoch data. A&A, **439**, 863–876, doi:10.1051/0004-6361:20041961.

536 Ilbert, O., Tresse, L., Arnouts, S., Zucca, E., Bardelli, S., Zamorani, G., Adami, C., Cappi, A., Garilli, B., Le Fèvre, O., Maccagni, D., Meneux, B., Scaramella, R., Scodeggio, M., Vettolani, G., and Zanichelli, A. (2004) Bias in the estimation of global luminosity functions. MNRAS, **351**, 541–551, doi:10.1111/j.1365-2966.2004.07796.x.

537 Binggeli, B., Sandage, A., and Tammann, G.A. (1988) The luminosity function of galaxies. ARA&A, **26**, 509–560, doi:10.1146/annurev.aa.26.090188.002453.

538 Eisenhardt, P.R.M., Brodwin, M., Gonzalez, A.H., Stanford, S.A., Stern, D., Barmby, P., Brown, M.J.I., Dawson, K., Dey, A., Doi, M., Galametz, A., Jannuzi, B.T., Kochanek, C.S., Meyers, J., Morokuma, T., and Moustakas, L.A. (2008) Clusters of galaxies in the first half of the universe from the IRAC shallow survey. ApJ, **684**, 905–932, doi:10.1086/590105.

539 Adami, C., Ilbert, O., Pelló, R., Cuillandre, J.C., Durret, F., Mazure, A., Picat, J.P., and Ulmer, M.P. (2008) Photometric redshifts as a tool for studying the Coma cluster galaxy populations. A&A, **491**, 681–692, doi:10.1051/0004-6361:200809845.

540 Cortese, L., Gavazzi, G., and Boselli, A. (2008) The ultraviolet luminosity function and star formation rate of the Coma cluster. MNRAS, **390**, 1282–1296, doi:10.1111/j.1365-2966.2008.13838.x.

541 De Propris, R., Colless, M., Driver, S.P., Couch, W., Peacock, J.A., Baldry, I.K., Baugh, C.M., Bland-Hawthorn, J., Bridges, T., Cannon, R., Cole, S., Collins, C., Cross, N., Dalton, G.B., Efstathiou, G., Ellis, R.S., Frenk, C.S., Glazebrook, K., Hawkins, E., Jackson, C., Lahav, O., Lewis, I., Lumsden, S., Maddox, S., Madgwick, D.S., Norberg, P., Percival, W., Peterson, B., Sutherland, W., and Taylor, K. (2003) The 2dF galaxy redshift survey: the luminosity function of cluster galaxies. MNRAS, **342**, 725–737, doi:10.1046/j.1365-8711.2003.06510.x.

542 Sandage, A., Binggeli, B., and Tammann, G.A. (1985) Studies of the Virgo cluster – part five – luminosity functions of Virgo cluster galaxies. AJ, **90**, 1759–1771, doi:10.1086/113875.

543 Schechter, P. (1976) An analytic expression for the luminosity function for galaxies. ApJ, **203**, 297–306, doi:10.1086/154079.

544 Saunders, W., Rowan-Robinson, M., Lawrence, A., Efstathiou, G., Kaiser, N., Ellis, R.S., and Frenk, C.S. (1990) The 60-micron and far-infrared luminosity functions of IRAS galaxies. MNRAS, **242**, 318–337.

545 Takeuchi, T.T., Yoshikawa, K., and Ishii, T.T. (2003) The luminosity function of IRAS point source catalog redshift survey galaxies. ApJ, **587**, L89–L92, doi:10.1086/375181.

546 Avni, Y. and Bahcall, J.N. (1980) On the simultaneous analysis of several complete samples – The V/V_{max} and V_e/V_a variables, with applications to quasars. ApJ, **235**, 694–716, doi:10.1086/157673.

547 Takeuchi, T.T. (2010) Constructing a bivariate distribution function with given marginals and correlation: application to the galaxy luminosity function. MNRAS, **406**, 1830–1840, doi:10.1111/j.1365-2966.2010.16778.x.

548 Avni, Y., Soltan, A., Tananbaum, H., and Zamorani, G. (1980) A method for determining luminosity functions incorporating both flux measurements and flux upper limits, with applications to the average X-ray to optical luminosity ratio for quasars. ApJ, **238**, 800–807, doi:10.1086/158040.

549 Hummel, E. (1981) The radio continuum properties of spiral galaxies. A&A, **93**, 93–105.

550 Gavazzi, G. and Boselli, A. (1999) On the local radio luminosity function of

galaxies. I. The Virgo cluster. *A&A*, **343**, 86–92.

551 Ueda, Y., Akiyama, M., Ohta, K., and Miyaji, T. (2003) Cosmological evolution of the hard X-ray active galactic nucleus luminosity function and the origin of the hard X-ray background. *ApJ*, **598**, 886–908, doi:10.1086/378940.

552 Georgantopoulos, I., Basilakos, S., and Plionis, M. (1999) The X-ray luminosity function of local galaxies. *MNRAS*, **305**, L31–L34, doi:10.1046/j.1365-8711.1999.02629.x.

553 Norman, C., Ptak, A., Hornschemeier, A., Hasinger, G., Bergeron, J., Comastri, A., Giacconi, R., Gilli, R., Glazebrook, K., Heckman, T., Kewley, L., Ranalli, P., Rosati, P., Szokoly, G., Tozzi, P., Wang, J., Zheng, W., and Zirm, A. (2004) The X-ray-derived cosmological star formation history and the galaxy x-ray luminosity functions in the Chandra deep fields north and south. *ApJ*, **607**, 721–738, doi:10.1086/383487.

554 Ranalli, P., Comastri, A., and Setti, G. (2005) The X-ray luminosity function and number counts of spiral galaxies. *A&A*, **440**, 23–37, doi:10.1051/0004-6361:20042598.

555 Wyder, T.K., Treyer, M.A., Milliard, B., Schiminovich, D., Arnouts, S., Budavári, T., Barlow, T.A., Bianchi, L., Byun, Y., Donas, J., Forster, K., Friedman, P.G., Heckman, T.M., Jelinsky, P.N., Lee, Y., Madore, B.F., Malina, R.F., Martin, D.C., Morrissey, P., Neff, S.G., Rich, R.M., Siegmund, O.H.W., Small, T., Szalay, A.S., and Welsh, B.Y. (2005) The ultraviolet galaxy luminosity function in the local universe from GALEX data. *ApJ*, **619**, L15–L18, doi:10.1086/424735.

556 Cole, S., Norberg, P., Baugh, C.M., Frenk, C.S., Bland-Hawthorn, J., Bridges, T., Cannon, R., Colless, M., Collins, C., Couch, W., Cross, N., Dalton, G., De Propris, R., Driver, S.P., Efstathiou, G., Ellis, R.S., Glazebrook, K., Jackson, C., Lahav, O., Lewis, I., Lumsden, S., Maddox, S., Madgwick, D., Peacock, J.A., Peterson, B.A., Sutherland, W., and Taylor, K. (2001) The 2dF galaxy redshift survey: near-infrared galaxy luminosity functions. *MNRAS*, **326**, 255–273, doi:10.1046/j.1365-8711.2001.04591.x.

557 Blanton, M.R., Lupton, R.H., Schlegel, D.J., Strauss, M.A., Brinkmann, J., Fukugita, M., and Loveday, J. (2005) The properties and luminosity function of extremely low luminosity galaxies. *ApJ*, **631**, 208–230, doi:10.1086/431416.

558 Arnouts, S., Schiminovich, D., Ilbert, O., Tresse, L., Milliard, B., Treyer, M., Bardelli, S., Budavari, T., Wyder, T.K., Zucca, E., Le Fèvre, O., Martin, D.C., Vettolani, G., Adami, C., Arnaboldi, M., Barlow, T., Bianchi, L., Bolzonella, M., Bottini, D., Byun, Y., Cappi, A., Charlot, S., Contini, T., Donas, J., Forster, K., Foucaud, S., Franzetti, P., Friedman, P.G., Garilli, B., Gavignaud, I., Guzzo, L., Heckman, T.M., Hoopes, C., Iovino, A., Jelinsky, P., Le Brun, V., Lee, Y., Maccagni, D., Madore, B.F., Malina, R., Marano, B., Marinoni, C., McCracken, H.J., Mazure, A., Meneux, B., Merighi, R., Morrissey, P., Neff, S., Paltani, S., Pellò, R., Picat, J.P., Pollo, A., Pozzetti, L., Radovich, M., Rich, R.M., Scaramella, R., Scodeggio, M., Seibert, M., Siegmund, O., Small, T., Szalay, A.S., Welsh, B., Xu, C.K., Zamorani, G., and Zanichelli, A. (2005) The GALEX VIMOS-VLT deep survey measurement of the evolution of the 1500 Å luminosity function. *ApJ*, **619**, L43–L46, doi:10.1086/426733.

559 Zucca, E., Ilbert, O., Bardelli, S., Tresse, L., Zamorani, G., Arnouts, S., Pozzetti, L., Bolzonella, M., McCracken, H.J., Bottini, D., Garilli, B., Le Brun, V., Le Fèvre, O., Maccagni, D., Picat, J.P., Scaramella, R., Scodeggio, M., Vettolani, G., Zanichelli, A., Adami, C., Arnaboldi, M., Cappi, A., Charlot, S., Ciliegi, P., Contini, T., Foucaud, S., Franzetti, P., Gavignaud, I., Guzzo, L., Iovino, A., Marano, B., Marinoni, C., Mazure, A., Meneux, B., Merighi, R., Paltani, S., Pellò, R., Pollo, A., Radovich, M., Bondi, M., Bongiorno, A., Busarello, G., Cucciati, O., Gregorini, L., Lamareille, F., Mathez, G., Mellier, Y., Merluzzi, P., Ripepi, V., and Rizzo, D. (2006) The VIMOS VLT deep survey. Evolution of the luminosity functions by galaxy type up to $z = 1.5$

from first epoch data. *A&A*, **455**, 879–890, doi:10.1051/0004-6361:20053645.

560 Faber, S.M., Willmer, C.N.A., Wolf, C., Koo, D.C., Weiner, B.J., Newman, J.A., Im, M., Coil, A.L., Conroy, C., Cooper, M.C., Davis, M., Finkbeiner, D.P., Gerke, B.F., Gebhardt, K., Groth, E.J., Guhathakurta, P., Harker, J., Kaiser, N., Kassin, S., Kleinheinrich, M., Konidaris, N.P., Kron, R.G., Lin, L., Luppino, G., Madgwick, D.S., Meisenheimer, K., Noeske, K.G., Phillips, A.C., Sarajedini, V.L., Schiavon, R.P., Simard, L., Szalay, A.S., Vogt, N.P., and Yan, R. (2007) Galaxy luminosity functions to $z \sim 1$ from DEEP2 and COMBO-17: implications for red galaxy formation. *ApJ*, **665**, 265–294, doi:10.1086/519294.

561 Kochanek, C.S., Pahre, M.A., Falco, E.E., Huchra, J.P., Mader, J., Jarrett, T.H., Chester, T., Cutri, R., and Schneider, S.E. (2001) The K-band galaxy luminosity function. *ApJ*, **560**, 566–579, doi:10.1086/322488.

562 Cirasuolo, M., McLure, R.J., Dunlop, J.S., Almaini, O., Foucaud, S., Smail, I., Sekiguchi, K., Simpson, C., Eales, S., Dye, S., Watson, M.G., Page, M.J., and Hirst, P. (2007) The evolution of the near-infrared galaxy luminosity function and colour bimodality up to $z \simeq 2$ from the UKIDSS Ultra Deep Survey Early Data Release. *MNRAS*, **380**, 585–595, doi:10.1111/j.1365-2966.2007.12038.x.

563 Boselli, A., Boissier, S., Cortese, L., and Gavazzi, G. (2008) The origin of dwarf ellipticals in the Virgo cluster. *ApJ*, **674**, 742–767, doi:10.1086/525513.

564 Lawrence, A., Walker, D., Rowan-Robinson, M., Leech, K.J., and Penston, M.V. (1986) Studies of IRAS sources at high galactic latitudes. II – results from a redshift survey at *B* greater than 60 deg: distribution in depth, luminosity function, and physical nature of IRAS galaxies. *MNRAS*, **219**, 687–701.

565 Soifer, B.T., Sanders, D.B., Madore, B.F., Neugebauer, G., Danielson, G.E., Elias, J.H., Lonsdale, C.J., and Rice, W.L. (1987) The IRAS bright galaxy sample. II – the sample and luminosity function. *ApJ*, **320**, 238–257, doi:10.1086/165536.

566 Pérez-González, P.G., Rieke, G.H., Egami, E., Alonso-Herrero, A., Dole, H., Papovich, C., Blaylock, M., Jones, J., Rieke, M., Rigby, J., Barmby, P., Fazio, G.G., Huang, J., and Martin, C. (2005) Spitzer view on the evolution of star-forming galaxies from $z = 0$ to $z \sim 3$. *ApJ*, **630**, 82–107, doi:10.1086/431894.

567 Condon, J.J. (1989) The 1.4 GHz luminosity function and its evolution. *ApJ*, **338**, 13–23, doi:10.1086/167176.

568 Machalski, J. and Godlowski, W. (2000) 1.4 GHz luminosity function of galaxies in the Las Campanas redshift survey and its evolution. *A&A*, **360**, 463–471.

569 Magliocchetti, M., Maddox, S.J., Jackson, C.A., Bland-Hawthorn, J., Bridges, T., Cannon, R., Cole, S., Colless, M., Collins, C., Couch, W., Dalton, G., de Propris, R., Driver, S.P., Efstathiou, G., Ellis, R.S., Frenk, C.S., Glazebrook, K., Lahav, O., Lewis, I., Lumsden, S., Peacock, J.A., Peterson, B.A., Sutherland, W., and Taylor, K. (2002) The 2dF galaxy redshift survey: the population of nearby radio galaxies at the 1-mJy level. *MNRAS*, **333**, 100–120, doi:10.1046/j.1365-8711.2002.05386.x.

570 Sadler, E.M., Jackson, C.A., Cannon, R.D., McIntyre, V.J., Murphy, T., Bland-Hawthorn, J., Bridges, T., Cole, S., Colless, M., Collins, C., Couch, W., Dalton, G., De Propris, R., Driver, S.P., Efstathiou, G., Ellis, R.S., Frenk, C.S., Glazebrook, K., Lahav, O., Lewis, I., Lumsden, S., Maddox, S., Madgwick, D., Norberg, P., Peacock, J.A., Peterson, B.A., Sutherland, W., and Taylor, K. (2002) Radio sources in the 2dF galaxy redshift survey – II. Local radio luminosity functions for AGN and star-forming galaxies at 1.4 GHz. *MNRAS*, **329**, 227–245, doi:10.1046/j.1365-8711.2002.04998.x.

571 Mauch, T. and Sadler, E.M. (2007) Radio sources in the 6dFGS: local luminosity functions at 1.4 GHz for star-forming galaxies and radio-loud AGN. *MNRAS*, **375**, 931–950, doi:10.1111/j.1365-2966.2006.11353.x.

572 Sadler, E.M., Ricci, R., Ekers, R.D., Ekers, J.A., Hancock, P.J., Jackson, C.A.,

Kesteven, M.J., Murphy, T., Phillips, C., Reinfrank, R.F., Staveley-Smith, L., Subrahmanyan, R., Walker, M.A., Wilson, W.E., and de Zotti, G. (2006) The properties of extragalactic radio sources selected at 20 GHz. *MNRAS*, **371**, 898–914, doi:10.1111/j.1365-2966.2006.10729.x.

573 Sadler, E.M., Ricci, R., Ekers, R.D., Sault, R.J., Jackson, C.A., and de Zotti, G. (2008) The extragalactic radio-source population at 95 GHz. *MNRAS*, **385**, 1656–1672, doi:10.1111/j.1365-2966.2008.12955.x.

574 Madau, P., Ferguson, H.C., Dickinson, M.E., Giavalisco, M., Steidel, C.C., and Fruchter, A. (1996) High-redshift galaxies in the Hubble Deep Field: colour selection and star formation history to $z \sim 4$. *MNRAS*, **283**, 1388–1404.

575 Lilly, S.J., Le Fevre, O., Hammer, F., and Crampton, D. (1996) The Canada–France redshift survey: the luminosity density and star formation history of the universe to Z approximately 1. *ApJ*, **460**, L1, doi:10.1086/309975.

576 Madau, P., Pozzetti, L., and Dickinson, M. (1998) The star formation history of field galaxies. *ApJ*, **498**, 106, doi:10.1086/305523.

577 Hopkins, A.M. and Beacom, J.F. (2006) On the normalization of the cosmic star formation history. *ApJ*, **651**, 142–154, doi:10.1086/506610.

578 Wilkins, S.M., Trentham, N., and Hopkins, A.M. (2008) The evolution of stellar mass and the implied star formation history. *MNRAS*, **385**, 687–694, doi:10.1111/j.1365-2966.2008.12885.x.

579 Hopkins, P.F., Younger, J.D., Hayward, C.C., Narayanan, D., and Hernquist, L. (2010) Mergers, active galactic nuclei and "normal" galaxies: contributions to the distribution of star formation rates and infrared luminosity functions. *MNRAS*, **402**, 1693–1713, doi:10.1111/j.1365-2966.2009.15990.x.

580 Oemler, Jr., A. (1976) The structure of elliptical and cD galaxies. *ApJ*, **209**, 693–709, doi:10.1086/154769.

581 Mateo, M.L. (1998) Dwarf galaxies of the local group. *ARA&A*, **36**, 435–506, doi:10.1146/annurev.astro.36.1.435.

582 Kauffmann, G., Heckman, T.M., White, S.D.M., Charlot, S., Tremonti, C., Peng, E.W., Seibert, M., Brinkmann, J., Nichol, R.C., SubbaRao, M., and York, D. (2003) The dependence of star formation history and internal structure on stellar mass for 10^5 low-redshift galaxies. *MNRAS*, **341**, 54–69, doi:10.1046/j.1365-8711.2003.06292.x.

583 Tully, R.B. and Fisher, J.R. (1977) A new method of determining distances to galaxies. *A&A*, **54**, 661–673.

584 Dressler, A., Lynden-Bell, D., Burstein, D., Davies, R.L., Faber, S.M., Terlevich, R., and Wegner, G. (1987) Spectroscopy and photometry of elliptical galaxies. I – a new distance estimator. *ApJ*, **313**, 42–58, doi:10.1086/164947.

585 Kennicutt, Jr., R.C. (1990) Large scale star formation and the interstellar medium, in *The Interstellar Medium in Galaxies*, Astrophysics and Space Science Library, vol. 161 (eds H.A. Thronson, Jr. and J.M. Shull), pp. 405–435.

586 Cowie, L.L., Songaila, A., Hu, E.M., and Cohen, J.G. (1996) New insight on galaxy formation and evolution from Keck spectroscopy of the Hawaii deep fields. *AJ*, **112**, 839, doi:10.1086/118058.

587 Baum, W.A. (1959) Population inferences from star counts, surface brightness and colors. *PASP*, **71**, 106–117, doi:10.1086/127346.

588 Sandage, A. (1972) Absolute magnitudes of E and so galaxies in the Virgo and Coma clusters as a function of $U - B$ color. *ApJ*, **176**, 21, doi:10.1086/151606.

589 Visvanathan, N. and Sandage, A. (1977) The color-absolute magnitude relation for E and S0 galaxies. I – calibration and tests for universality using Virgo and eight other nearby clusters. *ApJ*, **216**, 214–226, doi:10.1086/155464.

590 Bower, R.G., Lucey, J.R., and Ellis, R.S. (1992) Precision photometry of early type galaxies in the Coma and Virgo clusters – a test of the universality of the colour/magnitude relation – part two – analysis. *MNRAS*, **254**, 601.

591 Bernardi, M., Sheth, R.K., Annis, J., Burles, S., Finkbeiner, D.P., Lupton, R.H., Schlegel, D.J., SubbaRao, M., Bah-

call, N.A., Blakeslee, J.P., Brinkmann, J., Castander, F.J., Connolly, A.J., Csabai, I., Doi, M., Fukugita, M., Frieman, J., Heckman, T., Hennessy, G.S., Ivezić, Ž., Knapp, G.R., Lamb, D.Q., McKay, T., Munn, J.A., Nichol, R., Okamura, S., Schneider, D.P., Thakar, A.R., and York, D.G. (2003) Early-type galaxies in the Sloan digital sky survey. IV. Colors and chemical evolution. *AJ*, **125**, 1882–1896, doi:10.1086/367795.

592 Tully, R.B., Mould, J.R., and Aaronson, M. (1982) A color-magnitude relation for spiral galaxies. *ApJ*, **257**, 527–537, doi:10.1086/160009.

593 Gavazzi, G. (1993) Colors, luminosities, and masses of disk galaxies. *ApJ*, **419**, 469, doi:10.1086/173500.

594 Sandage, A. and Visvanathan, N. (1978) The color-absolute magnitude relation for E and S0 galaxies. II – new colors, magnitudes, and types for 405 galaxies. *ApJ*, **223**, 707–729, doi:10.1086/156305.

595 Baldry, I.K., Glazebrook, K., Brinkmann, J., Ivezić, Ž., Lupton, R.H., Nichol, R.C., and Szalay, A.S. (2004) Quantifying the bimodal color-magnitude distribution of galaxies. *ApJ*, **600**, 681–694, doi:10.1086/380092.

596 Gallazzi, A., Charlot, S., Brinchmann, J., and White, S.D.M. (2006) Ages and metallicities of early-type galaxies in the Sloan digital sky survey: new insight into the physical origin of the colour-magniude and the $Mg_2 - \sigma$ relations. *MNRAS*, **370**, 1106–1124, doi:10.1111/j.1365-2966.2006.10548.x.

597 Bressan, A., Chiosi, C., and Fagotto, F. (1994) Spectrophotometric evolution of elliptical galaxies. 1: ultraviolet excess and color-magnitude-redshift relations. *ApJS*, **94**, 63–115, doi:10.1086/192073.

598 Kauffmann, G. and Charlot, S. (1998) Chemical enrichment and the origin of the colour–magnitude relation of elliptical galaxies in a hierarchical merger model. *MNRAS*, **294**, 705, doi:10.1046/j.1365-8711.1998.01322.x.

599 De Lucia, G., Poggianti, B.M., Aragón-Salamanca, A., White, S.D.M., Zaritsky, D., Clowe, D., Halliday, C., Jablonka, P., von der Linden, A., Milvang-Jensen, B., Pelló, R., Rudnick, G., Saglia, R.P., and Simard, L. (2007) The build-up of the colour–magnitude relation in galaxy clusters since $z \sim 0.8$. *MNRAS*, **374**, 809–822, doi:10.1111/j.1365-2966.2006.11199.x.

600 Gavazzi, G. and Scodeggio, M. (1996) The mass dependence of the star formation history of disk galaxies. *A&A*, **312**, L29–L32.

601 Kodama, T., Arimoto, N., Barger, A.J., and Arag'on-Salamanca, A. (1998) Evolution of the colour–magnitude relation of early-type galaxies in distant clusters. *A&A*, **334**, 99–109.

602 Bernardi, M., Sheth, R.K., Nichol, R.C., Schneider, D.P., and Brinkmann, J. (2005) Colors, magnitudes, and velocity dispersions in early-type galaxies: implications for galaxy ages and metallicities. *AJ*, **129**, 61–72, doi:10.1086/426336.

603 Schawinski, K., Kaviraj, S., Khochfar, S., Yoon, S., Yi, S.K., Deharveng, J., Boselli, A., Barlow, T., Conrow, T., Forster, K., Friedman, P.G., Martin, D.C., Morrissey, P., Neff, S., Schiminovich, D., Seibert, M., Small, T., Wyder, T., Bianchi, L., Donas, J., Heckman, T., Lee, Y., Madore, B., Milliard, B., Rich, R.M., and Szalay, A. (2007) The effect of environment on the ultraviolet color-magnitude relation of early-type galaxies. *ApJS*, **173**, 512–523, doi:10.1086/516631.

604 Disney, M.J., Romano, J.D., Garcia-Appadoo, D.A., West, A.A., Dalcanton, J.J., and Cortese, L. (2008) Galaxies appear simpler than expected. *Nature*, **455**, 1082–1084, doi:10.1038/nature07366.

605 Iglesias-Páramo, J., Boselli, A., Gavazzi, G., and Zaccardo, A. (2004) Tracing the star formation history of cluster galaxies using the H_α/UV flux ratio. *A&A*, **421**, 887–897, doi:10.1051/0004-6361:20034572.

606 White, S.D.M. and Rees, M.J. (1978) Core condensation in heavy halos – a two-stage theory for galaxy formation and clustering. *MNRAS*, **183**, 341–358.

607 Silk, J. and Rees, M.J. (1998) Quasars and galaxy formation. *A&A*, **331**, L1–L4.

608 Efstathiou, G. (2000) A model of supernova feedback in galaxy formation. *MNRAS*, **317**, 697–719, doi:10.1046/j.1365-8711.2000.03665.x.

609 Lequeux, J., Peimbert, M., Rayo, J.F., Serrano, A., and Torres-Peimbert, S. (1979) Chemical composition and evolution of irregular and blue compact galaxies. A&A, **80**, 155–166.

610 Skillman, E.D., Kennicutt, R.C., and Hodge, P.W. (1989) Oxygen abundances in nearby dwarf irregular galaxies. ApJ, **347**, 875–882, doi:10.1086/168178.

611 Richer, M.G. and McCall, M.L. (1995) Oxygen abundances in diffuse ellipticals and the metallicity-luminosity relations for dwarf galaxies. ApJ, **445**, 642–659, doi:10.1086/175727.

612 Garnett, D.R. (2002) The luminosity–metallicity relation, effective yields, and metal loss in spiral and irregular galaxies. ApJ, **581**, 1019–1031, doi:10.1086/344301.

613 Savaglio, S., Glazebrook, K., Le Borgne, D., Juneau, S., Abraham, R.G., Chen, H., Crampton, D., McCarthy, P.J., Carlberg, R.G., Marzke, R.O., Roth, K., Jørgensenø, I., and Murowinski, R. (2005) The Gemini deep deep survey. VII. The redshift evolution of the mass–metallicity relation. ApJ, **635**, 260–279, doi:10.1086/497331.

614 Kobulnicky, H.A., Willmer, C.N.A., Phillips, A.C., Koo, D.C., Faber, S.M., Weiner, B.J., Sarajedini, V.L., Simard, L., and Vogt, N.P. (2003) The DEEP groth strip survey. VII. The metallicity of field galaxies at $0.26 < z < 0.82$ and the evolution of the luminosity-metallicity relation. ApJ, **599**, 1006–1030, doi:10.1086/379360.

615 Erb, D.K., Steidel, C.C., Shapley, A.E., Pettini, M., Reddy, N.A., and Adelberger, K.L. (2006) The stellar, gas, and dynamical masses of star-forming galaxies at $z \sim 2$. ApJ, **646**, 107–132, doi:10.1086/504891.

616 Maiolino, R., Nagao, T., Grazian, A., Cocchia, F., Marconi, A., Mannucci, F., Cimatti, A., Pipino, A., Ballero, S., Calura, F., Chiappini, C., Fontana, A., Granato, G.L., Matteucci, F., Pastorini, G., Pentericci, L., Risaliti, G., Salvati, M., and Silva, L. (2008) AMAZE. I. The evolution of the mass–metallicity relation at $z > 3$. A&A, **488**, 463–479, doi:10.1051/0004-6361:200809678.

617 Burstein, D., Davies, R.L., Dressler, A., Faber, S.M., and Lynden-Bell, D. (1988) Global stellar populations of elliptical galaxies. A – optical properties, in *Towards Understanding Galaxies at Large Redshift*, Astrophysics and Space Science Library, vol. 141 (eds R.G. Kron and A. Renzini), pp. 17–21.

618 Bender, R., Burstein, D., and Faber, S.M. (1993) Dynamically hot galaxies. II – global stellar populations. ApJ, **411**, 153–169, doi:10.1086/172815.

619 Trager, S.C., Faber, S.M., Worthey, G., and González, J.J. (2000) The stellar population histories of early-type galaxies. II. Controlling parameters of the stellar populations. AJ, **120**, 165–188, doi:10.1086/301442.

620 Jablonka, P., Martin, P., and Arimoto, N. (1996) The luminosity–metallicity relation for bulges of spiral galaxies. AJ, **112**, 1415, doi:10.1086/118109.

621 Grebel, E.K., Gallagher, III, J.S., and Harbeck, D. (2003) The progenitors of dwarf spheroidal galaxies. AJ, **125**, 1926–1939, doi:10.1086/368363.

622 Searle, L. and Sargent, W.L.W. (1972) Inferences from the composition of two dwarf blue galaxies. ApJ, **173**, 25, doi:10.1086/151398.

623 Edmunds, M.G. (1990) General constraints on the effect of gas flows in the chemical evolution of galaxies. MNRAS, **246**, 678.

624 Dalcanton, J.J. (2007) The metallicity of galaxy disks: infall versus outflow. ApJ, **658**, 941–959, doi:10.1086/508913.

625 Calura, F., Pipino, A., Chiappini, C., Matteucci, F., and Maiolino, R. (2009) The evolution of the mass–metallicity relation in galaxies of different morphological types. A&A, **504**, 373–388, doi:10.1051/0004-6361/200911756.

626 Buat, V., Boissier, S., Burgarella, D., Takeuchi, T.T., Le Floc'h, E., Marcillac, D., Huang, J., Nagashima, M., and Enoki, M. (2008) Star formation history of galaxies from $z = 0$ to $z = 0.7$. A backward approach to the evolution of star-forming galaxies. A&A, **483**, 107–119, doi:10.1051/0004-6361:20078263.

627 De Lucia, G., Kauffmann, G., and White, S.D.M. (2004) Chemical enrichment of the intracluster and intergalactic medium in a hierarchical galaxy formation model. *MNRAS*, **349**, 1101–1116, doi:10.1111/j.1365-2966.2004.07584.x.

628 Brooks, A.M., Governato, F., Booth, C.M., Willman, B., Gardner, J.P., Wadsley, J., Stinson, G., and Quinn, T. (2007) The origin and evolution of the mass–metallicity relationship for galaxies: results from cosmological N-body simulations. *ApJ*, **655**, L17–L20, doi:10.1086/511765.

629 Davé, R. and Oppenheimer, B.D. (2007) The enrichment history of baryons in the universe. *MNRAS*, **374**, 427–435, doi:10.1111/j.1365-2966.2006.11177.x.

630 de Rossi, M.E., Tissera, P.B., and Scannapieco, C. (2007) Clues for the origin of the fundamental metallicity relations – I. The hierarchical building up of the structure. *MNRAS*, **374**, 323–336, doi:10.1111/j.1365-2966.2006.11150.x.

631 Finlator, K. and Davé, R. (2008) The origin of the galaxy mass–metallicity relation and implications for galactic outflows. *MNRAS*, **385**, 2181–2204, doi:10.1111/j.1365-2966.2008.12991.x.

632 Roberts, M.S. and Haynes, M.P. (1994) Physical parameters along the Hubble sequence. *ARA&A*, **32**, 115–152, doi:10.1146/annurev.aa.32.090194.000555.

633 Bothwell, M.S., Kennicutt, R.C., and Lee, J.C. (2009) On the interstellar medium and star formation demographics of galaxies in the local universe. *MNRAS*, **400**, 154–167, doi:10.1111/j.1365-2966.2009.15471.x.

634 Feulner, G., Gabasch, A., Salvato, M., Drory, N., Hopp, U., and Bender, R. (2005) Specific star formation rates to redshift 5 from the FORS deep field and the GOODS-S field. *ApJ*, **633**, L9–L12, doi:10.1086/498109.

635 Juneau, S., Glazebrook, K., Crampton, D., McCarthy, P.J., Savaglio, S., Abraham, R., Carlberg, R.G., Chen, H., Le Borgne, D., Marzke, R.O., Roth, K., Jørgensen, I., Hook, I., and Murowinski, R. (2005) Cosmic star formation history and its dependence on galaxy stellar mass. *ApJ*, **619**, L135–L138, doi:10.1086/427937.

636 Binggeli, B., Sandage, A., and Tarenghi, M. (1984) Studies of the Virgo cluster. I – photometry of 109 galaxies near the cluster center to serve as standards. *AJ*, **89**, 64–82, doi:10.1086/113484.

637 Kormendy, J. (1985) Families of ellipsoidal stellar systems and the formation of dwarf elliptical galaxies. *ApJ*, **295**, 73–79, doi:10.1086/163350.

638 Ferguson, H.C. and Binggeli, B. (1994) Dwarf elliptical galaxies. *A&A Rev.*, **6**, 67–122, doi:10.1007/BF01208252.

639 Gavazzi, G., Donati, A., Cucciati, O., Sabatini, S., Boselli, A., Davies, J., and Zibetti, S. (2005) The structure of elliptical galaxies in the Virgo cluster. Results from the INT wide field survey. *A&A*, **430**, 411–419, doi:10.1051/0004-6361:20034571.

640 Aguerri, J.A.L., Iglesias-Páramo, J., Vílchez, J.M., Muñoz-Tuñón, C., and Sánchez-Janssen, R. (2005) Structural parameters of dwarf galaxies in the Coma cluster: on the origin of dS0 galaxies. *AJ*, **130**, 475–495, doi:10.1086/431360.

641 Misgeld, I., Mieske, S., and Hilker, M. (2008) The early-type dwarf galaxy population of the Hydra I cluster. *A&A*, **486**, 697–709, doi:10.1051/0004-6361:200810014.

642 Boselli, A., Boissier, S., Cortese, L., and Gavazzi, G. (2008) The origin of the $\mu - M_B$ and Kormendy relations in dwarf elliptical galaxies. *A&A*, **489**, 1015–1022, doi:10.1051/0004-6361:200809546.

643 Mathewson, D.S. and Ford, V.L. (1994) Large-scale streaming motions in the local universe. *ApJ*, **434**, L39–L42, doi:10.1086/187569.

644 Theureau, G., Hanski, M.O., Coudreau, N., Hallet, N., and Martin, J. (2007) Kinematics of the Local Universe. XIII. 21-cm line measurements of 452 galaxies with the Nançay radiotelescope, JHK Tully–Fisher relation, and preliminary maps of the peculiar velocity field. *A&A*, **465**, 71–85, doi:10.1051/0004-6361:20066187.

645 Fouque, P., Bottinelli, L., Gouguenheim, L., and Paturel, G. (1990) The extragalactic distance scale. II – the unbiased distance to the Virgo cluster from the

B-band Tully–Fisher relation. *ApJ*, **349**, 1–21, doi:10.1086/168288.

646 Yasuda, N., Fukugita, M., and Okamura, S. (1997) Study of the Virgo cluster using the B-band Tully–Fisher relation. *ApJS*, **108**, 417, doi:10.1086/312960.

647 Gavazzi, G., Boselli, A., Scodeggio, M., Pierini, D., and Belsole, E. (1999) The 3D structure of the Virgo cluster from H-band fundamental plane and Tully–Fisher distance determinations. *MNRAS*, **304**, 595–610, doi:10.1046/j.1365-8711.1999.02350.x.

648 Conselice, C.J., Bundy, K., Ellis, R.S., Brichmann, J., Vogt, N.P., and Phillips, A.C. (2005) Evolution of the near-infrared Tully–Fisher relation: constraints on the relationship between the stellar and total masses of disk galaxies since $z \sim 1$. *ApJ*, **628**, 160–168, doi:10.1086/430589.

649 Aaronson, M. and Mould, J. (1983) A distance scale from the infrared magnitude/HI velocity-width relation. IV – the morphological type dependence and scatter in the relation; the distances to nearby groups. *ApJ*, **265**, 1–17, doi:10.1086/160648.

650 Giovanelli, R., Haynes, M.P., Herter, T., Vogt, N.P., da Costa, L.N., Freudling, W., Salzer, J.J., and Wegner, G. (1997) The I band Tully–Fisher relation for cluster galaxies: a template relation, its scatter and bias corrections. *AJ*, **113**, 53–79, doi:10.1086/118234.

651 Courteau, S., Dutton, A.A., van den Bosch, F.C., MacArthur, L.A., Dekel, A., McIntosh, D.H., and Dale, D.A. (2007) Scaling relations of spiral galaxies. *ApJ*, **671**, 203–225, doi:10.1086/522193.

652 Masters, K.L., Springob, C.M., and Huchra, J.P. (2008) 2MTF. I. The Tully–Fisher Relation in the two micron all sky survey J, H, and K bands. *AJ*, **135**, 1738–1748, doi:10.1088/0004-6256/135/5/1738.

653 McGaugh, S.S., Schombert, J.M., Bothun, G.D., and de Blok, W.J.G. (2000) The baryonic Tully–Fisher relation. *ApJ*, **533**, L99–L102, doi:10.1086/312628.

654 McGaugh, S.S. (2005) The baryonic Tully–Fisher relation of galaxies with extended rotation curves and the stellar mass of rotating galaxies. *ApJ*, **632**, 859–871, doi:10.1086/432968.

655 Geha, M., Blanton, M.R., Masjedi, M., and West, A.A. (2006) The baryon content of extremely low mass dwarf galaxies. *ApJ*, **653**, 240–254, doi:10.1086/508604.

656 De Rijcke, S., Zeilinger, W.W., Hau, G.K.T., Prugniel, P., and Dejonghe, H. (2007) Generalizations of the Tully–Fisher relation for early- and late-type galaxies. *ApJ*, **659**, 1172–1175, doi:10.1086/512717.

657 Zaritsky, D., Zabludoff, A.I., and Gonzalez, A.H. (2008) Toward equations of galactic structure. *ApJ*, **682**, 68–80, doi:10.1086/529577.

658 Faber, S.M. and Jackson, R.E. (1976) Velocity dispersions and mass-to-light ratios for elliptical galaxies. *ApJ*, **204**, 668–683, doi:10.1086/154215.

659 Djorgovski, S. and Davis, M. (1987) Fundamental properties of elliptical galaxies. *ApJ*, **313**, 59–68, doi:10.1086/164948.

660 Jorgensen, I., Franx, M., and Kjaergaard, P. (1996) The fundamental plane for cluster E and S0 galaxies. *MNRAS*, **280**, 167–185.

661 Pahre, M.A., Djorgovski, S.G., and de Carvalho, R.R. (1998) Near-infrared imaging of early-type galaxies. III. The near-infrared fundamental plane. *AJ*, **116**, 1591–1605, doi:10.1086/300544.

662 Bender, R., Burstein, D., and Faber, S.M. (1992) Dynamically hot galaxies. I – structural properties. *ApJ*, **399**, 462–477, doi:10.1086/171940.

663 Pahre, M.A., Djorgovski, S.G., and de Carvalho, R.R. (1995) The near-infrared fundamental plane of elliptical galaxies. *ApJ*, **453**, L17, doi:10.1086/309740.

664 Scodeggio, M., Gavazzi, G., Belsole, E., Pierini, D., and Boselli, A. (1998) The tilt of the fundamental plane of early-type galaxies: wavelength dependence. *MNRAS*, **301**, 1001–1018, doi:10.1046/j.1365-8711.1998.02106.x.

665 Renzini, A. and Ciotti, L. (1993) Transverse dissections of the fundamental planes of elliptical galaxies and clusters of galaxies. *ApJ*, **416**, L49, doi:10.1086/187068.

666 Capelato, H.V., de Carvalho, R.R., and Carlberg, R.G. (1995) Mergers of dissipationless systems: clues about the fundamental plane. *ApJ*, **451**, 525, doi:10.1086/176240.

667 Hjorth, J. and Madsen, J. (1995) Small deviations from the $R^{1/4}$ law, the fundamental plane, and phase densities of elliptical galaxies. *ApJ*, **445**, 55–61, doi:10.1086/175672.

668 Ciotti, L., Lanzoni, B., and Renzini, A. (1996) The tilt of the fundamental plane of elliptical galaxies – I. Exploring dynamical and structural effects. *MNRAS*, **282**, 1–12.

669 Graham, A. and Colless, M. (1997) Some effects of galaxy structure and dynamics on the fundamental plane. *MNRAS*, **287**, 221–239.

670 Graves, G.J., Faber, S.M., and Schiavon, R.P. (2009) Dissecting the red sequence. I. Star-formation histories of quiescent galaxies: the color–magnitude versus the color–σ relation. *ApJ*, **693**, 486–506, doi:10.1088/0004-637X/693/1/486.

671 Graves, G.J., Faber, S.M., and Schiavon, R.P. (2009) Dissecting the red sequence. II. Star formation histories of early-type galaxies throughout the fundamental plane. *ApJ*, **698**, 1590–1608, doi:10.1088/0004-637X/698/2/1590.

672 Graves, G.J. and Faber, S.M. (2010) Dissecting the red sequence. III. Mass-to-light variations in three-dimensional fundamental plane space. *ApJ*, **717**, 803–824, doi:10.1088/0004-637X/717/2/803.

673 Zaritsky, D., Gonzalez, A.H., and Zabludoff, A.I. (2006) The fundamental manifold of spheroids. *ApJ*, **638**, 725–738, doi:10.1086/498672.

674 Burstein, D., Bender, R., Faber, S., and Nolthenius, R. (1997) Global relationships among the physical properties of stellar systems. *AJ*, **114**, 1365, doi:10.1086/118570.

675 Pierini, D., Gavazzi, G., Franzetti, P., Scodeggio, M., and Boselli, A. (2002) 1.65-μm (H-band) surface photometry of galaxies – VIII. The near-IR κ space at $z = 0$. *MNRAS*, **332**, 422–434, doi:10.1046/j.1365-8711.2002.05323.x.

676 Haehnelt, M.G., Natarajan, P., and Rees, M.J. (1998) High-redshift galaxies, their active nuclei and central black holes. *MNRAS*, **300**, 817–827, doi:10.1046/j.1365-8711.1998.01951.x.

677 Monaco, P., Salucci, P., and Danese, L. (2000) Joint cosmological formation of QSOs and bulge-dominated galaxies. *MNRAS*, **311**, 279–296, doi:10.1046/j.1365-8711.2000.03043.x.

678 Haehnelt, M.G. and Kauffmann, G. (2000) The correlation between black hole mass and bulge velocity dispersion in hierarchical galaxy formation models. *MNRAS*, **318**, L35–L38, doi:10.1046/j.1365-8711.2000.03989.x.

679 Robertson, B., Hernquist, L., Cox, T.J., Di Matteo, T., Hopkins, P.F., Martini, P., and Springel, V. (2006) The evolution of the M_{BH}–σ relation. *ApJ*, **641**, 90–102, doi:10.1086/500348.

680 Di Matteo, T., Colberg, J., Springel, V., Hernquist, L., and Sijacki, D. (2008) Direct cosmological simulations of the growth of black holes and galaxies. *ApJ*, **676**, 33–53, doi:10.1086/524921.

681 Sutter, P.M. and Ricker, P.M. (2010) Examining subgrid models of supermassive black holes in cosmological simulation. *ApJ*, **723**, 1308–1318, doi:10.1088/0004-637X/723/2/1308.

682 Magorrian, J., Tremaine, S., Richstone, D., Bender, R., Bower, G., Dressler, A., Faber, S.M., Gebhardt, K., Green, R., Grillmair, C., Kormendy, J., and Lauer, T. (1998) The demography of massive dark objects in galaxy centers. *AJ*, **115**, 2285–2305, doi:10.1086/300353.

683 Ferrarese, L. and Merritt, D. (2000) a fundamental relation between supermassive black holes and their host galaxies. *ApJ*, **539**, L9–L12, doi:10.1086/312838.

684 Tremaine, S., Gebhardt, K., Bender, R., Bower, G., Dressler, A., Faber, S.M., Filippenko, A.V., Green, R., Grillmair, C., Ho, L.C., Kormendy, J., Lauer, T.R., Magorrian, J., Pinkney, J., and Richstone, D. (2002) The slope of the black hole mass versus velocity dispersion correlation. *ApJ*, **574**, 740–753, doi:10.1086/341002.

685 Marconi, A. and Hunt, L.K. (2003) The relation between black hole mass, bulge mass, and near-infrared luminosity. *ApJ*, **589**, L21–L24, doi:10.1086/375804.

686 Gültekin, K., Richstone, D.O., Gebhardt, K., Lauer, T.R., Tremaine, S., Aller, M.C., Bender, R., Dressler, A., Faber, S.M., Filippenko, A.V., Green, R., Ho, L.C., Kormendy, J., Magorrian, J., Pinkney, J., and Siopis, C. (2009) The M–σ and M–L relations in galactic bulges, and determinations of their intrinsic scatter. ApJ, **698**, 198–221, doi:10.1088/0004-637X/698/1/198.

687 Kauffmann, G. and Haehnelt, M. (2000) A unified model for the evolution of galaxies and quasars. MNRAS, **311**, 576–588, doi:10.1046/j.1365-8711.2000.03077.x.

688 Adams, F.C., Graff, D.S., and Richstone, D.O. (2001) A theoretical model for the M_{BH}–σ relation for supermassive black holes in galaxies. ApJ, **551**, L31–L35, doi:10.1086/319828.

689 Di Matteo, T., Croft, R.A.C., Springel, V., and Hernquist, L. (2003) Black hole growth and activity in a Λ cold dark matter universe. ApJ, **593**, 56–68, doi:10.1086/376501.

690 Croton, D.J. (2009) A simple model to link the properties of quasars to the properties of dark matter haloes out to high redshift. MNRAS, **394**, 1109–1119, doi:10.1111/j.1365-2966.2009.14429.x.

691 Ferrarese, L. (2002) Beyond the bulge: A fundamental relation between supermassive black holes and dark matter halos. ApJ, **578**, 90–97, doi:10.1086/342308.

692 Baes, M., Buyle, P., Hau, G.K.T., and Dejonghe, H. (2003) Observational evidence for a connection between supermassive black holes and dark matter haloes. MNRAS, **341**, L44–L48, doi:10.1046/j.1365-8711.2003.06680.x.

693 Bandara, K., Crampton, D., and Simard, L. (2009) A relationship between supermassive black hole mass and the total gravitational mass of the host galaxy. ApJ, **704**, 1135–1145, doi:10.1088/0004-637X/704/2/1135.

694 Graham, A.W., Erwin, P., Caon, N., and Trujillo, I. (2001) A correlation between galaxy light concentration and supermassive black hole mass. ApJ, **563**, L11–L14, doi:10.1086/338500.

695 Graham, A.W. and Driver, S.P. (2007) A log-quadratic relation for predicting supermassive black hole masses from the host bulge Sérsic index. ApJ, **655**, 77–87, doi:10.1086/509758.

696 Stocke, J.T., Morris, S.L., Weymann, R.J., and Foltz, C.B. (1992) The radio properties of the broad-absorption-line QSOs. ApJ, **396**, 487–503, doi:10.1086/171735.

697 Snellen, I.A.G., Lehnert, M.D., Bremer, M.N., and Schilizzi, R.T. (2003) Fundamental galaxy parameters for radio-loud active galactic nuclei and the black hole-radio power connection. MNRAS, **342**, 889–900, doi:10.1046/j.1365-8711.2003.06629.x.

698 Merloni, A., Heinz, S., and di Matteo, T. (2003) A fundamental plane of black hole activity. MNRAS, **345**, 1057–1076, doi:10.1046/j.1365-2966.2003.07017.x.

699 Gültekin, K., Cackett, E.M., Miller, J.M., Di Matteo, T., Markoff, S., and Richstone, D.O. (2009) The fundamental plane of accretion onto black holes with dynamical masses. ApJ, **706**, 404–416, doi:10.1088/0004-637X/706/1/404.

700 Benson, A.J., Bower, R.G., Frenk, C.S., Lacey, C.G., Baugh, C.M., and Cole, S. (2003) What shapes the luminosity function of galaxies. ApJ, **599**, 38–49, doi:10.1086/379160.

701 Croton, D.J., Springel, V., White, S.D.M., De Lucia, G., Frenk, C.S., Gao, L., Jenkins, A., Kauffmann, G., Navarro, J.F., and Yoshida, N. (2006) The many lives of active galactic nuclei: cooling flows, black holes and the luminosities and colours of galaxies. MNRAS, **365**, 11–28, doi:10.1111/j.1365-2966.2005.09675.x.

702 Ferrara, A. and Tolstoy, E. (2000) The role of stellar feedback and dark matter in the evolution of dwarf galaxies. MNRAS, **313**, 291–309, doi:10.1046/j.1365-8711.2000.03209.x.

703 Elmegreen, B.G. and Scalo, J. (2004) Interstellar turbulence I: observations and processes. ARA&A, **42**, 211–273, doi:10.1146/annurev.astro.41.011802.094859.

704 McKee, C.F. and Ostriker, E.C. (2007) Theory of star formation. *ARA&A*, **45**, 565–687, doi:10.1146/annurev.astro.45.051806.110602.

705 Kennicutt, Jr., R.C. (1998) The global Schmidt law in star-forming galaxies. *ApJ*, **498**, 541, doi:10.1086/305588.

706 Schmidt, M. (1959) The rate of star formation. *ApJ*, **129**, 243, doi:10.1086/146614.

707 Bigiel, F., Leroy, A., Walter, F., Brinks, E., de Blok, W.J.G., Madore, B., and Thornley, M.D. (2008) The star formation law in nearby galaxies on sub-kpc scales. *AJ*, **136**, 2846–2871, doi:10.1088/0004-6256/136/6/2846.

708 Kennicutt, Jr., R.C. (1989) The star formation law in galactic disks. *ApJ*, **344**, 685–703, doi:10.1086/167834.

709 Martin, C.L. and Kennicutt, Jr., R.C. (2001) Star formation thresholds in galactic disks. *ApJ*, **555**, 301–321, doi:10.1086/321452.

710 Wong, T. and Blitz, L. (2002) The relationship between gas content and star formation in molecule-rich spiral galaxies. *ApJ*, **569**, 157–183, doi:10.1086/339287.

711 Boissier, S., Prantzos, N., Boselli, A., and Gavazzi, G. (2003) The star formation rate in disk galaxies: thresholds and dependence on gas amount. *MNRAS*, **346**, 1215–1230, doi:10.1111/j.1365-2966.2003.07170.x.

712 Boissier, S., Gil de Paz, A., Boselli, A., Madore, B.F., Buat, V., Cortese, L., Burgarella, D., Muñoz-Mateos, J.C., Barlow, T.A., Forster, K., Friedman, P.G., Martin, D.C., Morrissey, P., Neff, S.G., Schiminovich, D., Seibert, M., Small, T., Wyder, T.K., Bianchi, L., Donas, J., Heckman, T.M., Lee, Y., Milliard, B., Rich, R.M., Szalay, A.S., Welsh, B.Y., and Yi, S.K. (2007) Radial variation of attenuation and star formation in the largest late-type disks observed with GALEX. *ApJS*, **173**, 524–537, doi:10.1086/516642.

713 Kennicutt, Jr., R.C., Calzetti, D., Walter, F., Helou, G., Hollenbach, D.J., Armus, L., Bendo, G., Dale, D.A., Draine, B.T., Engelbracht, C.W., Gordon, K.D., Prescott, M.K.M., Regan, M.W., Thornley, M.D., Bot, C., Brinks, E., de Blok, E., de Mello, D., Meyer, M., Moustakas, J., Murphy, E.J., Sheth, K., and Smith, J.D.T. (2007) Star formation in NGC 5194 (M51a). II. The spatially resolved star formation law. *ApJ*, **671**, 333–348, doi:10.1086/522300.

714 Goddard, Q.E., Kennicutt, R.C., and Ryan-Weber, E.V. (2010) On the nature of star formation at large galactic radii. *MNRAS*, **405**, 2791–2809, doi:10.1111/j.1365-2966.2010.16661.x.

715 Leroy, A.K., Walter, F., Brinks, E., Bigiel, F., de Blok, W.J.G., Madore, B., and Thornley, M.D. (2008) The star formation efficiency in nearby galaxies: measuring where gas forms stars effectively. *AJ*, **136**, 2782–2845, doi:10.1088/0004-6256/136/6/2782.

716 Krumholz, M.R., McKee, C.F., and Tumlinson, J. (2009) The atomic-to-molecular transition in galaxies. II: HI and H_2 column densities. *ApJ*, **693**, 216–235, doi:10.1088/0004-637X/693/1/216.

717 Krumholz, M.R., McKee, C.F., and Tumlinson, J. (2009) The star formation law in atomic and molecular gas. *ApJ*, **699**, 850–856, doi:10.1088/0004-637X/699/1/850.

718 Madore, B.F. (1977) Numerical simulations of the rate of star formation in external galaxies. *MNRAS*, **178**, 1–9.

719 Elmegreen, B.G. (2002) Star formation from galaxies to globules. *ApJ*, **577**, 206–220, doi:10.1086/342177.

720 Bouché, N., Cresci, G., Davies, R., Eisenhauer, F., Förster Schreiber, N.M., Genzel, R., Gillessen, S., Lehnert, M., Lutz, D., Nesvadba, N., Shapiro, K.L., Sternberg, A., Tacconi, L.J., Verma, A., Cimatti, A., Daddi, E., Renzini, A., Erb, D.K., Shapley, A., and Steidel, C.C. (2007) Dynamical properties of $z \sim 2$ star-forming galaxies and a universal star formation relation. *ApJ*, **671**, 303–309, doi:10.1086/522221.

721 Daddi, E., Elbaz, D., Walter, F., Bournaud, F., Salmi, F., Carilli, C., Dannerbauer, H., Dickinson, M., Monaco, P., and Riechers, D. (2010) Different star formation laws for disks versus starbursts at low and high redshifts. *ApJ*, **714**, L118–L122, doi:10.1088/2041-8205/714/1/L118.

722 Genzel, R., Tacconi, L.J., Gracia-Carpio, J., Sternberg, A., Cooper, M.C., Shapiro, K., Bolatto, A., Bouché, N., Bournaud, F., Burkert, A., Combes, F., Comerford, J., Cox, P., Davis, M., Schreiber, N.M.F., Garcia-Burillo, S., Lutz, D., Naab, T., Neri, R., Omont, A., Shapley, A., and Weiner, B. (2010) A study of the gas-star formation relation over cosmic time. MNRAS, 407, 2091–2108, doi:10.1111/j.1365-2966.2010.16969.x.

723 Cattaneo, A., Faber, S.M., Binney, J., Dekel, A., Kormendy, J., Mushotzky, R., Babul, A., Best, P.N., Brüggen, M., Fabian, A.C., Frenk, C.S., Khalatyan, A., Netzer, H., Mahdavi, A., Silk, J., Steinmetz, M., and Wisotzki, L. (2009) The role of black holes in galaxy formation and evolution. Nature, 460, 213–219, doi:10.1038/nature08135.

724 King, A. (2003) Black holes, galaxy formation, and the $M_{BH}-\sigma$ relation. ApJ, 596, L27–L29, doi:10.1086/379143.

725 Murray, N., Quataert, E., and Thompson, T.A. (2005) On the maximum luminosity of galaxies and their central black holes: feedback from momentum-driven winds. ApJ, 618, 569–585, doi:10.1086/426067.

726 Fabian, A.C., Celotti, A., and Erlund, M.C. (2006) Radiative pressure feedback by a quasar in a galactic bulge. MNRAS, 373, L16–L20, doi:10.1111/j.1745-3933.2006.00234.x.

727 Springel, V., Di Matteo, T., and Hernquist, L. (2005) Modelling feedback from stars and black holes in galaxy mergers. MNRAS, 361, 776–794, doi:10.1111/j.1365-2966.2005.09238.x.

728 Monaco, P., Fontanot, F., and Taffoni, G. (2007) The MORGANA model for the rise of galaxies and active nuclei. MNRAS, 375, 1189–1219, doi:10.1111/j.1365-2966.2006.11253.x.

729 Bower, R.G., Benson, A.J., Malbon, R., Helly, J.C., Frenk, C.S., Baugh, C.M., Cole, S., and Lacey, C.G. (2006) Breaking the hierarchy of galaxy formation. MNRAS, 370, 645–655, doi:10.1111/j.1365-2966.2006.10519.x.

730 Cattaneo, A., Dekel, A., Devriendt, J., Guiderdoni, B., and Blaizot, J. (2006) Modelling the galaxy bimodality: shutdown above a critical halo mass. MNRAS, 370, 1651–1665, doi:10.1111/j.1365-2966.2006.10608.x.

731 Menci, N., Fontana, A., Giallongo, E., Grazian, A., and Salimbeni, S. (2006) The abundance of distant and extremely red galaxies: the role of AGN feedback in hierarchical models. ApJ, 647, 753–762, doi:10.1086/505528.

732 Schawinski, K., Khochfar, S., Kaviraj, S., Yi, S.K., Boselli, A., Barlow, T., Conrow, T., Forster, K., Friedman, P.G., Martin, D.C., Morrissey, P., Neff, S., Schiminovich, D., Seibert, M., Small, T., Wyder, T.K., Bianchi, L., Donas, J., Heckman, T., Lee, Y., Madore, B., Milliard, B., Rich, R.M., and Szalay, A. (2006) Suppression of star formation in early-type galaxies by feedback from supermassive black holes. Nature, 442, 888–891, doi:10.1038/nature04934.

733 Somerville, R.S., Hopkins, P.F., Cox, T.J., Robertson, B.E., and Hernquist, L. (2008) A semi-analytic model for the co-evolution of galaxies, black holes and active galactic nuclei. MNRAS, 391, 481–506, doi:10.1111/j.1365-2966.2008.13805.x.

734 Tenorio-Tagle, G. and Bodenheimer, P. (1988) Large-scale expanding superstructures in galaxies. ARA&A, 26, 145–197, doi:10.1146/annurev.aa.26.090188.001045.

735 Larson, R.B. (1974) Effects of supernovae on the early evolution of galaxies. MNRAS, 169, 229–246.

736 Dekel, A. and Silk, J. (1986) The origin of dwarf galaxies, cold dark matter, and biased galaxy formation. ApJ, 303, 39–55, doi:10.1086/164050.

737 Recchi, S. and Hensler, G. (2007) The effect of clouds in a galactic wind on the evolution of gas-rich dwarf galaxies. A&A, 476, 841–852, doi:10.1051/0004-6361:20078211.

738 Mayer, L., Mastropietro, C., Wadsley, J., Stadel, J., and Moore, B. (2006) Simultaneous ram pressure and tidal stripping; how dwarf spheroidals lost their gas. MNRAS, 369, 1021–1038, doi:10.1111/j.1365-2966.2006.10403.x.

739 Sarazin, C.L. (1986) X-ray emission from clusters of galaxies. *Rev. Mod. Phys.*, **58**, 1–115, doi:10.1103/RevModPhys.58.1.

740 Dressler, A. (2004) Star-forming galaxies in clusters, in *Clusters of Galaxies: Probes of Cosmological Structure and Galaxy Evolution*, pp. 206.

741 Dressler, A., Oemler, Jr., A., Couch, W.J., Smail, I., Ellis, R.S., Barger, A., Butcher, H., Poggianti, B.M., and Sharples, R.M. (1997) Evolution since $Z = 0.5$ of the morphology–density relation for clusters of galaxies. *ApJ*, **490**, 577, doi:10.1086/304890.

742 Cucciati, O., Marinoni, C., Iovino, A., Bardelli, S., Adami, C., Mazure, A., Scodeggio, M., Maccagni, D., Temporin, S., Zucca, E., De Lucia, G., Blaizot, J., Garilli, B., Meneux, B., Zamorani, G., Le Fèvre, O., Cappi, A., Guzzo, L., Bottini, D., Le Brun, V., Tresse, L., Vettolani, G., Zanichelli, A., Arnouts, S., Bolzonella, M., Charlot, S., Ciliegi, P., Contini, T., Foucaud, S., Franzetti, P., Gavignaud, I., Ilbert, O., Lamareille, F., McCracken, H.J., Marano, B., Merighi, R., Paltani, S., Pellò, R., Pollo, A., Pozzetti, L., Vergani, D., and Pérez-Montero, E. (2010) The VIMOS-VLT deep survey: the group catalogue. *A&A*, **520**, A42, doi:10.1051/0004-6361/200911831.

743 Abell, G.O. (1958) The distribution of rich clusters of galaxies. *ApJS*, **3**, 211, doi:10.1086/190036.

744 Zwicky, F., Herzog, E., and Wild, P. (1968) *Catalogue of Galaxies and of Clusters of Galaxies*.

745 Hickson, P. (1997) Compact groups of galaxies. *ARA&A*, **35**, 357–388, doi:10.1146/annurev.astro.35.1.357.

746 Ebeling, H., Voges, W., Bohringer, H., Edge, A.C., Huchra, J.P., and Briel, U.G. (1996) Properties of the X-ray-brightest Abell-type clusters of galaxies (XBACs) from ROSAT All-Sky Survey data – I. The sample. *MNRAS*, **281**, 799–829.

747 Vikhlinin, A., McNamara, B.R., Forman, W., Jones, C., Quintana, H., and Hornstrup, A. (1998) A catalog of 200 galaxy clusters serendipitously detected in the ROSAT PSPC pointed observations. *ApJ*, **502**, 558, doi:10.1086/305951.

748 Finoguenov, A., Guzzo, L., Hasinger, G., Scoville, N.Z., Aussel, H., Böhringer, H., Brusa, M., Capak, P., Cappelluti, N., Comastri, A., Giodini, S., Griffiths, R.E., Impey, C., Koekemoer, A.M., Kneib, J., Leauthaud, A., Le Fèvre, O., Lilly, S., Mainieri, V., Massey, R., McCracken, H.J., Mobasher, B., Murayama, T., Peacock, J.A., Sakelliou, I., Schinnerer, E., Silverman, J.D., Smolčić, V., Taniguchi, Y., Tasca, L., Taylor, J.E., Trump, J.R., and Zamorani, G. (2007) The XMM-Newton wide-field survey in the COSMOS field: statistical properties of clusters of galaxies. *ApJS*, **172**, 182–195, doi:10.1086/516577.

749 Donahue, M., Scharf, C.A., Mack, J., Lee, Y.P., Postman, M., Rosati, P., Dickinson, M., Voit, G.M., and Stocke, J.T. (2002) Distant cluster hunting. II. A comparison of X-ray and optical cluster detection techniques and catalogs from the ROSAT optical X-ray survey. *ApJ*, **569**, 689–719, doi:10.1086/339401.

750 Popesso, P., Biviano, A., Böhringer, H., and Romaniello, M. (2007) RASS-SDSS galaxy cluster survey. V. The X-ray-underluminous Abell clusters. *A&A*, **461**, 397–410, doi:10.1051/0004-6361:20054493.

751 Gavazzi, R. and Soucail, G. (2007) Weak lensing survey of galaxy clusters in the CFHTLS Deep. *A&A*, **462**, 459–471, doi:10.1051/0004-6361:20065677.

752 Limousin, M., Cabanac, R., Gavazzi, R., Kneib, J., Motta, V., Richard, J., Thanjavur, K., Foex, G., Pello, R., Crampton, D., Faure, C., Fort, B., Jullo, E., Marshall, P., Mellier, Y., More, A., Soucail, G., Suyu, S., Swinbank, M., Sygnet, J., Tu, H., Valls-Gabaud, D., Verdugo, T., and Willis, J. (2009) A new window of exploration in the mass spectrum: strong lensing by galaxy groups in the SL2S. *A&A*, **502**, 445–456, doi:10.1051/0004-6361/200811473.

753 Sunyaev, R.A. and Zeldovich, Y.B. (1972) The observations of relic radiation as a test of the nature of X-ray radiation from the clusters of galaxies. *Comm. Astrophys. Space Phys.*, **4**, 173.

754 Sunyaev, R.A. and Zeldovich, I.B. (1980) Microwave background radiation as a probe of the contemporary structure and history of the universe. *ARA&A*, **18**, 537–560, doi:10.1146/annurev.aa. 18.090180.002541.

755 Deltorn, J., Le Fevre, O., Crampton, D., and Dickinson, M. (1997) A massive cluster of galaxies at $Z = 0.996$. *ApJ*, **483**, L21, doi:10.1086/310730.

756 Gladders, M.D. and Yee, H.K.C. (2000) A new method for galaxy cluster detection. I. The algorithm. *AJ*, **120**, 2148–2162, doi:10.1086/301557.

757 Butcher, H. and Oemler, Jr., A. (1984) The evolution of galaxies in clusters. V – a study of populations since Z approximately equal to 0.5. *ApJ*, **285**, 426–438, doi:10.1086/162519.

758 Materne, J. (1978) The structure of nearby clusters of galaxies – hierarchical clustering and an application to the Leo region. *A&A*, **63**, 401–409.

759 Tully, R.B. (1980) Nearby groups of galaxies. I – the NGC 1023 group. *ApJ*, **237**, 390–403, doi:10.1086/157881.

760 Huchra, J.P. and Geller, M.J. (1982) Groups of galaxies. I – nearby groups. *ApJ*, **257**, 423–437, doi:10.1086/160000.

761 Kepner, J., Fan, X., Bahcall, N., Gunn, J., Lupton, R., and Xu, G. (1999) An automated cluster finder: the adaptive matched filter. *ApJ*, **517**, 78–91, doi:10.1086/307160.

762 Miller, C.J., Nichol, R.C., Reichart, D., Wechsler, R.H., Evrard, A.E., Annis, J., McKay, T.A., Bahcall, N.A., Bernardi, M., Boehringer, H., Connolly, A.J., Goto, T., Kniazev, A., Lamb, D., Postman, M., Schneider, D.P., Sheth, R.K., and Voges, W. (2005) The C4 clustering algorithm: clusters of galaxies in the Sloan digital sky survey. *AJ*, **130**, 968–1001, doi:10.1086/431357.

763 Marinoni, C., Davis, M., Newman, J.A., and Coil, A.L. (2002) Three-dimensional identification and reconstruction of galaxy systems within flux-limited redshift surveys. *ApJ*, **580**, 122–143, doi:10.1086/343092.

764 Adami, C., Mazure, A., Ilbert, O., Cappi, A., Bottini, D., Garilli, B., Le Brun, V., Le Fèvre, O., Maccagni, D., Picat, J.P., Scaramella, R., Scodeggio, M., Tresse, L., Vettolani, G., Zanichelli, A., Arnaboldi, M., Arnouts, S., Bardelli, S., Bolzonella, M., Charlot, S., Ciliegi, P., Contini, T., Covone, G., Foucaud, S., Franzetti, P., Gavignaud, I., Guzzo, L., Iovino, A., Lauger, S., McCracken, H.J., Marano, B., Marinoni, C., Meneux, B., Merighi, R., Paltani, S., Pellò, R., Pollo, A., Pozzetti, L., Radovich, M., Zamorani, G., Zucca, E., Bondi, M., Bongiorno, A., Busarello, G., Gregorini, L., Mathez, G., Mellier, Y., Merluzzi, P., Ripepi, V., and Rizzo, D. (2005) The Vimos VLT deep survey: compact structures in the CDFS. *A&A*, **443**, 805–818, doi:10.1051/0004-6361:20053202.

765 Kauffmann, G., White, S.D.M., Heckman, T.M., Ménard, B., Brinchmann, J., Charlot, S., Tremonti, C., and Brinkmann, J. (2004) The environmental dependence of the relations between stellar mass, structure, star formation and nuclear activity in galaxies. *MNRAS*, **353**, 713–731, doi:10.1111/j.1365-2966.2004.08117.x.

766 Kovač, K., Lilly, S.J., Cucciati, O., Porciani, C., Iovino, A., Zamorani, G., Oesch, P., Bolzonella, M., Knobel, C., Finoguenov, A., Peng, Y., Carollo, C.M., Pozzetti, L., Caputi, K., Silverman, J.D., Tasca, L.A.M., Scodeggio, M., Vergani, D., Scoville, N.Z., Capak, P., Contini, T., Kneib, J., Le Fèvre, O., Mainieri, V., Renzini, A., Bardelli, S., Bongiorno, A., Coppa, G., de la Torre, S., de Ravel, L., Franzetti, P., Garilli, B., Guzzo, L., Kampczyk, P., Lamareille, F., Le Borgne, J., Le Brun, V., Maier, C., Mignoli, M., Pello, R., Perez Montero, E., Ricciardelli, E., Tanaka, M., Tresse, L., Zucca, E., Abbas, U., Bottini, D., Cappi, A., Cassata, P., Cimatti, A., Fumana, M., Koekemoer, A.M., Maccagni, D., Marinoni, C., McCracken, H.J., Memeo, P., Meneux, B., and Scaramella, R. (2010) The density field of the 10k zCOSMOS galaxies. *ApJ*, **708**, 505–533, doi:10.1088/0004-637X/708/1/505.

767 Kneib, J., Hudelot, P., Ellis, R.S., Treu, T., Smith, G.P., Marshall, P., Czoske, O., Smail, I., and Natarajan, P. (2003) A wide-field hubble space telescope study

of the cluster Cl 0024 + 1654 at $z = 0.4$. II. the cluster mass distribution. *ApJ*, **598**, 804–817, doi:10.1086/378633.

768 Kennicutt, Jr., R.C. (1983) On the evolution of the spiral galaxies in the Virgo cluster. *AJ*, **88**, 483–488, doi:10.1086/113334.

769 Lewis, I., Balogh, M., De Propris, R., Couch, W., Bower, R., Offer, A., Bland-Hawthorn, J., Baldry, I.K., Baugh, C., Bridges, T., Cannon, R., Cole, S., Colless, M., Collins, C., Cross, N., Dalton, G., Driver, S.P., Efstathiou, G., Ellis, R.S., Frenk, C.S., Glazebrook, K., Hawkins, E., Jackson, C., Lahav, O., Lumsden, S., Maddox, S., Madgwick, D., Norberg, P., Peacock, J.A., Percival, W., Peterson, B.A., Sutherland, W., and Taylor, K. (2002) The 2dF galaxy redshift survey: The environmental dependence of galaxy star formation rates near clusters. *MNRAS*, **334**, 673–683, doi:10.1046/j.1365-8711.2002.05558.x.

770 Gómez, P.L., Nichol, R.C., Miller, C.J., Balogh, M.L., Goto, T., Zabludoff, A.I., Romer, A.K., Bernardi, M., Sheth, R., Hopkins, A.M., Castander, F.J., Connolly, A.J., Schneider, D.P., Brinkmann, J., Lamb, D.Q., SubbaRao, M., and York, D.G. (2003) Galaxy star formation as a function of environment in the early data release of the Sloan digital sky survey. *ApJ*, **584**, 210–227, doi:10.1086/345593.

771 Blanton, M.R. and Moustakas, J. (2009) Physical properties and environments of nearby galaxies. *ARA&A*, **47**, 159–210, doi:10.1146/annurev-astro-082708-101734.

772 Giovanelli, R. and Haynes, M.P. (1985) Gas deficiency in cluster galaxies – a comparison of nine clusters. *ApJ*, **292**, 404–425, doi:10.1086/163170.

773 Karachentseva, V.E. (1973) The catalogue of isolated galaxies, *Astrofizicheskie Issledovaniia Izvestiya Spetsial'noj Astrofizicheskoj Observatorii*, **8**, 3–49.

774 Verdes-Montenegro, L., Sulentic, J., Lisenfeld, U., Leon, S., Espada, D., Garcia, E., Sabater, J., and Verley, S. (2005) The AMIGA project. I. Optical characterization of the CIG catalog. *A&A*, **436**, 443–455, doi:10.1051/0004-6361:20042280.

775 Boselli, A., Eales, S., Cortese, L., Bendo, G., Chanial, P., Buat, V., Davies, J., Auld, R., Rigby, E., Baes, M., Barlow, M., Bock, J., Bradford, M., Castro-Rodriguez, N., Charlot, S., Clements, D., Cormier, D., Dwek, E., Elbaz, D., Galametz, M., Galliano, F., Gear, W., Glenn, J., Gomez, H., Griffin, M., Hony, S., Isaak, K., Levenson, L., Lu, N., Madden, S., O'Halloran, B., Okamura, K., Oliver, S., Page, M., Panuzzo, P., Papageorgiou, A., Parkin, T., Perez-Fournon, I., Pohlen, M., Rangwala, N., Roussel, H., Rykala, A., Sacchi, N., Sauvage, M., Schulz, B., Schirm, M., Smith, M.W.L., Spinoglio, L., Stevens, J., Symeonidis, M., Vaccari, M., Vigroux, L., Wilson, C., Wozniak, H., Wright, G., and Zeilinger, W. (2010) The Herschel reference survey. *PASP*, **122**, 261–287, doi:10.1086/651535.

776 Vollmer, B., Cayatte, V., Boselli, A., Balkowski, C., and Duschl, W.J. (1999) Kinematics of the anemic cluster galaxy NGC 4548. Is stripping still active. *A&A*, **349**, 411–423.

777 Vollmer, B., Marcelin, M., Amram, P., Balkowski, C., Cayatte, V., and Garrido, O. (2000) The consequences of ram pressure stripping on the Virgo cluster spiral galaxy NGC 4522. *A&A*, **364**, 532–542.

778 Vollmer, B. (2003) NGC 4654: gravitational interaction or ram pressure stripping. *A&A*, **398**, 525–539, doi:10.1051/0004-6361:20021729.

779 Vollmer, B., Balkowski, C., Cayatte, V., van Driel, W., and Huchtmeier, W. (2004) NGC 4569: recent evidence for a past ram pressure stripping event. *A&A*, **419**, 35–46, doi:10.1051/0004-6361:20034552.

780 Malin, D.F. and Carter, D. (1983) A catalog of elliptical galaxies with shells. *ApJ*, **274**, 534–540, doi:10.1086/161467.

781 Dupraz, C. and Combes, F. (1986) Shells around galaxies – testing the mass distribution and the 3-D shape of ellipticals. *A&A*, **166**, 53–74.

782 Hernquist, L. and Quinn, P.J. (1988) Formation of shell galaxies. I – spherical potentials. *ApJ*, **331**, 682–698, doi:10.1086/166592.

783 Kojima, M. and Noguchi, M. (1997) Sinking satellite disk galaxies. I. Shell forma-

tion preceded by cessation of star formation. *ApJ*, **481**, 132, doi:10.1086/304021.

784 Baum, W.A. (1962) Photoelectric magnitudes and red-shifts, in *Problems of Extra-Galactic Research*, IAU Symposium, vol. 15 (ed. G.C. McVittie), pp. 390.

785 Steidel, C.C., Pettini, M., and Hamilton, D. (1995) Lyman imaging of high-redshift galaxies. III. New observations of four QSO fields. *AJ*, **110**, 2519, doi:10.1086/117709.

786 Steidel, C.C., Giavalisco, M., Pettini, M., Dickinson, M., and Adelberger, K.L. (1996) Spectroscopic confirmation of a population of normal star-forming galaxies at redshifts $Z > 3$. *ApJ*, **462**, L17, doi:10.1086/310029.

787 Bolzonella, M., Miralles, J., and Pelló, R. (2000) Photometric redshifts based on standard SED fitting procedures. *A&A*, **363**, 476–492.

788 Kotulla, R. and Fritze, U. (2009) Impact of subsolar metallicities on photometric redshifts. *MNRAS*, **393**, L55–L59, doi:10.1111/j.1745-3933.2008.00598.x.

789 Benítez, N. (2000) Bayesian photometric redshift estimation. *ApJ*, **536**, 571–583, doi:10.1086/308947.

790 Dahlen, T., Mobasher, B., Dickinson, M., Ferguson, H.C., Giavalisco, M., Grogin, N.A., Guo, Y., Koekemoer, A., Lee, K., Lee, S., Nonino, M., Riess, A.G., and Salimbeni, S. (2010) A detailed study of photometric redshifts for GOODS-south galaxies. *ApJ*, **724**, 425–447, doi:10.1088/0004-637X/724/1/425.

791 Brammer, G.B., van Dokkum, P.G., and Coppi, P. (2008) EAZY: A fast, public photometric redshift code. *ApJ*, **686**, 1503–1513, doi:10.1086/591786.

792 Csabai, I., Budavári, T., Connolly, A.J., Szalay, A.S., Győry, Z., Benítez, N., Annis, J., Brinkmann, J., Eisenstein, D., Fukugita, M., Gunn, J., Kent, S., Lupton, R., Nichol, R.C., and Stoughton, C. (2003) The application of photometric redshifts to the SDSS early data release. *AJ*, **125**, 580–592, doi:10.1086/345883.

793 Feldmann, R., Carollo, C.M., Porciani, C., Lilly, S.J., Capak, P., Taniguchi, Y., Le Fèvre, O., Renzini, A., Scoville, N., Ajiki, M., Aussel, H., Contini, T., McCracken, H., Mobasher, B., Murayama, T., Sanders, D., Sasaki, S., Scarlata, C., Scodeggio, M., Shioya, Y., Silverman, J., Takahashi, M., Thompson, D., and Zamorani, G. (2006) The Zurich extragalactic bayesian redshift analyzer and its first application: COSMOS. *MNRAS*, **372**, 565–577, doi:10.1111/j.1365-2966.2006.10930.x.

794 Ilbert, O., Arnouts, S., McCracken, H.J., Bolzonella, M., Bertin, E., Le Fèvre, O., Mellier, Y., Zamorani, G., Pellò, R., Iovino, A., Tresse, L., Le Brun, V., Bottini, D., Garilli, B., Maccagni, D., Picat, J.P., Scaramella, R., Scodeggio, M., Vettolani, G., Zanichelli, A., Adami, C., Bardelli, S., Cappi, A., Charlot, S., Ciliegi, P., Contini, T., Cucciati, O., Foucaud, S., Franzetti, P., Gavignaud, I., Guzzo, L., Marano, B., Marinoni, C., Mazure, A., Meneux, B., Merighi, R., Paltani, S., Pollo, A., Pozzetti, L., Radovich, M., Zucca, E., Bondi, M., Bongiorno, A., Busarello, G., de La Torre, S., Gregorini, L., Lamareille, F., Mathez, G., Merluzzi, P., Ripepi, V., Rizzo, D., and Vergani, D. (2006) Accurate photometric redshifts for the CFHT legacy survey calibrated using the VIMOS VLT deep survey. *A&A*, **457**, 841–856, doi:10.1051/0004-6361:20065138.

795 Lahav, O., Naim, A., Buta, R.J., Corwin, H.G., de Vaucouleurs, G., Dressler, A., Huchra, J.P., van den Bergh, S., Raychaudhury, S., Sodre, Jr., L., and Storrie-Lombardi, M.C. (1995) Galaxies, human eyes, and artificial neural networks. *Science*, **267**, 859–862, doi:10.1126/science.267.5199.859.

796 Collister, A.A. and Lahav, O. (2004) ANNz: estimating photometric redshifts using artificial neural networks. *PASP*, **116**, 345–351, doi:10.1086/383254.

797 Collister, A., Lahav, O., Blake, C., Cannon, R., Croom, S., Drinkwater, M., Edge, A., Eisenstein, D., Loveday, J., Nichol, R., Pimbblet, K., de Propris, R., Roseboom, I., Ross, N., Schneider, D.P., Shanks, T., and Wake, D. (2007) MegaZ-LRG: a photometric redshift catalogue of one million SDSS luminous red galaxies. *MNRAS*, **375**, 68–76, doi:10.1111/j.1365-2966.2006.11305.x.

798 Oyaizu, H., Lima, M., Cunha, C.E., Lin, H., Frieman, J., and Sheldon, E.S. (2008) A galaxy photometric redshift catalog for the Sloan digital sky survey data release 6. *ApJ*, **674**, 768–783, doi:10.1086/523666.

799 Csabai, I., Dobos, L., Trencséni, M., Herczegh, G., Józsa, P., Purger, N., Budavári, T., and Szalay, A.S. (2007) Multidimensional indexing tools for the virtual observatory. *Astron. Nachr.*, **328**, 852, doi:10.1002/asna.200710817.

800 Li, I.H. and Yee, H.K.C. (2008) Finding galaxy groups in photometric-redshift space: the probability friends-of-friends algorithm. *AJ*, **135**, 809–822, doi:10.1088/0004-6256/135/3/809.

801 Wolf, C. (2009) Bayesian photometric redshifts with empirical training sets. *MNRAS*, **397**, 520–533, doi:10.1111/j.1365-2966.2009.14953.x.

802 Gerdes, D.W., Sypniewski, A.J., McKay, T.A., Hao, J., Weis, M.R., Wechsler, R.H., and Busha, M.T. (2010) ArborZ: photometric redshifts using boosted decision trees. *ApJ*, **715**, 823–832, doi:10.1088/0004-637X/715/2/823.

803 Carliles, S., Budavári, T., Heinis, S., Priebe, C., and Szalay, A.S. (2010) Random forests for photometric redshifts. *ApJ*, **712**, 511–515, doi:10.1088/0004-637X/712/1/511.

804 Hildebrandt, H., Arnouts, S., Capak, P., Moustakas, L.A., Wolf, C., Abdalla, F.B., Assef, R.J., Banerji, M., Benítez, N., Brammer, G.B., Budavári, T., Carliles, S., Coe, D., Dahlen, T., Feldmann, R., Gerdes, D., Gillis, B., Ilbert, O., Kotulla, R., Lahav, O., Li, I.H., Miralles, J., Purger, N., Schmidt, S., and Singal, J. (2010) PHAT: photo-z accuracy testing. *A&A*, **523**, A31, doi:10.1051/0004-6361/201014885.

805 Blain, A.W., Smail, I., Ivison, R.J., Kneib, J., and Frayer, D.T. (2002) Submillimeter galaxies. *Phys. Rep.*, **369**, 111–176, doi:10.1016/S0370-1573(02)00134-5.

806 Smail, I., Ivison, R.J., Blain, A.W., and Kneib, J. (2002) The nature of faint submillimetre-selected galaxies. *MNRAS*, **331**, 495–520, doi:10.1046/j.1365-8711.2002.05203.x.

807 Chapman, S.C., Blain, A.W., Smail, I., and Ivison, R.J. (2005) A redshift survey of the submillimeter galaxy population. *ApJ*, **622**, 772–796, doi:10.1086/428082.

808 Carilli, C.L. and Yun, M.S. (1999) The radio-to-submillimeter spectral index as a redshift indicator. *ApJ*, **513**, L13–L16, doi:10.1086/311909.

809 Carilli, C.L. and Yun, M.S. (2000) The scatter in the relationship between redshift and the radio-to-submillimeter spectral index. *ApJ*, **530**, 618–624, doi:10.1086/308418.

810 Barger, A.J., Cowie, L.L., and Richards, E.A. (2000) Mapping the evolution of high-redshift dusty galaxies with submillimeter observations of a radio-selected sample. *AJ*, **119**, 2092–2109, doi:10.1086/301341.

811 Dunne, L., Clements, D.L., and Eales, S.A. (2000) Constraining the radio-submillimetre redshift indicator using data from the SCUBA local universe galaxy survey. *MNRAS*, **319**, 813–820, doi:10.1046/j.1365-8711.2000.03882.x.

812 Rengarajan, T.N. and Takeuchi, T.T. (2001) The radio-to-submillimeter flux density ratio of galaxies as a measure of redshift. *PASJ*, **53**, 433–437.

813 Yun, M.S. and Carilli, C.L. (2002) Radio-to-far-infrared spectral energy distribution and photometric redshifts for dusty starburst galaxies. *ApJ*, **568**, 88–98, doi:10.1086/338924.

814 Blain, A.W., Barnard, V.E., and Chapman, S.C. (2003) Submillimetre and far-infrared spectral energy distributions of galaxies: the luminosity-temperature relation and consequences for photometric redshifts. *MNRAS*, **338**, 733–744, doi:10.1046/j.1365-8711.2003.06086.x.

815 Aretxaga, I., Hughes, D.H., Coppin, K., Mortier, A.M.J., Wagg, J., Dunlop, J.S., Chapin, E.L., Eales, S.A., Gaztañaga, E., Halpern, M., Ivison, R.J., van Kampen, E., Scott, D., Serjeant, S., Smail, I., Babbedge, T., Benson, A.J., Chapman, S., Clements, D.L., Dunne, L., Dye, S., Farrah, D., Jarvis, M.J., Mann, R.G., Pope, A., Priddey, R., Rawlings, S., Seigar, M., Silva, L., Simpson, C., and Vaccari, M. (2007)

The SCUBA half degree extragalactic survey – IV. Radio-mm-FIR photometric redshifts. *MNRAS*, **379**, 1571–1588, doi:10.1111/j.1365-2966.2007.12036.x.

816 Pérez-González, P.G., Egami, E., Rex, M., Rawle, T.D., Kneib, J., Richard, J., Johansson, D., Altieri, B., Blain, A.W., Bock, J.J., Boone, F., Bridge, C.R., Chung, S.M., Clément, B., Clowe, D., Combes, F., Cuby, J., Dessauges-Zavadsky, M., Dowell, C.D., Espino-Briones, N., Fadda, D., Fiedler, A.K., Gonzalez, A., Horellou, C., Ilbert, O., Ivison, R.J., Jauzac, M., Lutz, D., Pelló, R., Pereira, M.J., Rieke, G.H., Rodighiero, G., Schaerer, D., Smith, G.P., Valtchanov, I., Walth, G.L., van der Werf, P., Werner, M.W., and Zemcov, M. (2010) Improving the identification of high-z Herschel sources with position priors and optical/NIR and FIR/mm photometric redshifts. *A&A*, **518**, L15, doi:10.1051/0004-6361/201014593.

817 Oke, J.B. and Sandage, A. (1968) Energy distributions, K corrections, and the Stebbins–Whitford effect for giant elliptical galaxies. *ApJ*, **154**, 21, doi:10.1086/149737.

818 Hogg, D.W., Baldry, I.K., Blanton, M.R., and Eisenstein, D.J. (2002) The K correction. *ArXiv Astrophysics e-prints*.

819 Blanton, M.R. and Roweis, S. (2007) K-corrections and filter transformations in the ultraviolet, optical, and near-infrared. *AJ*, **133**, 734–754, doi:10.1086/510127.

820 Mannucci, F., Basile, F., Poggianti, B.M., Cimatti, A., Daddi, E., Pozzetti, L., and Vanzi, L. (2001) Near-infrared template spectra of normal galaxies: k-corrections, galaxy models and stellar populations. *MNRAS*, **326**, 745–758, doi:10.1046/j.1365-8711.2001.04628.x.

821 O'Mill, A.L., Duplancic, F., García Lambas, D., and Sodré, Jr., L. (2011) Photometric redshifts and k-corrections for the Sloan digital sky survey data release 7. *MNRAS*, pp. 201, doi:10.1111/j.1365-2966.2011.18222.x.

822 Blanton, M.R., Brinkmann, J., Csabai, I., Doi, M., Eisenstein, D., Fukugita, M., Gunn, J.E., Hogg, D.W., and Schlegel, D.J. (2003) Estimating fixed-frame galaxy magnitudes in the Sloan digital sky survey. *AJ*, **125**, 2348–2360, doi:10.1086/342935.

823 Roche, N., Bernardi, M., and Hyde, J. (2009) Spectral-based k-corrections and implications for the colour–magnitude relation of E/S0s and its evolution. *MNRAS*, **398**, 1549–1562, doi:10.1111/j.1365-2966.2009.15222.x.

824 Chilingarian, I.V., Melchior, A., and Zolotukhin, I.Y. (2010) Analytical approximations of K-corrections in optical and near-infrared bands. *MNRAS*, **405**, 1409–1420, doi:10.1111/j.1365-2966.2010.16506.x.

825 Westra, E., Geller, M.J., Kurtz, M.J., Fabricant, D.G., and Dell'Antonio, I. (2010) Empirical optical k-corrections for redshifts > 0.7. *PASP*, **122**, 1258–1284, doi:10.1086/657452.

826 Oke, J.B. (1974) Absolute spectral energy distributions for white dwarfs. *ApJS*, **27**, 21, doi:10.1086/190287.

827 Oke, J.B. and Gunn, J.E. (1983) Secondary standard stars for absolute spectrophotometry. *ApJ*, **266**, 713–717, doi:10.1086/160817.

828 Bessell, M.S., Castelli, F., and Plez, B. (1998) Model atmospheres broad-band colors, bolometric corrections and temperature calibrations for O–M stars. *A&A*, **333**, 231–250.

829 Cohen, M., Walker, R.G., Barlow, M.J., and Deacon, J.R. (1992) Spectral irradiance calibration in the infrared. I – ground-based and IRAS broad-band calibrations. *AJ*, **104**, 1650–1657, doi:10.1086/116349.

830 Cohen, M., Wheaton, W.A., and Megeath, S.T. (2003) Spectral irradiance calibration in the infrared. XIV. The absolute calibration of 2MASS. *AJ*, **126**, 1090–1096, doi:10.1086/376474.

831 Frei, Z. and Gunn, J.E. (1994) Generating colors and K corrections from existing catalog data. *AJ*, **108**, 1476–1485, doi:10.1086/117172.

Index

a

absorption cross-section 67
absorption line 26, 35, 37, 40, 57, 91, 99, 103, 106–107, 110, 123, 158, 185, 188, 220, 224
accretion disk 2, 5, 17, 21, 25, 31, 81, 135, 231, 239
AGN 2, 13, 17, 20, 25, 31, 45, 49, 51, 65, 69, 71, 74, 77, 86, 95, 99, 135, 153, 155, 157, 187, 197, 205, 207, 213, 218, 225, 231, 235, 239
AKARI 5, 67, 84
ALMA 51
α/Fe 104
antenna temperature 117
artificial neural network 257, 260
asymmetry index 161, 168, 253
atomic hydrogen 37, 45, 55, 107–108, 110, 150, 226, 235
attenuation
– internal 9, 91, 125, 132, 138, 141, 227, 256
– law 135, 139

b

Balmer
– decrement 95, 133, 156
– lines 33, 37, 92, 104, 106, 157–158
β grain emissivity index 43, 66, 68
β parameter 141
binary system 5, 7, 17–19, 104, 155, 240
birthrate parameter 146, 222
BL Lac 92
black body 5, 19, 21, 26, 44, 52, 54
black hole 18, 20, 31, 155, 242
– intermediate mass 17
– supermassive 17, 20, 91, 161, 187, 190, 231, 233, 235, 239–240
blue sequence 216

Brackett
– lines 33, 157
Bremsstrahlung 19, 21, 52, 250
bulge 1, 17, 162, 167, 180, 220, 228, 231, 239–240
bulge to disk ratio 167
bulge to total ratio 167
burst
– instantaneous 84, 159, 172

c

C4 density algorithm 247
Calzetti's law 133, 136, 139
case B 33, 49, 133
[CII] 47, 157
cirrus 136, 152
close box 220
clumpiness index 168–169
cluster 2, 4, 18, 21, 62, 92, 158, 161–162, 171, 177, 187, 203, 207, 215, 218, 231, 241, 243, 245–247, 249–250
– A1367 247, 250
– Coma 247, 250
– Virgo 164, 203, 205, 223, 225, 254
CO 5, 7, 44, 47, 51, 54, 72, 107, 115, 117, 119, 123, 180, 237
– conversion factor 239
COBE 126
color–color diagram 83, 89, 216
color–magnitude relation 166, 195, 203, 216, 218, 250
concentration index 161, 168, 233
confusion limited 7, 196, 201, 258
cosmic microwave background 4, 246
curve of growth 111–112

d

damped Lyman-α system 57, 114

A Panchromatic View of Galaxies, First Edition. Alessandro Boselli.
© 2012 WILEY-VCH Verlag GmbH & Co. KGaA. Published 2012 by WILEY-VCH Verlag GmbH & Co. KGaA.

Index

dark matter 171, 178, 180–181, 186, 200, 207, 219, 225, 230, 233, 242, 245, 249
density
 – pseudo-isothermal sphere 182
density contrast 245, 247
density profile
 – general 181
 – isothermal sphere 241
 – NFW 181
distance indicator 5, 215, 218, 255
$D_n(4000)$ 38, 92, 158, 256, 260
Doppler parameter 111–112
downsizing effect 216, 218
dust
 – attenuation 41
 – extinction 12, 33, 37, 49, 79, 95, 125, 132, 134, 137, 139–140, 147, 151–152, 156, 172, 211, 235, 256
 – grain 41–42, 46, 54, 65, 68, 125, 235, 241
 – mass 41, 63, 65, 68–69, 200
dust extinction 68
dynamical mass 12, 37, 61–62, 115, 155, 171, 179–180, 185–187, 216, 224, 227, 233
 – exponential disk approximation 179
 – flat-disk approximation 179
 – homogeneous sphere 178
 – homogeneous spheroid approximation 178
 – Keplerian approximation 178, 187
 – mass point approximation 178

e

Eddington luminosity 20, 240
effective radius 161, 166, 224, 228, 230, 233
effective surface brightness 166, 223–224, 228, 230
electron
 – density 34, 62, 101, 154, 157
 – relativistic 5, 7, 19, 51, 53, 74, 153
 – temperature 35, 71, 101
emission line
 – collisionally excited 29, 35, 48, 57, 122
 – hydrogen recombination 29, 32, 49, 71, 133, 150, 152–153, 156–157
equivalent width 37, 103, 106, 111, 115, 146
escape fraction 125, 132, 149
extinction
 – atmospheric 125
 – curve 44, 127, 136, 139
 – Galactic 125–128, 134, 195
 – LMC 128
 – SMC 128

f

Faber–Jackson relation 228
far-IR luminosity 45, 63, 65, 152, 154
feedback
 – AGN 17, 207, 218, 225, 231, 239, 241
 – supernova 207, 218, 225, 235, 242
FIR/radio correlation 75, 153, 258
flux density 43, 52, 61, 63, 65, 68, 71, 73, 77, 85, 88, 115, 125, 139, 152, 158, 195, 198, 201, 204, 258, 263, 265
forbidden line 39, 47, 56, 92, 115, 156
fragmentation 143
free–bound emission 22
free–free emission 5, 21–22, 52, 54, 71–72, 75
friend of a friend density algorithm 247
fundamental mainfold of spheroids 230
fundamental plane 161, 166, 215, 225, 228, 230, 233
 – tilt 229

g

galaxy
 – BCD 1, 81, 94, 224
 – cD 162, 215, 249
 – dwarf 1, 4, 13
 – dwarf elliptical 1, 82, 185, 207, 223
 – dwarf spheroidal 1, 82, 215, 220, 223
 – elliptical 1, 4, 10, 13, 18, 25, 29, 37–38, 41, 61–62, 74, 77, 79, 81, 89, 104, 110, 123, 151, 161–162, 169, 172, 177, 184–185, 187, 198, 203, 206, 215–216, 218, 220, 223–224, 228, 230–231, 239–240, 245–246, 250, 257, 260
 – HII 95
 – irregular 1, 10, 13, 39, 81, 90, 156, 169, 182, 206, 250
 – lenticular 1, 4, 10, 13, 25, 29, 38, 74, 79, 206, 231, 250
 – low surface brightness 1, 182, 207
 – Lyman break 256
 – merger 1, 12, 158, 161, 171, 196, 213, 218
 – pair 250
 – passively evolving 93
 – post-star-forming 93
 – poststarburst 92–93
 – spiral 1, 4, 10, 13, 18, 29, 39, 45, 65–66, 75, 79, 81, 90, 94, 97, 107, 110, 115, 120, 123, 136, 140–141, 151–153, 156, 161–162, 166, 177, 180, 182, 187, 206, 208, 213, 215–216, 220–221, 224–225, 227, 231, 237, 239, 250, 256

– star-forming   17–18, 25, 41, 74, 79, 82, 87, 91, 93, 95, 97, 99, 135, 137, 140, 142, 149, 151–152, 154, 156, 169, 198, 205–206, 208, 218, 220, 224, 250
– starburst   17, 41, 45, 49, 65–66, 69, 74–75, 79, 84, 89–90, 93–94, 96, 101, 116, 120, 122–123, 132, 135–136, 139, 141, 149, 151–152, 157, 172, 198, 238, 256
– submillimetre   201, 258
GALEX   5, 7, 136, 141, 151, 205, 216
gas depletion   147, 237
gas to dust ratio   68, 116, 121, 126
giant molecular cloud   4, 39, 45, 115–116, 120, 143, 177, 235
Gini coefficient   170
globular cluster   185, 231
grain opacity   66
gravitational interaction   4, 47, 166, 245, 249, 253
group   18, 62, 171, 231, 245, 247, 249–250
– compact   2, 161, 177, 245–246
– local   215, 220
– loose   245, 247

h

H_2 39, 44, 47, 55, 99, 107, 115–116, 119–123, 221, 235, 237
H_α 7, 34, 71, 92, 95, 97, 102, 133–134, 146–147, 150, 152, 155, 211, 222
harassment   249
H_β 34, 72, 92, 95, 102, 106, 133–134, 147, 150, 156, 158
HCN   55, 72
H_δ 33, 104, 106, 134, 158
Herschel   7, 54, 64, 67, 84–85, 88, 258
H_γ 33, 104, 106, 134, 158
HI   7, 33, 49, 51, 237
– absorption   37, 57, 109
– emission   55, 107
– mass   108
HI-deficiency parameter   110, 251
hierarchical density algorithm   247
hierarchical model   171, 207, 249
HII region   1, 5, 7, 29, 39, 45, 47, 49, 51, 67, 75, 91, 95, 101, 135, 146, 150, 156, 169, 218, 236
HST   164, 188, 190
Humphreys
– lines   33
hydrostatic equilibrium   143, 186
HyLIRG   64

i

infall   19–20, 220, 239, 245

initial mass function   28, 34, 143–144, 146, 150–151, 154, 156–157, 171, 212, 230, 235, 238, 256
interferometric filter   148
intergalactic medium   4, 18, 21, 57, 112, 187, 196, 231, 236, 243, 246, 249, 251, 256
interstellar medium   5, 13, 21, 38–39, 41, 44–45, 48, 51, 54–56, 65–66, 69, 85, 107, 115–122, 132, 146–147, 157
interstellar radiation field   5, 7, 41, 44, 47, 63, 65, 77, 84, 136, 152
inverse Compton   246
ionizing radiation   5, 49, 97, 99, 125, 132, 147, 152
IRAS   5, 12, 63, 65–67, 84, 88, 126, 208
ISO   64, 84, 122
isothermal sphere   62, 186, 241

j

jet   25, 90, 233, 240

k

K-correction   12, 153, 196, 201, 203, 259
k-space   230
Kormendy relation   161, 166, 223–224

l

Lick/IDS system   38, 103–104, 220
light profile   161, 223
– central   162
– core-Sersic   162
– de Vaucouleurs   162, 167
– double-Sersic   163
– exponential   162, 167, 179, 182, 226
– Nuker   163
– radial   161, 166, 168, 180, 233
– scale height   166
– scale length   162, 183
– Sersic   162, 167, 223, 233
– vertical   166
line diagnostic   91, 95, 97, 99
LINER   31, 45, 91–92, 95, 97, 99
LIRG   64, 152–153
low surface brightness   207
LTE   48
luminosity density   209
luminosity function   11–12, 200, 203, 205–206, 209, 219, 242
– bivariate   204
– C^- 202
– cluster   203, 207
– Schechter   196, 203, 205, 210
– STY   202

- SWML 202
- V_{max} 201

Lyα 37, 49, 56, 112, 150

Lyman
- break 113, 256
- continuum 25, 75, 114, 132, 150
- escape fraction 125
- lines 33, 47, 49, 56, 112, 150, 157

Lyman-α forest 114

Lyman limit system 113

m

magnetic field 5, 35, 51, 71, 74, 77, 153, 186, 236, 242

magnitude 2, 79, 125, 141, 161, 166, 182, 195, 200, 203–204, 263, 265
- AB system 264
- STMAG system 264
- Vega system 264

Magorrian relation 231

main beam 119
- efficiency 117
- temperature 117

maser 188

2MASS 5, 206, 216, 223, 226, 260

mass function 11, 195, 200, 207, 242

mass gas relation 220

mass–metallicity relation 218, 220

mass–star formation relation 222

mass to light ratio 171–172, 178–180, 189, 227

matter cycle 4, 12, 107, 235

metallicity 12, 19, 26, 37, 45, 47, 65, 68, 77, 83, 91, 101, 103, 116–117, 121, 127, 132, 150–153, 156, 158–159, 172, 215–216, 218, 230, 256
- solar 102

Mg$_2$ 220

morphological type 2, 10, 13, 79, 83, 88, 110, 121, 145, 171, 182, 215, 222, 226

morphology
- CAS classification 168
- Gini coefficient 168

morphology segregation effect 250

n

[NII] 95, 101–102, 134, 146, 148, 151

number counts 11–12, 126, 195, 197, 201, 203

o

[OI] 47, 95, 97, 134

[OII] 29, 35, 91, 93, 97, 101, 134, 156, 212

[OIII] 30, 35, 92, 95, 97, 101, 103, 134, 156

optical spectral index 37–38, 103, 220

outflow 17, 220

p

PAH 7, 42, 44, 65, 68–69, 79, 84, 99, 153

Paschen
- lines 33

Pfund
- lines 33

photodissociation region 39, 46–47

photoionization 46, 147, 241
- model 95, 101

photometric redshift 12, 201, 203, 247, 255, 260

photometric system 12, 263

planetary nebulae 18, 29

population synthesis model 28, 34, 37, 68, 83, 95, 103, 106, 140, 145, 150, 158, 171, 180, 216, 220, 239, 256

pseudo-isothermal sphere 182

q

QSO 2, 31, 37, 57, 80, 90, 92, 110, 114, 198, 201

quasi soft X-ray source 19

r

radiative
- cascade 39
- de-excitation 48
- decay 35
- excitation 48, 122
- lifetime 56
- pressure 241
- process 12, 17, 47–48
- transfer model 134

radio galaxy 2, 79, 110, 190, 246

radio luminosity 74, 92, 154, 204

ram pressure 241, 243, 249

recessional velocity 185, 247

recombination line 21, 29, 32, 34, 49, 71, 133, 147, 150, 152–153, 156–157

recycled gas 235
- fraction 146
- rate 61

red sequence 216, 223, 242, 250

reddening 41, 125, 133, 216

Roberts time 147

ROSAT 5

rotation curve 178–180, 182, 190, 227, 233, 238
- Keplerian approximation 178, 187, 227
- Polyex model 183
- universal 182

s

sandwich model   137
scaling relation   4, 12, 161, 166, 215, 220, 223–224, 231, 241, 252
scattering   41, 125
 – Compton   241
 – electron   20
 – resonant   49, 150
 – Thomson   240
Schmidt law   236, 238
Schwarzschild radius   188
SDSS   5, 31, 95, 97, 101, 174, 205–206, 216, 219, 223, 230, 247, 250, 260
self-absorption   57, 90, 111
self-gravity   26, 236
semianalytical model   207, 218, 225, 231, 241, 257
Seyfert   2, 17, 31, 69, 81, 87, 91–92, 95–99, 122, 206
shock   19, 30, 39, 47, 49, 61, 91, 99, 122, 241
[SII]   35, 91, 95, 101, 103, 134
silicate   42, 66, 68, 99
SKA   51, 216
specific star formation rate   146, 250
spectral classification   93
spectral energy distribution   7, 12, 68, 71, 77, 79, 83, 85, 134, 136–137, 143, 158, 172, 200–201, 255, 260
spectroscopic redshift   203, 247, 255
sphere of influence   187, 231
spiral arm   1, 7, 18, 170
Spitzer   7, 64–65, 67, 69, 84, 88, 99, 122, 208, 258
star
 – AGB   83, 166
 – cluster   143–144, 177, 188
 – main sequence   26, 83, 144, 166
 – mass loss   18, 172
 – mass loss rate   61
 – massive   4–5, 7, 27, 51, 63, 77, 132, 134, 144, 153, 156, 220, 242
 – neutron   18, 20, 155, 242
 – OB   49, 83, 99, 140, 147, 236
 – old   30, 77
 – TP-AGB   25, 172
 – white dwarf   19–20, 26, 30, 83, 242
 – Wolf–Rayet   99, 236, 242
 – young   5, 7, 25, 51, 77, 134–135, 148
star formation
 – activity   12, 17–18, 34, 37, 51, 75, 93, 134, 141, 143–146, 148, 151–152, 154, 156–158, 198, 207, 216, 218, 222, 240, 243, 245, 247

 – density   201, 211
 – efficiency   147, 237
 – history   12, 25, 68, 83–84, 92, 103–104, 136, 143, 158–159, 171–172, 196, 211, 217, 220, 222, 230–231, 256
 – rate   49, 144, 148, 150–158, 211, 222, 236, 239, 243, 250
stationarity   145, 152, 158
stellar evolutionary track   28, 83
stellar library   104, 106
stellar mass   83, 146, 171–172, 180, 209, 213, 216, 218, 220, 222, 239, 241, 250
stellar wind   18, 21, 41, 47, 77, 146, 235, 242
structural parameter   161, 166
Sunyaev–Zeldovich effect   246
super soft X-ray source   19
supercluster   2, 245, 250
supermassive black hole   2, 239
supernova   21, 41, 47, 75, 143, 218, 220, 225, 235, 242
 – remnant   5, 7, 17, 19, 29, 49, 51, 75, 77, 153, 155, 242
 – Type Ia   104, 242
 – Type Ib   242
 – Type II   104, 242
surface brightness-absolute magnitude relation   223
synchrotron   5, 7, 19, 25, 51, 53–54, 71–72, 74, 77, 153
 – spectral index   53, 74, 80, 90

t

three-dimensional adaptive matched filter density algorithm   247
tidal tail   1, 253
transition
 – electronic   39
 – rotational   39–40, 44, 47, 54, 115, 120, 122
 – vibrational   39–40, 44, 115, 122
Tully–Fisher relation   215, 225, 227
turbulence   143, 186, 189, 235

u

ULIRG   12, 45, 64–65, 69, 74, 85, 99, 120, 152–153, 157, 201, 238, 258
ultra-luminous X-ray source   19
underlying absorption   148
UV bump   44, 127, 136, 140

v

Vega system   172, 174
velocity dispersion   62, 104, 175, 184, 187, 220, 228, 230–231, 241, 247, 249

virial
- equilibrium   116
- radius   249
- theorem   175–176, 228
viscous friction   19–20, 239
viscous stripping   249
void   2
Voronoi–Delaunay density algorithm   247

w
WMAP   4

x
X conversion factor   55, 116, 119–120
X-ray binary   17, 19, 155
- high-mass X-ray binary   18
- low-mass X-ray binary   18
X-ray emission   17
X-ray luminosity   17, 61, 155, 204, 233

y
yield   220